T0211940

Communications in Computer and Information Science 1109

Commenced Publication in 2007
Founding and Former Series Editors:
Phoebe Chen, Alfredo Cuzzocrea, Xiaoyong Du, Orhun Kara, Ting Liu,
Krishna M. Sivalingam, Dominik Ślęzak, Takashi Washio, Xiaokang Yang,
and Junsong Yuan

More information about this series at http://www.springer.com/series/7899

Alexander Dudin · Anatoly Nazarov ·
Alexander Moiseev (Eds.)

Information Technologies
and Mathematical Modelling

Queueing Theory and Applications

18th International Conference, ITMM 2019
Named after A.F. Terpugov, Saratov, Russia, June 26–30, 2019
Revised Selected Papers

 Springer

Editors
Alexander Dudin 🆔
Faculty of Applied Mathematics
and Computer Science
Belarusian State University
Minsk, Belarus

Anatoly Nazarov 🆔
Institute of Applied Mathematics
and Computer Science
Tomsk State University
Tomsk, Russia

Alexander Moiseev 🆔
Institute of Applied Mathematics
and Computer Science
Tomsk State University
Tomsk, Russia

ISSN 1865-0929   ISSN 1865-0937 (electronic)
Communications in Computer and Information Science
ISBN 978-3-030-33387-4   ISBN 978-3-030-33388-1 (eBook)
https://doi.org/10.1007/978-3-030-33388-1

This Springer imprint is published by the registered company Springer Nature Switzerland AG
The registered company address is: Gewerbestrasse 11, 6330 Cham, Switzerland

Preface

The series of scientific conferences Information Technologies and Mathematical Modelling (ITMM) was started in 2002. In 2012, the series acquired an international status, and selected revised papers have been published in *Communication in Computer and Information Science* since 2014. The conference series was named after Alexander Terpugov, one of the first organizers of the conference, an outstanding scientist of the Tomsk State University and a leader of the famous Siberian school on applied probability, queueing theory, and applications.

Traditionally, the conferences have about ten sections in various fields of mathematical modelling and information technologies. Throughout the years, the sections on probabilistic methods and models, queueing theory, and communication networks have been the most popular ones at the conference. These sections gather many scientists from different countries. Many foreign participants come to this Siberia conference every year because of our warm welcome and serious scientific discussions. This year, the ITMM conference was held in Saratov, one of the biggest research and engineering centers of the Russian Federation.

This volume presents selected papers from the 18th ITMM conference. The papers are devoted to new results in queueing theory and its applications. Its target audience includes specialists in probabilistic theory, random processes, mathematical modeling, as well as engineers engaged in logical and technical design and operational management of data processing systems, communication, and computer networks.

June 2019

Alexander Dudin
Anatoly Nazarov
Alexander Moiseev

Organization

The conference was organized by the National Research Tomsk State University, Peoples Friendship University of Russia (RUDN University), Trapeznikov Institute of Control Sciences of Russian Academy of Sciences, and National Research Saratov State University.

International Program Committee Chairs

Alexander Dudin	Belarusian State University, Belarus
Anatoly Nazarov	Tomsk State University, Russia

International Program Committee

Khalid Al-Begain	University of South Wales, UK
Ivan Atencia	University of Malaga, Spain
Pedro Cabral	Universidade Nova de Lisboa, Portugal
Pau Fonseca i Casas	Universitat Politècnica de Catalunya, Spain
Srinivas Chakravarthy	Kettering University, USA
Bong Dae Choi	National Institute for Mathematical Sciences, South Korea
Tadeusz Czachórski	Institute of Theoretical and Applied Informatics, Polish Academy of Sciences, Poland
Rui Dinis	Universidade Nova de Lisboa, Portugal
Dmitry Efrosinin	Johannes Kepler University Linz, Austria
Mais Farhadov	Institute of Control Sciences, Russian Academy of Sciences, Russia
Yulia Gaydamaka	Peoples Friendship University of Russia (RUDN University), Russia
Erol Gelenbe	Imperial College, UK
Alexander Gortsev	Tomsk State University, Russia
Viktor Ivnitskii	Railway Research Institute, Russia
Bara Kim	Korea University, South Korea
Che Soong Kim	Sangji University, South Korea
Alexander Kirpichnikov	Kazan National Research Technological University, Russia
Udo Krieger	Universität Bamberg, Germany
B. Krishna Kumar	Anna University, Chennai, India
Achyutha Krishnamoorthy	Cochin University of Science and Technology, India
Quan-Lin Li	Yan Shan University, China
Yury Malinkovsky	Francisk Skorina Gomel State University, Belarus
Gennady Medvedev	Belarusian State University, Belarus

Agassi Melikov	National Aviation Academy of Azerbaijan, Azerbaijan
Alexander Moiseev	Tomsk State University, Russia
Svetlana Moiseeva	Tomsk State University, Russia
Paulo Montezuma-Carvalho	Universidade Nova de Lisboa, Portugal
Evsey Morozov	Institute of Applied Mathematical Research, Karelian Research Centre of Russian Academy of Sciences, Russia
Rein Nobel	Vrije Universiteit Amsterdam, The Netherlands
Michele Pagano	Pisa University, Italy
Tuan Phung-Duc	University of Tsukuba, Japan
Thomas B. Preußer	Technische Universität Dresden, Germany
Jacques Resing	Eindhoven University of Technology, The Netherlands
Vladimir Rykov	Gubkin Russian State University of Oil and Gas, Russia
Konstantin Samouylov	Peoples Friendship University of Russia (RUDN University), Russia
Sergey Suschenko	Tomsk State University, Russia
Daniel Stamate	Goldsmiths College, UK
János Sztrik	University of Debrecen, Hungary
Henk Tijms	Vrije Universiteit Amsterdam, The Netherlands
Oleg Tikhonenko	Cardinal Stefan Wyszynski University in Warsaw, Poland
Gurami Tsitsiashvili	Institute of Applied Mathematics, Far Eastern Branch of Russian Academy of Sciences, Russia
Vladimir Vishnevsky	Institute of Control Sciences, Russian Academy of Sciences, Russia
Vladimir Zadorozhny	Omsk State Technical University, Russia

Local Organizing Committee

Svetlana Moiseeva (Chair)	Anna Morozova
Dmitry Andreichenko (Co-chair)	Oleg Osipov
Mikhail Abrosimov	Svetlana Paul
Inna Batraeva	Ekaterina Rogachko
Valentina Broner	Svetlana Rozhkova
Elena Danilyuk	Daria Semenova
Vitaliy Dolgov	Maria Shklennik
Antonina Fedorova	Alexey Shkurkin
Ekaterina Fedorova	Igor Shmyrin
Nadezhda Fokina	Elena Stankevich
Elena Glukhova	Igor Tananko
Ivan Lapatin	Konstantin Voytikov
Ekaterina Lisovskaya	Anton Voitishek
Mikhail Matalytski	Andrey Zorin

Contents

On the Busy Period in a Finite-Source Retrial Queue with Outgoing Calls

Velika Dragieva[1]([✉]) and Tuan Phung-Duc[2]

[1] University of Forestry, 10 Kliment Ohridsky, 1756 Sofia, Bulgaria
dragievav@yahoo.com
[2] University of Tsukuba, 1-1-1 Tennodai, Tsukuba, Ibaraki 305-8573, Japan
tuan@sk.tsukuba.ac.jp

Abstract. The main purpose of present paper is to study the busy period in one single-server, finite-source retrial queue with outgoing calls. The specific feature of this system is that the outgoing calls do not affect the customers in the system. This allows the model to be considered as a queue with two types of customers. The service times of incoming and the outgoing calls follow two distinct arbitrary distributions. We derive formulas for computing the Laplace-Stieltjes transform of the distribution of the busy period length and its first moment.

Keywords: Finite queues · Retrials · Outgoing calls · Busy period

1 Introduction

This paper deals with a single-server queue with two types of customers (units, users). The customers of the first type form a quasi-random input of demands, and if the service cannot start at the time of arrival, the customer joins a virtual waiting room, called orbit and after some exponentially distributed random time repeats his/her attempt for service. The customers of the second type arrive according to a Poisson flow. When their service can not start at the moment of arrival they do not join the orbit, they are lost. This behavior of the second type customers allows the model to be considered as a single-server, finite-source retrial queue where the server makes outgoing calls after some exponentially distributed idle time.

In the queueing models with quasi-random input (also called closed queueing models or queues with finite source) it is assumed that the server/servers serve a finite number of customers as it is in most of the real situations. Each of these customers produces its flow of demands which means that the generalized input flow depends on the number of customers able to produce demands. These models have been used to analyze the performance of telephone, computer, communication and other systems (see [6, 7, 20, 21]).

Retrial queues have been used to model problems in telephone, computer, communication system, etc. The most obvious example of a retrial queue appears

© Springer Nature Switzerland AG 2019
A. Dudin et al. (Eds.): ITMM 2019, CCIS 1109, pp. 1–13, 2019.
https://doi.org/10.1007/978-3-030-33388-1_1

in a telephone call, when a person phones and finds the line engaged. Usually in such a situation the subscriber repeats his attempts until he is satisfied. Other examples arrise in diverse real situations including telephone switching systems, telecommunication and computer networks, call centers, cellular and local area networks, etc. (see [1,3,8,9,31,32]). A systematic account of the fundamental methods and the latest results, as well as an classified bibliography on this topic can be found, for example in [3,15,17,23] and references therein.

Single-server retrial queues with quasi-random input are useful in modeling magnetic disk memory systems (see [29]), a star-like local area networks (see [22,26]), a local area networks with non-persistent CSMA/CD protocol (see [25]) and in other real life situations. These models have been studied in a number of articles by a number of authors: Ohmura and Takahashi [29], Falin and Artalejo [18], Amador [2], Dragieva [10]. Recently, such models, extended with additional features of the service regime and/or the customers behaviour, have been extensively analyzed. This includes models with an unreliable server (in [33,35]), with two phases of the service times (in [34]), models with collisions (in [27]), with random access (in [19]), models with two-way communication (in [13,24,28]).

In various real life situations, especially in systems with human servers such as call centers, mobile phone, etc., the server not only accepts and serves incoming demands (calls) but, while being idle he/she himself/herself initiates outgoing calls. Such situations are usually modelled by queueing systems, recently known as two-way communication queues or queues with outgoing calls. Ones of the first results about such models are obtained by Falin in his paper [16], where the author analyzes a single server queue with outgoing and incoming calls whose service times have the same arbitrary distribution. Single and multiple-servers retrial models with incoming and outgoing calls that follow exponential distribution with distinct parameters are analyzed by Artalejo and Phung-Duc [4]. The corresponding $M/G/1$ queue where the service times of incoming and outgoing calls follow two distinct arbitrary distributions is investigated by the same authors in their paper [5]. Sakurai and Phung-Duc [30] consider two-way communication retrial queues with multiple types of outgoing calls. A two-way communication $M/M/1$ retrial queue with server-orbit interaction is studied by Dragieva and Phung-Duc [12].

As stated above, a single-server, finite-source retrial queues with two-way communication are analyzed by Dragieva and Phung-Duc [13], Nazarov et al. [28], Kuki et al. [24]. In these works it is assumed that the outgoing calls are directed to the customers in the system: in the orbit or outside it (the customers in free state), while the corresponding model with outgoing calls directed to some customers outside the system is studied by Dragieva and Phung-Duc [14]. In the present paper we extend this analysis, considering the system performance in non-stationary regime. Applying the discrete transformations method, common it the investigation of queues with finite source (see [11,17,18,21,29,33–35]) we derive formulas for computing the Laplace-Stieltjes transform of the distribution function of the system busy period. We also derive convenient recursive formulas for computing the mathematical expectation of the busy period length.

The assumption that outgoing calls are directed to some customers outside the system allows to consider the model under investigation as a model with vacations (or breakdowns and repairs), made by the server during its idle time, or as a model with two types of customers. The motivation for studying such model are many real situations like call centers, repair centers, or medical centers. Usually in these centers along with the regular customers there is and a special group of subscribed customers, or customers (patients) under special care whose service consists mainly of preventive activities, initiated by the server (operator) when being idle.

Further the paper is organized as follows. In Sect. 2 we describe the model in detail and introduce the necessary notations. Section 3 is devoted to the busy period investigation and contains the main results of the paper. Section 4 closes the paper and presents some topics for possible further works.

2 Model Description and Notations

We consider queueing system with one server that serves K customers. Each of these customers in its free state that means not being served or waiting for service, produces a Poisson flow of demands (calls) with intensity λ_1, i.e. a customer in free state calls for service with rate λ_1 which will be also called primary intensity. If the server is idle at the time of a customer arrival, the service of this customer starts. Otherwise it enters the orbit and repeats its attempts for service until it finds the server idle. The intervals between repetitions are exponentially distributed with parameter μ, the secondary intensity. The customers in the orbit are called secondary incoming customers or repeated customers, while those that are outside it - primary customers or customers in free state. The service duration of primary and secondary incoming customers follows the same arbitrary law with common probability distribution function $B_1(x)$, hazard rate function $b_1(x) = B_1'(x)[1 - B_1(x)]^{-1}$, Laplace-Stieltjes transform $\beta_1(s)$ and mean $1/\nu_1$. After the service all customers (primary or secondary) move to a free state, i.e. can produce a Poisson flow of demands with intensity λ_1.

On the other hand, the server produces outgoing calls after some idle time having exponential distribution with parameter λ_2 (the intensity of outgoing calls, or just outgoing intensity): The duration of outgoing calls follows an arbitrary distribution with probability distribution function $B_2(x)$, hazard rate function $b_2(x)$, Laplace-Stieltjes transform $\beta_2(s)$ and mean $1/\nu_2$.

Thus, after each service the next server occupation is determined by a competition between three flows of demands - demands of the primary and secondary incoming customers, and demands of the server outgoing calls. More precisely, if at a time moment t the server is idle and there are n customers in the orbit, the rates of these flows are equal to $(K - n)\lambda_1$, $n\mu$ and λ_2, respectively.

3 Busy Period

Let us assume that the busy period starts at time $t_0 = 0$ at which all customers are in free state and one of them generates a call. It ends at the first epoch at which the server is idle and there are no customers in the orbit. We assume that the length of the busy period is ζ, its distribution function, $P\{\zeta \leq x\}$ is $H(x)$ with Laplace – Stieltjes transform $\eta(s)$. For each $t \geq 0$ we consider the probabilities (densities)

$$P_{i,n}(t,x)dx = P\{\zeta > t, C(t) = i, R(t) = n, x \leq z(t) < x + dx\},$$
$$P_{i,n}(t) = P\{\zeta > t, C(t) = i, R(t) = n\} = \int P_{i,n}(t,x)dx, \ i = 1,2,$$
$$P_{0,n}(t) = P\{\zeta > t, C(t) = 0, R(t) = n\}, 0 \leq n \leq K,$$

with $P_{1,K}(t,x) = 0$,

$$P_{0,n}(0) = P_{2,n}(0,x) = 0, \ P_{1,n}(0,x) = \delta(x)\delta_{0,n}, \tag{1}$$

and Laplace transforms $\overline{P}_{i,n}(s,x)$ $(i = 1,2)$, $\overline{P}_{i,n}(s)$ $(i = 0,1,2)$. Here $C(t)$ denotes the server state at time t (0, 1 or 2, according to the server is idle, busy with an incoming or with an outgong call, respectively), $R(t)$ is the number of repeated customers at time t, $z(t)$ is a supplementary variable, equal to the elapsed service time, $\delta(x)$ is Dirac delta and $\delta_{i,j}$ is Kronecker's delta.

Remark 1. In our definition of a busy period we have not included the case when this period is initiated by an outgoing call. Since a client can arrive during this outgoing call, thus starting a new busy period, such a situation is interesting and should be stated as a possible future topic for investigation. The same refers to the case when we consider the model as a queue with two types of customers.

The Kolmogorov's equations for these transient probabilities are:

$$\frac{d}{dt}P_{0,n}(t) = -[(K-n)\lambda_1 + \lambda_2 + n\mu]P_{0,n}(t) + \int_0^t \sum_{i=1}^{2} P_{i,n}(t,x)b_i(x)dx,$$

$$\frac{\partial}{\partial t}P_{2,n}(t,x) = -\left[(K-n)\lambda_1 + b_2(x) + \frac{\partial}{\partial x}\right]P_{2,n}(t,x)+$$
$$+ (1 - \delta_{n,1})(K-n+1)\lambda_1 P_{2,n-1}(t,x),$$
$$P_{2,n}(t,0) = \lambda_2 P_{0,n}(t),$$
$$1 \leq n \leq K,$$

$$\frac{\partial}{\partial t}P_{1,n}(t,x) = -\left[(K-n-1)\lambda_1 + b_1(x) + \frac{\partial}{\partial x}\right]P_{1,n}(t,x)+$$
$$+ (1 - \delta_{n,1})(K-n)\lambda_1 P_{1,n-1}(t,x),$$
$$P_{1,n}(t,0) = (1 - \delta_{0,n})(K-n)\lambda_1 P_{0,n}(t) + (n+1)\mu P_{0,n+1}(t),$$
$$0 \leq n \leq K-1,$$

with initial conditions (1).

In addition, it holds:

$$\frac{d}{dt}H(t) = \int\limits_0^\infty P_{1,0}(t,x)b_1(x)dx,$$

$$\sum_{n=1}^K \left(P_{0,n}(t) + \int_0^\infty P_{2,n}(t,x)dx \right) + \sum_{n=0}^{K-1} \int_0^\infty P_{1,n}(t,x)dx = 1 - H(t).$$

Applying in these equations Laplace transform, we get

$$[(K-n)\lambda_1 + \lambda_2 + n\mu + s]\overline{P}_{0,n}(s) = \int_0^\infty \left(\overline{P}_{1,n}(s,x)b_1(x) + \overline{P}_{2,n}(s,x)b_2(x) \right) dx, \tag{2}$$

$$\left[(K-n)\lambda_1 + b_2(x) + s + \frac{\partial}{\partial x} \right] \overline{P}_{2,n}(s,x) = (1-\delta_{n,1})(K-n+1)\lambda_1\overline{P}_{2,n-1}(s,x), \tag{3}$$

$$\overline{P}_{2,n}(s,0) = \lambda_2\overline{P}_{0,n}(s), \quad 1 \le n \le K, \tag{4}$$

$$\left[(K-n-1)\lambda_1 + b_1(x) + s + \frac{\partial}{\partial x} \right] \overline{P}_{1,n}(s,x) = \delta(x)\delta_{n,0} + (1-\delta_{n,0})(K-n)\lambda_1\overline{P}_{1,n-1}(s,x), \tag{5}$$

$$\overline{P}_{1,n}(s,0) = (1-\delta_{n,0})(K-n)\lambda_1\overline{P}_{0,n}(s) + (n+1)\mu\overline{P}_{0,n+1}(s), \quad 0 \le n \le K-1, \tag{6}$$

$$\eta(s) = \int\limits_0^\infty \overline{P}_{1,0}(s,x)b_1(x)dx, \tag{7}$$

$$\sum_{n=1}^K \left(\overline{P}_{0,n}(s) + \int_0^\infty \overline{P}_{2,n}(s,x)dx \right) + \sum_{n=0}^{K-1} \int_0^\infty \overline{P}_{1,n}(s,x)dx = \begin{cases} \frac{1-\eta(s)}{s}, & \text{if } s \ne 0 \\ E[\zeta], & \text{if } s = 0 \end{cases}. \tag{8}$$

The matrix form of Eqs. (3) and (5) is

$$\left(\overline{\theta}_i I - A \right) \overline{P}_i(s,x) = D_i(x), \tag{9}$$

$i = 1, 2$, respectively, where θ_i are defined as,

$$\theta_i = b_i(x) + s + \frac{\partial}{\partial x},$$

I is the identity matrix of order K, A is obtained by (3) and (5) in the usual way,

$$A = \begin{pmatrix} -(K-1)\lambda_1 & 0 & ... & 0 & 0 & 0 \\ (K-1)\lambda_1 & -(K-2)\lambda_1 & ... & 0 & 0 & 0 \\ 0 & (K-2)\lambda_1 & ... & 0 & 0 & 0 \\ . & . & . & . & . & . \\ 0 & 0 & ... & 2\lambda_1 & -\lambda_1 & 0 \\ 0 & 0 & ... & 0 & \lambda_1 & 0 \end{pmatrix},$$

and

$$\overline{P}_i(s,x) = \left(\overline{P}_{i,i-1}(s,x), ..., \overline{P}_{i,K-2+i}(s,x)\right)^T,$$
$$D_1(x) = (\delta(x), 0, ..., 0)^T, \quad D_2(x) = 0.$$

It is not difficult to verify that applying in Eq. (9) the transformations

$$\overline{P}_i(s,x) = Y\overline{Q}_i(s,x),$$

these equations get a simpler form

$$\theta_i \overline{Q}_i(s,x) - \Lambda \overline{Q}_i(s,x) = Y^{-1}D_i, \qquad (10)$$

where

$$Y = \begin{pmatrix} 0 & 0 & 0 & ... & 0 & 1 \\ 0 & 0 & 0 & ... & 1 & -\binom{K-1}{K-2} \\ 0 & 0 & 0 & ... & -\binom{K-2}{K-3} & \binom{K-1}{K-3} \\ ... & ... & ... & . & ... & ... \\ 0 & 1 & -2 & ... & (-1)^{K-3}\binom{K-2}{1} & (-1)^{K-2}\binom{K-2}{1} \\ 1 & -1 & 1 & ... & (-1)^{K-2} & (-1)^{K-1} \end{pmatrix},$$

$$Y^{-1} = \begin{pmatrix} 1 & 1 & ... & 1 & 1 & 1 \\ \binom{K-1}{1} & \binom{K-2}{1} & ... & \binom{2}{1} & 1 & 0 \\ ... & ... & ... & ... & ... & ... \\ \binom{K-1}{K-3} & \binom{K-2}{K-3} & ... & 0 & 0 & 0 \\ \binom{K-1}{K-2} & 1 & ... & 0 & 0 & 0 \\ 1 & 0 & ... & 0 & 0 & 0 \end{pmatrix},$$

$\Lambda = diag\{0, -\lambda_1, ..., -(K-1)\lambda_1\}$.

Transformations of this type are known as discrete transformations and, as stated in the Introduction are widely used in the investigation of queueing systems with finite number of customers. In fact, the diagonal elements of the matrix Λ are the eigenvalues of the matrix A, and the columns of the matrix Y are the corresponding eigenvectors.

Thus, transforming (2)–(8) according to the formulas

$$\overline{P}_{1,n}(s,x) = \sum_{m=0}^{n} (-1)^{n-m} \binom{K-1-m}{n-m} \overline{Q}_{1,K-1-m}(s,x), \qquad (11)$$
$$0 \le n \le K-1,$$

$$\overline{P}_{2,n}(s,x) = \sum_{m=0}^{n-1} (-1)^{n-m-1} \binom{K-1-m}{n-m-1} \overline{Q}_{2,K-m}(s,x),$$
$$1 \leq n \leq K, \tag{12}$$

we can simplify and solve Eqs. (5) and (3), and get formulas for calculating $\overline{P}_{i,n}(s,x)$ $(i = 1,2)$, $\overline{P}_{0,n}(s)$, $\overline{P}_{i,n}(s) = \int_0^\infty \overline{P}_{i,n}(s,x)dx$ $(i = 1,2)$ and $\eta(s)$. These formulas are given in the next theorem.

Theorem 1. *The Laplace transforms $\overline{P}_{i,n}(s,x)$ $(i = 1,2)$ and $\overline{P}_{i,n}(s)$ $(i = 0,1,2)$ of the system state distribution during the busy period, $P_{i,n}(t,x)$ and $P_{i,n}(t)$, respectively, and the Laplace – Stieltjes transform, $\eta(s)$, of the busy period distribution function, $H(x)$ can be calculated by the formulas*

$$\overline{P}_{1,n}(s,x) = (1 - B_1(x)) \sum_{m=0}^{n} (-1)^{n-m} e^{-[(K-m-1)\lambda_1 + s]x} \times$$
$$\binom{K-1-m}{n-m} \left(\overline{Q}_{1,K-1-m}(s,0) + \binom{K-1}{m} \right), \tag{13}$$

$$\overline{P}_{1,n}(s) =$$
$$\sum_{m=0}^{n} (-1)^{n-m} \widetilde{\beta}_{1,m}(s) \binom{K-1-m}{n-m} \left(\overline{Q}_{1,K-1-m}(s,0) + \binom{K-1}{m} \right), \tag{14}$$
$$0 \leq n \leq K-1,$$

$$\overline{P}_{2,n}(s,x) = (1 - B_2(x)) \times$$
$$\sum_{m=0}^{n-1} (-1)^{n-m-1} \binom{K-1-m}{n-m-1} e^{-[(K-m-1)\lambda_2 + s]x} \overline{Q}_{2,K-m}(s,0), \tag{15}$$

$$\overline{P}_{2,n}(s) = \sum_{m=0}^{n-1} (-1)^{n-m-1} \binom{K-m-1}{n-m-1} \widetilde{\beta}_{2,m}(s) \overline{Q}_{2,K-m}(s,0), \tag{16}$$

$$\overline{P}_{0,n}(s) = \frac{\overline{P}_{2,n}(s,0)}{\lambda_2} =$$
$$\overline{P}_{0,n}(s) = \frac{\overline{P}_{2,n}(s,0)}{\lambda_2} = \frac{1}{\lambda_2} \sum_{m=0}^{n-1} (-1)^{n-m-1} \binom{K-m-1}{n-m-1} \overline{Q}_{2,K-m}(s,0), \tag{17}$$
$$1 \leq n \leq K,$$

$$\eta(s) = \int_0^\infty \overline{P}_{1,0}(s,x) b_1(x) dx = \overline{\beta}_{1,K-1}(s) \left(\overline{Q}_{1,K-1}(s,0) + 1 \right). \tag{18}$$

Here

$$\widetilde{\beta}_{i,k}(s) = \begin{cases} \frac{1-\overline{\beta}_{i,K-k-1}(s)}{(K-k-1)\lambda_i + s}, & \text{if } (K-k-1)\lambda_i + s \neq 0 \\ \frac{1}{\nu_i}, & \text{if } (K-k-1)\lambda_i + s = 0 \end{cases},$$
$$\overline{\beta}_{i,k}(s) = \beta_i (k\lambda_i + s), \quad \overline{\beta}_{i,k} = \overline{\beta}_{i,k}(0),$$

and initial conditions $\overline{Q}_{i,k}(s,0)$ satisfy the following system of linear equations

$$
\sum_{m=0}^{n} \lambda_2 \left(1 - \delta_{n,K}\right) (-1)^{n-m} \overline{\beta}_{1,K-m-1}(s) \binom{K-1-m}{n-m} \overline{Q}_{1,K-1-m}(s,0) +
$$

$$
\sum_{m=0}^{n-1} (-1)^{n-m} \binom{K-1-m}{n-m-1} \times
$$

$$
\left[(K-n)\lambda_1 + \lambda_2 \left(1 - \overline{\beta}_{2,K-m-1}(s)\right) + n\mu + s\right] \overline{Q}_{2,K-m}(s,0) = \tag{19}
$$

$$
\lambda_2 \left(1 - \delta_{n,K}\right) \sum_{m=0}^{n} (-1)^{n-m-1} \overline{\beta}_{1,K-m-1}(s) \binom{K-1-m}{n-m} \binom{K-1}{m},
$$

$$
1 \le n \le K,
$$

$$
\lambda_2 \sum_{m=0}^{n} (-1)^{n-m} \binom{K-1-m}{n-m} \overline{Q}_{1,K-1-m}(s,0) + \delta_{n,0}\lambda_2 +
$$

$$
\sum_{m=0}^{n} (-1)^{n-m} \binom{K-1-m}{n-m} \left[(n-m)\lambda_1 - (n+1)\mu\right] \overline{Q}_{2,K-m}(s,0) = 0 \tag{20}
$$

$$
0 \le n \le K-1,
$$

$$
\overline{Q}_{2,1}(s,0) \left(\widetilde{\beta}_{2,K-1}(s) + \frac{1}{\lambda_2}\right) + \widetilde{\beta}_{1,K-1}(s) \left(\overline{Q}_{1,0}(s,0) + 1\right) =
$$

$$
\begin{cases} \dfrac{1 - \overline{\beta}_{1,K-1}(s)\left(\overline{Q}_{1,K-1}(s,0)+1\right)}{s}, & \text{if } s \neq 0 \\ E[\zeta], & \text{if } s = 0 \end{cases}. \tag{21}
$$

Proof. Transforming Eqs. (5) and (3) according to (9) we obtain them in the form

$$
\frac{\partial}{\partial x}\overline{Q}_{1,m}(s,x) + \left[m\lambda_1 + b_1(x) + s\right]\overline{Q}_{1,m}(s,x) = \binom{K-1}{m}\delta(x),
$$

$$
0 \le m \le K-1,
$$

$$
\frac{\partial}{\partial x}\overline{Q}_{2,m}(s,x) + \left[(m-1)\lambda_2 + b_2(x) + s\right]\overline{Q}_{2,m}(s,x) = 0,
$$

$$
1 \le m \le K,
$$

whose solutions are

$$
\overline{Q}_{1,m}(s,x) = (1 - B_1(x))\, e^{-(m\lambda_1+s)x} \left(\overline{Q}_{1,m}(s,0) + \binom{K-1}{m}\right),
$$

$$
\overline{Q}_{2,m}(s,x) = (1 - B_2(x))\, e^{-[(m-1)\lambda_2+s]x}\overline{Q}_{2,m}(s,0).
$$

Substituting with these expressions in (11), (12) and (4) we obtain formulas (13)–(17) for the quantities $\overline{P}_{i,n}(s,x), \overline{P}_{i,n}(s)$ $(i = 1,2)$ and $\overline{P}_{0,n}(s)$. Formula (18) for $\eta(s)$ follows from (7) and (13).

Further, from Eqs. (2) and (6) we derive relations between the initial conditions $\overline{Q}_{1,m}(s,0)$ and $\overline{Q}_{2,m}(s,0)$. First, substituting in (2) $\overline{P}_{0,n}(s)$ according to (4) and then using (13) and (15) we obtain

$$[(K-n)\lambda_1 + \lambda_2 + n\mu + s] \sum_{m=0}^{n-1} (-1)^{n-m-1} \binom{K-m-1}{n-m-1} \overline{Q}_{2,K-m}(s,0) =$$

$$\lambda_2 (1-\delta_{n,K}) \sum_{m=0}^{n} (-1)^{n-m} \overline{\beta}_{1,K-m-1}(s) \binom{K-1-m}{n-m} \times$$

$$\left(\overline{Q}_{1,K-1-m}(s,0) + \binom{K-1}{m}\right) +$$

$$\lambda_2 \sum_{m=0}^{n-1} (-1)^{n-m-1} \binom{K-1-m}{n-m-1} \overline{\beta}_{2,K-m-1}(s) \overline{Q}_{2,K-m}(s,0),$$

$$1 \le n \le K.$$

These equations can be transformed to the form (19).

In a similar way, substituting in (6) $\overline{P}_{0,n}(s)$ and $\overline{P}_{0,n+1}(s)$ according to (4), after some rearrangements we get the following relations between the initial conditions $\overline{Q}_{1,m}(s,0)$ and $\overline{Q}_{2,m}(s,0)$:

$$\lambda_2 \sum_{m=0}^{n} (-1)^{n-m} \left(\binom{K-1-m}{n-m} \overline{Q}_{1,K-1-m}(s,0) + \binom{K-1}{n}\binom{n}{m}\right) =$$

$$(1-\delta_{n,0}) (K-n)\lambda_1 \sum_{m=0}^{n-1} (-1)^{n-m-1} \binom{K-1-m}{n-m-1} \overline{Q}_{2,K-m}(s,0) +$$

$$(n+1)\mu \sum_{m=0}^{n} (-1)^{n-m} \binom{K-1-m}{n-m} \overline{Q}_{2,K-m}(s,0),$$

$$0 \le n \le K-1.$$

Last relations lead to formulas (20).

Finally we use the normalizing condition (8). From (16) and (17), after some transformations we have

$$\sum_{n=1}^{K} \left(\overline{P}_{0,n}(s) + \int_0^\infty \overline{P}_{2,n}(s,x)dx\right) = \sum_{n=1}^{K} \left(\overline{P}_{0,n}(s) + \overline{P}_{2,n}(s)\right) =$$

$$\overline{Q}_{2,1}(s,0) \left(\tilde{\beta}_{2,K-1}(s) + \frac{1}{\lambda_2}\right).$$

Similarly, from (14) it follows that

$$\sum_{n=0}^{K-1} \int_0^\infty \overline{P}_{1,n}(s,x)dx = \sum_{n=0}^{K-1} \overline{P}_{1,n}(s) = \tilde{\beta}_{1,K-1}(s) \left(\overline{Q}_{1,0}(s,0) + 1\right).$$

These two expressions and formula (18) show that Eq. (8) can be presented in the form (21) which ends the proof of the theorem.

Thus, to calculate $\eta(s)$ we have to find the solutions $\overline{Q}_{1k}(s,0)$ of the linear system (19)–(21). Further, upon suitable differentiations in (18)–(20) we can obtain formulas for computing the moments of the busy period length, ζ. Besides this way, the mean busy period can be calculated with the help of formula (21) for $s = 0$,

$$E[\zeta] = \overline{Q}_{2,1}(0,0)\left(\widetilde{\beta}_{2,K-1}(0) + \frac{1}{\lambda_2}\right) + \widetilde{\beta}_{1,K-1}(0)\left(\overline{Q}_{1,0}(0,0) + 1\right).$$

Having in mind that

$$\widetilde{\beta}_{i,K-1}(0) = \frac{1}{\nu_i},$$

and that $\overline{Q}_{i,i-1}(0,0)$, $(i = 1,2)$ can be calculated from formulas (19)–(20) and (18) for $s = 0$ we prove the following corollary.

Corollary 1. *The mean busy period can be calculated by the formula*

$$E[\zeta] = \frac{\overline{Q}_{1,0}(0,0)}{\nu_1} + \overline{Q}_{2,1}(0,0)\frac{(\lambda_2 + \nu_2)}{\lambda_2\nu_2} + \frac{1}{\nu_1},$$

where $\overline{Q}_{i,i-1}(0,0)$ $(i = 1,2)$ can be calculated solving the system

$$\sum_{m=0}^{n}\lambda_2\left(1 - \delta_{n,K}\right)(-1)^{n-m}\overline{\beta}_{1,K-m-1}\binom{K-1-m}{n-m}\overline{Q}_{1,K-1-m}(0,0)+$$

$$\sum_{m=0}^{n-1}(-1)^{n-m}\binom{K-1-m}{n-m-1}[(K-n)\lambda_1 + \lambda_2\left(1 - \overline{\beta}_{2,K-m-1}\right) + n\mu]\overline{Q}_{2,K-m}(0,0) = \quad (22)$$

$$\lambda_2\left(1 - \delta_{n,K}\right)\sum_{m=0}^{n}(-1)^{n-m-1}\overline{\beta}_{1,K-m-1}\binom{K-1-m}{n-m}\binom{K-1}{m},$$

$$1 \le n \le K,$$

$$\lambda_2\sum_{m=0}^{n}(-1)^{n-m}\binom{K-1-m}{n-m}\overline{Q}_{1,K-1-m}(0,0) + \delta_{n,0}\lambda_2+$$

$$\sum_{m=0}^{n}(-1)^{n-m}\binom{K-1-m}{n-m}[(n-m)\lambda_1 - (n+1)\mu]\overline{Q}_{2,K-m}(0,0) = 0, \quad (23)$$

$$0 \le n \le K-1,$$

$$\overline{\beta}_{1,K-1}\left(\overline{Q}_{1,K-1}(0,0) + 1\right) = 1. \quad (24)$$

The solution of system (22)–(24) can be computed recursively, starting with the value of $\overline{Q}_{1,K-1}(0,0)$ that we find from (24),

$$\overline{\beta}_{1,K-1}\overline{Q}_{1,K-1}(0,0) = 1 - \overline{\beta}_{1,K-1}.$$

Then from Eq. (23) for $n = 0$ we calculate $\overline{Q}_{2,K}(0,0)$,

$$\mu\overline{Q}_{2,K}(0,0) = \lambda_2 + \lambda_2\overline{Q}_{1,K-1}(0,0),$$

and from Eq. (22) for $n = 1$ we calculate $\overline{Q}_{1,K-2}(0,0)$,

$$\lambda_2\overline{\beta}_{1,K-2}\overline{Q}_{1,K-2}(0,0) = \lambda_2\left(K-1\right)\left(\overline{\beta}_{1,K-1} - \overline{\beta}_{1,K-2}\right) +$$
$$\lambda_2\overline{\beta}_{1,K-1}\left(K-1\right)\overline{Q}_{1,K-1}(0,0) + [(K-1)\lambda_1 + \lambda_2\left(1 - \overline{\beta}_{2,K-1}\right) + \mu]\overline{Q}_{2,K}(0,0).$$

Further, once we have calculated all quantities $\overline{Q}_{1,K-k-1}(0,0)$ ($k = 0,..,n$, $n = 0,...,K-2$) and $\overline{Q}_{2,K-k}(0,0)$ ($k = 0,..,n-1$, $n = 1,...,K-2$), from (23) for n we calculate $\overline{Q}_{2,K-n}(0,0)$,

$$(n+1)\mu\overline{Q}_{2,K-n}(0,0) = \lambda_2 \sum_{m=0}^{n} (-1)^{n-m} \binom{K-1-m}{n-m}\overline{Q}_{1,K-1-m}(0,0)+ $$
$$\sum_{m=0}^{n-1} (-1)^{n-m} \binom{K-1-m}{n-m} [(n-m)\lambda_1 - (n+1)\mu] \overline{Q}_{2,K-m}(0,0), \tag{25}$$

and then from (22) for $n+1$ we calculate $\overline{Q}_{1,K-n-2}(0,0)$,

$$\lambda_2\overline{\beta}_{1,K-n-2}\overline{Q}_{1,K-n-2}(0,0) = -\lambda_2\overline{\beta}_{1,K-n-2}\binom{K-1}{n+1} + \sum_{m=0}^{n} \overline{Q}_{2,K-m}(0,0)\times$$
$$(-1)^{n-m} \binom{K-1-m}{n-m}[(K-n-1)\lambda_1 + \lambda_2\left(1 - \overline{\beta}_{2,K-m-1}\right) + (n+1)\mu]+$$
$$\sum_{m=0}^{n} \lambda_2(-1)^{n-m}\overline{\beta}_{1,K-m-1}\binom{K-1-m}{n+1-m}\left[\overline{Q}_{1,K-1-m}(0,0) + \binom{K-1}{m}\right].$$

The quantity $\overline{Q}_{2,1}(0,0)$ can be calculated by formula (25) for $n = K-1$ (which comes from Eq. (23) for $n = K-1$) or from Eq. (22) for $n = K$,

$$K\mu\overline{Q}_{2,1}(0,0) = \sum_{m=0}^{K-2} (-1)^{K-m} \left[\lambda_2\left(1 - \overline{\beta}_{2,K-m-1}\right) + K\mu\right] \overline{Q}_{2,K-m}(0,0).$$

4   Conclusions

In this paper we analyze the busy period in an $M/G/1//K$ retrial queue with outgoing calls, directed outside the customers in the system. This assumption means that the model under consideration can be thought of as a model with server vacation or as a model with two types of customers. The duration of outgoing calls follows an arbitrary distribution, other than the distribution of incoming calls. Applying the discrete transformations method we derive formulas for computing the Laplace-Stieltjes transform of the busy period distribution as well as the first moment of this distribution.

The results, obtained in present paper can be applied for further investigation of the busy period, like deriving formulas for computation of the second moment of the busy period length. We also plan to extend our analysis assuming that the busy period can be initiated not only by an incoming call, as it is assumed here, but also by an outgoing call (or by a customer of second type if we consider the model as a queue with two types of customers).

References

1. Aguir, S., Karaesmen, E., Aksin, O., Chauvet, F.: The impact of retrials on call center performance. OR Spectr. **26**, 353–376 (2004)
2. Amador, J.: On the distribution of the successful and blocked events in retrial queues with finite number of sources. In: Proceedings of the 5th International Conference on Queueing Theory and Network Applications, pp. 15–22 (2010)
3. Artalejo, J., Gómez-Corral, A.: Retrial Queueing Systems: A Computational Approach. Springer, Heidelberg (2008)
4. Artalejo, J., Phung-Duc, T.: Markovian retrial queues with two way communication. J. Ind. Manag. Optim. **8**, 781–806 (2012)
5. Artalejo, J., Phung-Duc, T.: Single server retrial queues with two way communication. Appl. Math. Model. **37**, 1811–1822 (2013)
6. Balazsfalvi, G., Sztrik, J.: A tool for modeling distributed protocols. PIK **31**(1), 39–44 (2008)
7. Biro, J., Bérczes, T., Korosi, A., Heszberger, Z., Sztrik, J.: Discriminatory processor sharing from optimization point of view. In: Dudin, A., De Turck, K. (eds.) ASMTA 2013. LNCS, vol. 7984, pp. 67–80. Springer, Heidelberg (2013). https://doi.org/10.1007/978-3-642-39408-9_6
8. Choi, B., Shin, Y.W., Ahn, W.C.: Retrial queues with collision arising from unslotted CSMA/CD protocol. Queueing Syst. **11**(4), 335–356 (1992)
9. Deslauriers, A., L'Ecuyer, P., Pichitlamken, J., Ingolfsson, A., Avramidis, A.: Markov chain models of a telephone call center with call blending. Comput. Oper. Res. **34**, 1616–1645 (2007)
10. Dragieva, V.: A finite source retrial queue: number of retrials. Commun. Stat. Theor. Methods **42**(5), 812–829 (2013)
11. Dragieva, V.: Steady state analysis of the M/G/1//N queue with orbit of blocked customers. Ann. Oper. Res. **247**, 121–140 (2016)
12. Dragieva, V., Phung-Duc, T.: Two-way communication M/M/1 retrial queue with server-orbit interaction. In: Proceedings of the 11th International Conference on Queueing Theory and Network Applications, 7 p. ACM Digital Library (2016). https://doi.org/10.1145/3016032.3016049
13. Dragieva, V., Phung-Duc, T.: Two-way communication M/M/1//N retrial queue. In: Thomas, N., Forshaw, M. (eds.) ASMTA 2017. LNCS, vol. 10378, pp. 81–94. Springer, Cham (2017). https://doi.org/10.1007/978-3-319-61428-1_6
14. Dragieva, V., Phung-Duc, T.: An M/G/1//K retrial queue with outgoing calls. Proceedings of the 13th International Conference on Queueing Theory and Network Applications, QTNA 2018, Tsukuba, Japan, 25–27 July, pp. 9–13 (2018)
15. Gómez-Corral, A., Phung-Duc, T.: Retrial queues and related models. Ann. Oper. Res. **247**(1), 1–2 (2016)
16. Falin, G.: Model of coupled switching in presence of recurrent calls. Eng. Cybern. Rev. **17**, 53–59 (1979)
17. Falin, G., Templeton, J.: Retrial Queues. Chapman and Hall, London (1997)
18. Falin, G., Artalejo, J.: A finite source retrial queue. Eur. J. Oper. Res. **108**, 409–424 (1998)
19. Fiems, D., Phung-Duc, T.: Light-traffic analysis of random access systems without collisions. Ann. Oper. Res. (2017). https://doi.org/10.1007/s10479-017-2636-7
20. Jain, R.: The Art of Computer Systems Performance Analysis. Wiley, New York (1991)
21. Jaiswal, N.: Priority Queues. Academic Press, New York (1969)

22. Janssens, G.: The quasi-random input queueing system with repeated attempts as a model for collision-avoidance star local area network. IEEE Trans. Commun. **45**, 360–364 (1997)

23. Kim, J., Kim, B.: A survey of retrial queueing systems. Ann. Oper. Res. **247**(1), 3–36 (2016)

24. Kuki, A., Sztrik, J., Tóth, Á., Bérczes, T.: A contribution to modeling two-way communication with retrial queueing systems. In: Dudin, A., Nazarov, A., Moiseev, A. (eds.) ITMM/WRQ -2018. CCIS, vol. 912, pp. 236–247. Springer, Cham (2018). https://doi.org/10.1007/978-3-319-97595-5_19

25. Li, H., Yang, T.: A single server retrial queue with server vacations and a finite number of input sources. Eur. Oper. Res. **85**, 149–160 (1995)

26. Mehmet-Ali, M., Elhakeem, A., Hayes, J.: Traffic analysis of a local area network with star topology. IEEE Trans. Commun. **36**, 703–712 (1988)

27. Nazarov, A., Sztrik, J., Kvach, A.: Some features of a finite-source M/GI/1 retrial queuing system with collisions of customers. In: Vishnevskiy, V.M., Samouylov, K.E., Kozyrev, D.V. (eds.) DCCN 2017. CCIS, vol. 700, pp. 186–200. Springer, Cham (2017). https://doi.org/10.1007/978-3-319-66836-9_16

28. Nazarov, A., Sztrik, J., Kvach, A.: Asymptotic sojourn time analysis of finite-source M/M/1 retrial queuing system with two-way communication. In: Dudin, A., Nazarov, A., Moiseev, A. (eds.) ITMM/WRQ -2018. CCIS, vol. 912, pp. 172–183. Springer, Cham (2018). https://doi.org/10.1007/978-3-319-97595-5_14

29. Ohmura, H., Takahashi, Y.: An analysis of repeated call model with a finite number of sources. Electron. Commun. Jpn. **68**, 112–121 (1985)

30. Sakurai, H., Phung-Duc, T.: Two-way communication retrial queues with multiple types of outgoing calls. TOP **23**, 466–492 (2015)

31. Tran-Gia, P., Mandjes, M.: Modeling of customer retrial phenomenon in cellular mobile networks. IEEE J. Sel. Areas Commun. **15**, 1406–1414 (1997)

32. Van Do, T., Wochner, P., Berches, T., Sztrik, J.: A new finite-source queueing model for mobile cellular networks applying spectrum renting. Asia-Pac. J. Oper. Res. **31**, 14400004 (2014)

33. Wang, J., Zhao, L., Zhang, F.: Analysis of the finite source retrial queues with server breakdowns and repairs. J. Ind. Manag. Optim. **7**(3), 655–676 (2011)

34. Wang, J., Wang, F., Sztrick, J., Kuki, A.: Finite source retrial queue with two phase service. Int. J. Oper. Res. **3**(4), 421–440 (2017)

35. Zhang, F., Wang, J.: Performance analysis of the retrial queues with finite number of sources and service interruption. J. Korean Stat. Soc. **42**, 117–131 (2013)

An Algorithmic Approach for the Analysis of Finite-Source M/GI/1 Retrial Queueing Systems with Collisions and Server Subject to Breakdowns and Repairs

Anatoly Nazarov[1], János Sztrik[2(✉)], Anna Kvach[1], and Attila Kuki[2]

[1] National Research Tomsk State University,
36 Lenina Avenue, Tomsk 634050, Russia
nazarov.tsu@gmail.com, kvach_as@mail.ru
[2] University of Debrecen, Debrecen, Hungary
{sztrik.janos,kuki.attila}@inf.unideb.hu

Abstract. In this paper retrial queuing systems with a finite number of sources and collisions of the customers is considered, where the server is subjects to random breakdowns and repairs depending on whether it is idle or busy. The novelty of this system comparing to the previous ones is that the service time is assumed to follow a general distribution while the source times, retrial times, servers lifetime and repair time are supposed to be exponentially distributed. A new numerical algorithm for finding the joint probability distribution of the number of customers in the system and the server's state is proposed. Several numerical examples and Figures show the effect of different input parameters on the main steady state performance measures, such as mean response and waiting time of the customers, probability of collision and retrials.

Keywords: Finite-source queuing system · Closed queuing systems · Retrial queue · Collision · Server breakdowns and repairs · Unreliable server · Asymptotic analysis · Method of residual service time · Method of elapsed service time

1 Introduction

Finite-source retrial queues are very useful and effective stochastic systems to model several problems arising in telephone switching systems, telecommunication networks, computer networks and computer systems, call centers, wireless communication systems, etc. To see their importance the interested reader is referred to the following works and references cited in them, for example [3,5–7,10]. Searching the scientific databases we have noticed that relatively just a small number of papers have been devoted to systems when the arriving calls (primary or secondary) causes collisions to the request under service and both

© Springer Nature Switzerland AG 2019
A. Dudin et al. (Eds.): ITMM 2019, CCIS 1109, pp. 14–27, 2019.
https://doi.org/10.1007/978-3-030-33388-1_2

go to the orbit, see for example [1,9,11,21]. It should be noted that collisions decreases the effectiveness of the system performance and that is why new protocols should be developed to avoid the collision but unfortunately it cannot be neglected, see [4,8,16]. This fact shows the importance of the mathematical modeling of such systems.

Nazarov and his research group developed a very effective asymptotic method [20] by the help of which various systems have been investigated. Concerning to finite-source retrial systems with collision we should mention the following papers [12–15,18].

Sztrik and his research group have been dealing with systems with unreliable server/s as can be seen, for example in [2,23] and that is why it was understandable that the two research groups started cooperation in 2017.

The primary aim of the present paper is to give a new numerical algorithm for finding the joint probability distribution of the number of customers in the system and the server's state. The method of supplementary variable is used by introducing the residual service time to derive the system of steady state Kolmogorov equations. An effective algorithmic approach is proposed to get the solution of these equations resulting the steady state distribution of the underlying process. Several numerical examples and Figures show the effect of different input parameters on the main steady state performance measures, such as mean response and waiting time of the customers, probability of collision and retrials. The present model is a generalization of the $M/G/1//N$ retrial system treated in [14] where the server was reliable and the $M/M/1//N$ system with unreliable server analyzed in [17].

The rest of the paper is organized as follows. In Sect. 2 description of the model is given, the corresponding multi-dimensional non-Markov process is defined. Then in Sect. 3 by the help of the residual service time technique the corresponding steady state Kolmogorov equations are derived. Section 4 is devoted to the solution of these equations by proposing and new algorithmic approach and important performance measures are defined. In Sect. 5 several numerical examples and Figures show the effect of different input parameters on the main steady state performance measures and some comments are made. Finally, the paper ends with a Conclusion and some future plans are highlighted.

2   Model Description and Notation

Let us consider a closed retrial queuing system of type $M/GI/1//N$ with collision of the customers and an unreliable server. The number of sources is N and each of them can generate a primary request with rate λ/N. A source cannot generate a new call until the end of the successful service of this customer. If a primary customer finds the server idle and not failed, he enters into service immediately, in which the required service time has a probability distribution function $B(x)$. Let us denote its service rate function by $\mu(y) = B^{'}(y)(1 - B(y))^{-1}$ and its Laplace -Stieltjes transform by $B^*(y)$, respectively. Otherwise, if the server is busy, an arriving (primary or repeated) customer involves into collision with the customer under service and they both moves into the orbit. The retrial times of requests are exponentially distributed with rate σ/N. We assume that the server is unreliable, that is its lifetime is supposed to be exponentially distributed with failure rate γ_0 if the server is idle and with rate γ_1 if it is busy. When the server breaks down, it is immediately sent for repair and the recovery time is assumed to be exponentially distributed with rate γ_2. We deal with the case when the server is down all sources continue generation of customers and send it to the server, similarly customers may retry from the orbit to the server but all arriving customers immediately go into the orbit. Furthermore, in this unreliable model we suppose that the interrupted request goes to the orbit immediately and its next service is independent of the interrupted one. The explanation of using λ/N, and σ/N is that in a consecutive paper we would like to investigate the same system by means of asymptotic methods as N tends to infinity and we would like to compare the asymptotic results to the exact ones. All random variables involved in the model construction are assumed to be independent of each other. Let $Q(t)$ be the number of customers in the system at time t, that is, the total number of customers in the orbit and in service. Similarly, let $C(t)$ be the server's state at time t, that is

$$C(t) = \begin{cases} 0, & \text{if the server is idle,} \\ 1, & \text{if the server is busy,} \\ 2, & \text{if the server is down} \\ & \text{(under repair).} \end{cases}$$

Thus, we will investigate the process $\{C(t), Q(t)\}$, which is not a Markov-type unless the required service time is exponentially distributed. Using the supplementary variable method let us introduce the random process $Z(t)$, equal to the residual service time, that is the time interval from moment t until the end of the successful service. It should be noted that the other standard method is to introduce the elapsed service time as the continuous component, see for example [6,25,26] where the resulting Kolmogorov equations are solved by the help of so-called discrete transform. This approach is more common but in our case the residual service time method is more effective as we will show it later on.

As we can see $\{C(t), Q(t), Z(t)\}$ is a three-dimensional Markov process, which has variable number of components, depending on the server's state, since the component $Z(t)$ is determined only in those moments when the server is busy, that is $C(t) = 1$.

3   Kolmogorov Equations for the Probability Distribution

Let us define the following probabilities

$$P_k(j, t) = P\{C(t) = k, Q(t) = j\}, \qquad k = 0, 2$$
$$P_1(j, z, t) = P\{C(t) = 1, Q(t) = j, Z(t) < z\}.$$

Since the introduction of the residual service time is not so standard as the elapsed service time approach we derive the Kolmogorov equations in more details, namely we can write

$$P_0(0, t + \Delta t) = P_0(0, t)(1 - \lambda \Delta t)(1 - \gamma_0 \Delta t) + \gamma_2 \Delta t P_2(0, t) + P_1(1, \Delta t, t) + o(\Delta t), \tag{1}$$

$$P_1(1, z - \Delta t, t + \Delta t) = \left[P_1(1, z, t) - P_1(1, \Delta t, t) \right] \left(1 - \lambda \frac{N - 1}{N} \Delta t \right)(1 - \gamma_1 \Delta t) \tag{2}$$

$$+ P_0(0, t) \lambda B(z) \Delta t + P_0(1, t) \frac{\sigma}{N} B(z) \Delta t + o(\Delta t),$$

$$P_2(0, t + \Delta t) = P_2(0, t)(1 - \lambda \Delta t)(1 - \gamma_2 \Delta t) + \gamma_0 \Delta t P_0(0, t) + o(\Delta t), \tag{3}$$

$$P_0(j, t + \Delta t) = P_0(j, t)(1 - \lambda \frac{N - j}{N} \Delta t)(1 - \gamma_0 \Delta t)\left(1 - \frac{j}{N} \sigma \Delta t \right) + P_1(j + 1, \Delta t, t) \tag{4}$$

$$+ P_1(j - 1, t)\lambda \frac{N - j + 1}{N} \Delta t + P_1(j, t) \frac{(j - 1)\sigma}{N} \Delta t + P_2(j, t)\gamma_2 \Delta t + o(\Delta t),$$

$$P_1(j, z - \Delta t, t + \Delta t) = \tag{5}$$

$$\left[P_1(j, z, t) - P_1(j, \Delta t, t) \right] \left(1 - \lambda \frac{N - j}{N} \Delta t \right)(1 - \gamma_1 \Delta t)\left(1 - \frac{j - 1}{N} \sigma \Delta t \right)$$

$$+ P_0(j - 1, t)\lambda \frac{N - j + 1}{N} B(z) \Delta t + P_0(j, t) \frac{j\sigma}{N} B(z) \Delta t + o(\Delta t),$$

$$P_2(j, t + \Delta t) = P_2(j, t)\left(1 - \lambda \frac{N - j}{N} \Delta t \right)(1 - \gamma_2 \Delta t) + \gamma_0 \Delta t P_0(j, t) \tag{6}$$

$$+ P_2(j - 1, t)\lambda \frac{N - j + 1}{N} \Delta t + P_1(j, t)\gamma_1 \Delta t + o(\Delta t)$$

Assuming that system is operating in steady state, then from the above relations it is not difficult to get the system of equations for the stationary probability distribution $P_0(j), P_1(j, z), P_2(j)$, $j = 0, ..., N$ in a shorter form, namely we have

$$- \left[\lambda \frac{N - j}{N} + \sigma \frac{j}{N} + \gamma_0 \right] P_0(j) + \frac{\partial P_1(j + 1, 0)}{\partial z} + \lambda \frac{N - j + 1}{N} P_1(j - 1)$$

$$+ \frac{j - 1}{N} \sigma P_1(j) + \gamma_2 P_2(j) = 0 ,$$

$$\frac{\partial P_1(j, z)}{\partial z} - \frac{\partial P_1(j, 0)}{\partial z} - \left[\lambda \frac{N - j}{N} + \sigma \frac{j - 1}{N} + \gamma_1 \right] P_1(j, z) \tag{7}$$

$$+ \lambda \frac{N - j + 1}{N} B(z) P_0(j - 1) + \frac{j}{N} \sigma B(z) P_0(j) = 0 ,$$

$$- \left[\lambda \frac{N - j}{N} + \gamma_2 \right] P_2(j) + \lambda \frac{N - j + 1}{N} P_2(j - 1) + \gamma_0 P_0(j) + \gamma_1 P_1(j) = 0 .$$

where the meaningless probabilities are zero.

4 Numerical Algorithm for Finding the Probability Distribution of the System State and Performance Measures

4.1 Algorithmic Approach for the Steady State Distribution

In order to find the probability distribution of the number of customers in the system, we will solve system (7) numerically. We first obtain some very important equalities used later on.

Let us consider the second equation of system (7) for case $j = 1$, that is

$$
\frac{\partial P_1(1,z)}{\partial z} - \frac{\partial P_1(1,0)}{\partial z} - \left[\lambda \frac{N-1}{N} + \gamma_1 \right] P_1(1,z)
$$
$$
+ \lambda B(z) P_0(0) + \frac{\sigma}{N} B(z) P_0(1) = 0. \tag{8}
$$

The solution of this equation can be written in the form

$$
P_1(1,z) = e^{\left[\lambda \frac{N-1}{N} + \gamma_1 \right] z} \int_0^z e^{-\left[\lambda \frac{N-1}{N} + \gamma_1 \right] y} \left\{ \frac{\partial P_1(1,0)}{\partial z} \right.
$$
$$
\left. - \left[\lambda P_0(0) + \frac{\sigma}{N} P_0(1) \right] B(y) \right\} dy. \tag{9}
$$

Then by carrying out the limiting transition at $z \to \infty$ we obtain that the first factor of the right part of equality (9) in a limiting condition tends to infinity, therefore we can conclude that the second factor will be equal to zero, that is

$$
\int_0^\infty e^{-\left[\lambda \frac{N-1}{N} + \gamma_1 \right] y} \left\{ \frac{\partial P_1(1,0)}{\partial z} - \left[\lambda P_0(0) + \frac{\sigma}{N} P_0(1) \right] B(y) \right\} dy = 0,
$$

from which it is not difficult to obtain that

$$
\frac{\partial P_1(1,0)}{\partial z} = \left[\lambda P_0(0) + \frac{\sigma}{N} P_0(1) \right] B^* \left(\lambda \frac{N-1}{N} + \gamma_1 \right). \tag{10}
$$

We can perform similar transformations for the second equation of system (7) for the general case and, as a result we obtain

$$
\frac{\partial P_1(j,0)}{\partial z} = \left[\lambda \frac{N-j+1}{N} P_0(j-1) + \frac{j}{N} \sigma P_0(j) \right] B^* \left(\lambda \frac{N-j}{N} + \frac{j-1}{N} \sigma + \gamma_1 \right). \tag{11}
$$

In Eq. (8) let us execute the limiting transition at $z \to \infty$ then we get

$$
\frac{\partial P_1(1,0)}{\partial z} \left[\lambda \frac{N-1}{N} + \gamma_1 \right] P_1(1) = \lambda P_0(0) + \frac{\sigma}{N} P_0(1). \tag{12}
$$

Similarly, for the general case, that is for j, using the second equation of system (7), we can obtain

$$\frac{\partial P_1(j,0)}{\partial z} \left[\lambda \frac{N-j}{N} + \frac{j-1}{N}\sigma + \gamma_1 \right] P_1(j) = \lambda \frac{N-j+1}{N} P_0(j-1)$$
$$+ \frac{j}{N}\sigma P_0(j). \tag{13}$$

Let us write down the system of Eq. (7) for the case $j = 0$ then we get

$$\frac{\partial P_1(1,0)}{\partial z} = [\lambda + \gamma_0] P_0(0) - \gamma_2 P_2(0),$$
$$\tag{14}$$
$$- [\lambda + \gamma_2] P_2(0) + \gamma_0 P_0(0) = 0.$$

Hence combining equations of the system (7) for case $j = 1$ by using Eqs. (10) and (12) we obtain

$$\frac{\partial P_1(1,0)}{\partial z} = \left[\lambda P_0(0) + \frac{\sigma}{N} P_0(1) \right] B^* \left(\lambda \frac{N-1}{N} + \gamma_1 \right),$$

$$\frac{\partial P_1(1,0)}{\partial z} + \left[\lambda \frac{N-1}{N} + \gamma_1 \right] P_1(1) = \lambda P_0(0) + \frac{\sigma}{N} P_0(1),$$

$$\left[\lambda \frac{N-1}{N} + \gamma_2 \right] P_2(1) = \gamma_0 P_0(1) + \gamma_1 P_1(1) + \lambda P_2(0), \tag{15}$$

$$\frac{\partial P_1(2,0)}{\partial z} = \left[\lambda \frac{N-1}{N} + \gamma_0 + \frac{\sigma}{N} \right] P_0(1) - \gamma_2 P_2(1) - \lambda P_1(0).$$

Similarly, using the equations of system (7) and the equalities (11), (13) obtained earlier we can write down the extended system of equations for $2 \leq j \leq N$ as follows

$$\left[\lambda \frac{N-j}{N} + \sigma \frac{j}{N} + \gamma_0 \right] P_0(j) = \frac{\partial P_1(j+1,0)}{\partial z} + \lambda \frac{N-j+1}{N} P_1(j-1)$$
$$+ \frac{j-1}{N}\sigma P_1(j) + \gamma_2 P_2(j),$$

$$\left[\lambda \frac{N-j}{N} + \gamma_2 \right] P_2(j) = \lambda \frac{N-j+1}{N} P_2(j-1) + \gamma_0 P_0(j) + \gamma_1 P_1(j),$$

$$\frac{\partial P_1(j,0)}{\partial z} = \left[\lambda \frac{N-j+1}{N} P_0(j-1) + \frac{j}{N}\sigma P_0(j) \right] B^* \left(\lambda \frac{N-j}{N} + \frac{j-1}{N}\sigma + \gamma_1 \right),$$

$$\frac{\partial P_1(j,0)}{\partial z} + \left[\lambda \frac{N-j}{N} + \frac{j-1}{N}\sigma + \gamma_1 \right] P_1(j) = \lambda \frac{N-j+1}{N} P_0(j-1) + + \frac{j}{N}\sigma P_0(j). \tag{16}$$

The joint stationary probability distribution $\Pi_k(j)$ of the server's state and the number of customers in the system is the normalized solution of systems (14)–(16). Starting with $P_0(0) = 1$ using the algorithmic steps after normalization we obtain the probability distribution $\Pi_k(j)$. Thus our proposed algorithmic solution consists of the following steps

1. Put $P_0(0) = 1$.
2. From the second equation of system (14) we get

$$P_2(0) = \frac{\gamma_0}{\lambda + \gamma_2} P_0(0).$$

3. From the first equations of systems (14), (15) we obtain

$$P_0(1) = \frac{N}{\sigma B^* \left(\lambda \frac{N-1}{N} + \gamma_1\right)} \{-\gamma_2 P_2(0) +$$

$$+ \left(\lambda \left[1 - B^* \left(\lambda \frac{N-1}{N} + \gamma_1\right)\right] + \gamma_0\right) P_0(0)\}.$$

4. From the first equation of system (14) and second equation of system (15) we have

$$P_1(1) = \frac{1}{\lambda \frac{N-1}{N} + \gamma_1} \left\{\frac{\sigma}{N} P_0(1) - \gamma_0 P_0(0) + \gamma_2 P_2(0)\right\}.$$

5. From the third equation of system (15) we determine

$$P_2(1) = \frac{1}{\lambda \frac{N-1}{N} + \gamma_2} \{\gamma_0 P_0(1) + \gamma_1 P_1(1) + \lambda P_2(0)\}.$$

6. For general case, that is for $2 \leq j \leq N$, from system (16) it is not difficult to obtain formulas for calculating $P_k(j)$ in the form

$$P_0(j) = \frac{1}{j\sigma B^* \left(\lambda \frac{N-j}{N} + \frac{j-1}{N}\sigma + \gamma_1\right)} \left\{ -\lambda(N - j + 2)P_1(j - 2)\right.$$

$$+ \left(\lambda(N - j + 1)\left[1 - B^* \left(\lambda \frac{N-j}{N} + \frac{j-1}{N}\sigma + \gamma_1\right)\right] + (j - 1)\sigma\right.$$

$$\left. + \gamma_0 N\right) P_0(j - 1) - (j - 2)\sigma P_1(j - 1) - \gamma_2 N P_2(j - 1)\right\},$$

$$P_1(j) = \frac{1}{\lambda(N - j) + \sigma(j - 1) + \gamma_1 N} \{-[\sigma(j - 1) + \gamma_0 N] P_0(j - 1)$$

$$+ j\sigma P_0(j) + \lambda(N - j + 2)P_1(j - 2) + (j - 2)\sigma P_1(j - 1) + \gamma_2 N P_2(j - 1)\},$$

$$P_2(j) = \frac{1}{\lambda(N - j) + \gamma_2 N} [\gamma_0 N P_0(j) + \gamma_1 N P_1(j) + \lambda(N - j + 1)P_2(j - 1)],$$

$P_1(0) = 0$ by convention.

7. The solution obtained in the previous steps does not satisfy the normalization condition. For the normalizing constant let us calculate the sum

$$d = \sum_{j=0}^{N} [P_0(j) + P_1(j) + P_2(j)],$$

where $P_k(j)$ is the quantities obtained in the previous steps.

8. To calculate the two-dimensional probability distribution $\Pi_k(j)$ carry on the normalization, that is

$$\Pi_k(j) = \frac{1}{d} P_k(j), \quad k = 0, 1, 2, \quad j = 0, 1, ..., N.$$

9. The marginal distribution of the number of customers in the system $\Pi(j)$, and the server's state Π_k, respectively can be calculated as follows

$$\Pi(j) = \Pi_0(j) + \Pi_1(j) + \Pi_2(j), j = 0, ..., N, \qquad \Pi_k = \sum_{j=0}^{N} \Pi_k(j), k = 0, 1, 2.$$

4.2 Performance Measures

To show the effect of the input parameters on the operation of the system let us define the most important characteristics which can be determine directly from the steady state probabilities. Unfortunately only mean values are obtained but our intention is to continue the research to get the distribution of the response and waiting time of the customers, distribution of the number of retrials just to mention some.

– *Mean number of customers in the system* $\overline{Q}$

$$\overline{Q} = \sum_{j=0}^{N} j \Pi(j), \tag{17}$$

– *Mean arrival rate* $\overline{\lambda}$

$$\overline{\lambda} = (N - \overline{Q}) \frac{\lambda}{N}, \tag{18}$$

– *Mean response time* $\overline{T}$ can be obtained by the Little-formula

$$\overline{T} = \frac{\overline{Q}}{\overline{\lambda}}, \tag{19}$$

– *Mean number of customers in the orbit* $\overline{O}$

$$\overline{O} = \overline{Q} - \Pi_1, \tag{20}$$

– *Mean waiting time in the orbit* $\overline{W}$

$$\overline{W} = \frac{\overline{O}}{\overline{\lambda}}, \tag{21}$$

– *Mean total service time* $E(T_S)$

$$E(T_S) = \overline{T} - \overline{W}, \tag{22}$$

– *Probability of collision of a customer arriving from the source (Primary Customer)* P_{PC}

$$P_{PC} = \frac{\sum_{k=1}^{N}(N-k)\frac{\lambda}{N}\Pi_1(k)}{\sum_{j=0}^{N}(N-j)\frac{\lambda}{N}(\Pi_0(j) + \Pi_1(j))}, \tag{23}$$

– *Probability of collision of a customer arriving from the orbit (Secondary Customer)* P_{SC}

$$P_{SC} = \frac{\sum_{j=1}^{N}(j-1)\frac{\sigma}{N}\Pi_1(j)}{\sum_{j=0}^{N}j\frac{\sigma}{N}\Pi_0(j) + \sum_{j=1}^{N}(j-1)\frac{\sigma}{N}\Pi_1(j)}, \tag{24}$$

– *Probability of collision* P_C

$$P_C = \tag{25}$$
$$\frac{\sum_{j=1}^{N}\left[(N-j)\frac{\lambda}{N} + (j-1)\frac{\sigma}{N}\right]\Pi_1(j)}{\sum_{j=0}^{N}\left[(N-j)\frac{\lambda}{N} + j\frac{\sigma}{N}\right]\Pi_0(j) + \sum_{j=0}^{N}(N-j)\frac{\lambda}{N}\Pi_1(j) + \sum_{j=1}^{N}(j-1)\frac{\sigma}{N}\Pi_1(j)},$$

– *Probability of retrial* P_R

$$P_R = \tag{26}$$
$$\frac{\sum_{j=0}^{N}(N-j)\frac{\lambda}{N}\left(\Pi_1(j) + \Pi_2(j)\right) + \sum_{j=1}^{N}(j-1)\frac{\sigma}{N}\Pi_1(j) + \sum_{j=1}^{N}j\frac{\sigma}{N}\Pi_2(j) + \gamma_1\Pi_1}{\sum_{j=0}^{N}(N-j)\frac{\lambda}{N}\Pi(j) + \sum_{j=1}^{N}(j-1)\frac{\sigma}{N}\Pi_1(j) + \sum_{j=1}^{N}j\frac{\sigma}{N}\left(\Pi_0(j) + \Pi_2(j)\right) + \gamma_1\Pi_1}.$$

5   Numerical Examples

The proposed algorithm has been tested by the numerical results of [13] where the service time is exponentially distributed and the server is reliable, [14] in which the service time is generally distributed and the server is reliable, and [19] where the service time is exponentially distributed and the server is subject to breakdowns and repairs. Finally, in [24] the present model has been analyzed by means of stochastic simulation.

Table 1. Numerical values of model parameters

Case studies								
No.	N	λ	σ	α	β	γ_0	γ_1	γ_2
Fig. 1	30	2	10	0.5	0.5	0.1	0.2	1
Fig. 2	30	2	10	0.5, 1, 2	0.5, 1, 2	0.1	0.2	1
Fig. 3	100	0.01..10	1	0.5, 1, 2	0.5, 1, 2	0.1	0.1	1
Fig. 4	100	0.01..0.2	1	0.5, 1, 2	0.5, 1, 2	0.1	0.1	1
Fig. 5	100	0.01..0.2	1	0.5, 1, 2	0.5, 1, 2	0.1	0.1	1
Fig. 6	100	0.01..0.25	0.1	0.5	0.5	0.5	0.5	1
Fig. 7	100	0.01..0.15	0.1	1	1	0.5	0.5	1
Fig. 8	100	1	1	1	1	0.05..1	0.05..1	2

In our examples we will choose gamma distributed service time S with a shape parameter α and scale parameter β, with Laplace-Stieltjes transform $B^*(\delta)$ of the form

$$B^*(\delta) = \left(1 + \frac{\delta}{\beta}\right)^{-\alpha},$$

in the case when $\alpha = \beta$, that is when the average service time will be equal to unit.

It can be shown that

$$\mathsf{E}(S) = \frac{\alpha}{\beta}, \quad Var(S) = \frac{\alpha}{\beta^2}, \quad V_S^2 = \frac{1}{\alpha},$$

where V_S^2 denotes the squared coefficient of variation of S. This distribution allows us to show the effect of the distribution on the main performance measures, because dealing with the same mean we can see the impact of the variance, too.

From the system probabilities the well known system characteristics are calculated. The most interesting performance characteristics obtained by these tools are graphically presented in this section. On the Figures the lines represent different working assumptions or cases (e.g. different parameters of the distribution of the service time). The input parameters are listed in Table 1.

Figures 1 and 2 display distributions of the steady-state system probabilities where values of x-axes represent the numbers of customers staying in the system, i.e. the states of the system. On the other Figures the effects of a running parameter are shown. In Table 1 a parameter running from n to m is denoted by $n..m$. If the effect of an other parameter is also considered, a separate curve is presented for each values of that parameter, and these values are listed in Table 1, as well.

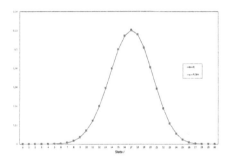

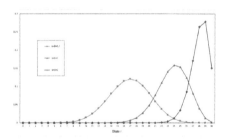

Fig. 1. Comparison of numerical and simulation results

Fig. 2. Comparison the distributions for different α and β parameters

On Fig. 1 the numerical and simulation results for the steady-state probabilities are compared to each other. As we can see the values are very close to each other, so the two curves are identical illustrating that the numerical and simulation procedures operate correctly.

Figure 2 shows the effect of the different values of the shape α, and scale/rate β parameters. The curves represent the cases of $\alpha = \beta$ with values 0.5, 1, 2, respectively. Thus, the expected values of the service times are equal but the variances are different. For higher values of α, and β parameters, the standard deviation and the coefficient of variance will be smaller. For small values of parameters, i.e. high value of standard deviation, the distribution is more tailed than for higher values of α, and β.

On Fig. 3 the mean waiting time can be seen in different cases. For Case 1, 2, and 3 the values of α, and β are 0.5, 1, 2, respectively ($\alpha = \beta$ for all cases). A maximum point can be observed for this performance measure, as the arrival rate increases. In retrial systems this maximum feature is an unexpected and quite unique phenomenon. Many times there exists a combination of parameters, for which the response time, waiting time or queue length have a maximum point, see for example in [5, 6, 22, 25].

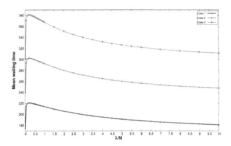

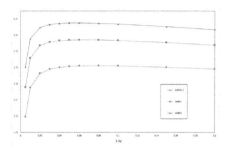

Fig. 3. Mean waiting time $\overline{W}$ vs arrival intensity from the source

Fig. 4. Probability of collision for primary customers P_{PC} vs arrival intensity from the source

The mentioned maximum feature can be observed on Fig. 4, as well. Here again the arrival rate is the running parameter. The different lines correspond to the different α, and β parameters as on Fig. 2 and 3. As mentioned above, this maximum point can be achieved only a specific set of parameters. With the parameters of Fig. 4, the probabilities of retrial are computed and displayed on Fig. 5. Here there are no maximum points for the probabilities. The P_R values increase with the increasing arrival rate. But when some parameters (retrial rate, failure rates) are modified, the following results can be obtained: for $\alpha = \beta = 0.5$ the Fig. 6, and for $\alpha = \beta = 1$ the Fig. 7. The maximum feature and the decreasing trend of the probabilities can be seen on both Figures. A similar Figure could be generated for $\alpha = \beta = 2$ case, too.

Finally, Fig. 8 displays the result of the effect of modification of failure rates. γ_0, and γ_1 are modified parallel, the same way, so for each point $\gamma_0 = \gamma_1$. The range of the parameters can be found in Table 1. The P_{PC}, P_C, and P_R probabilities are displayed, but only two lines are in the Figure. The values of P_{PC}, and P_C are so close (not identical) to each other, that only one line can be seen for these two parameters. The results show what is expected, that is as the failure rate increasing more and more requests are sent to the orbit causing retrials, but the chance of collision is decreasing since the server is broken.

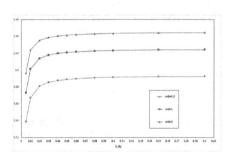

Fig. 5. Probability of retrial P_R vs arrival intensity from the source

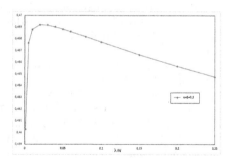

Fig. 6. Probability of retrial P_R vs arrival intensity from the source

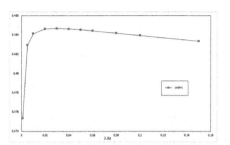

Fig. 7. Probability of retrial P_R vs arrival intensity from the source

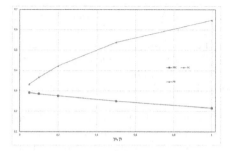

Fig. 8. Values of probabilities P_{PC}, P_C, and P_R vs failure rate

6 Conclusion

In this paper finite source $M/GI/1$ retrial queuing systems with collisions of the customers and an unreliable server were considered. Applying the method of residual service times as supplementary variable the steady state Kolmogorov equations were solved by means of a new algorithmic approach. The main performance measures were defined and several numerical sample examples illustrated the effect of the input parameters on these characteristics. In the near future, for the considered system we plan to investigate the distribution of the number of transitions of the customer into the orbit, distribution of the sojourn time of the customer in the system and other system performance descriptors.

Acknowledgments. The work/publication of J. Sztrik is supported by the EFOP-3.6.1-16-2016-00022 project. The project is co-financed by the European Union and the European Social Fund.

References

1. Ali, A.A., Wei, S.: Modeling of coupled collision and congestion in finite source wireless access systems. In: 2015 IEEE Conference on Wireless Communications and Networking Conference (WCNC), pp. 1113–1118. IEEE (2015)
2. Almási, B., Roszik, J., Sztrik, J.: Homogeneous finite-source retrial queues with server subject to breakdowns and repairs. Math. Comput. Modell. **42**(5–6), 673–682 (2005)
3. Artalejo, J., Corral, A.G.: Retrial Queueing Systems: A Computational Approach. Springer, Heidelberg (2008)
4. Cao, Y., Khosla, D., Chen, Y., Huber, D.J.: System and method for real-time collision detection. US Patent 9,934,437, 3 Apr 2018
5. Dragieva, V., Phung-Duc, T.: Two-way communication M/M/1//N retrial queue. In: Thomas, N., Forshaw, M. (eds.) ASMTA 2017. LNCS, vol. 10378, pp. 81–94. Springer, Cham (2017). https://doi.org/10.1007/978-3-319-61428-1_6
6. Falin, G., Artalejo, J.: A finite source retrial queue. Eur. J. Oper. Res. **108**, 409–424 (1998)
7. Gómez-Corral, A., Phung-Duc, T.: Retrial queues and related models. Ann. Oper. Res. **247**(1), 1–2 (2016)
8. Jinsoo, A., Kim, Y., Kwak, J., Son, J.: Wireless communication method for multiuser transmission scheduling, and wireless communication terminal using same. US Patent App. 15/736,968, 6 Sept 2018
9. Kim, J.S.: Retrial queueing system with collision and impatience. Commun. Korean Math. Soc. **25**(4), 647–653 (2010)
10. Kim, J., Kim, B.: A survey of retrial queueing systems. Ann. Oper. Res. **247**(1), 3–36 (2016)
11. Kumar, B.K., Vijayalakshmi, G., Krishnamoorthy, A., Basha, S.S.: A single server feedback retrial queue with collisions. Comput. Oper. Res. **37**(7), 1247–1255 (2010)
12. Kvach, A., Nazarov, A.: Sojourn time analysis of finite source Markov retrial queuing system with collision. In: Dudin, A., Nazarov, A., Yakupov, R. (eds.) ITMM 2015. CCIS, vol. 564, pp. 64–72. Springer, Cham (2015). https://doi.org/10.1007/978-3-319-25861-4_6

13. Kvach, A.: Numerical research of a Markov closed retrial queueing system without collisions and with the collision of the customers. In: Proceedings of Tomsk State University. A Series Of Physics and Mathematics. Tomsk. Materials of the II All-Russian Scientific Conference, vol. 295, pp. 105–112. TSU Publishing House (2014). (In Russian)

14. Kvach, A., Nazarov, A.: Numerical research of a closed retrial queueing system M/GI/1//N with collision of the customers. In: Proceedings of Tomsk State University. A Series of Physics and Mathematics. Tomsk. Materials of the III All-Russian Scientific Conference, vol. 297, pp. 65–70. TSU Publishing House (2015). (In Russian)

15. Kvach, A., Nazarov, A.: The research of a closed RQ-system M/GI/1//N with collision of the customers in the condition of an unlimited increasing number of sources. In: Probability Theory, Random Processes, Mathematical Statistics and Applications: Materials of the International Scientific Conference Devoted to the 80th Anniversary of Professor Gennady Medvedev, Doctor of Physical and Mathematical Sciences, pp. 65–70 (2015). (In Russian)

16. Kwak, B.J., Rhee, J.K., Kim, J., Kyounghye, K.: Random access method and terminal supporting the same. US Patent 9,954,754, 24 Apr 2018

17. Nazarov, A., Sztrik, J., Kvach, A., Bérczes, T.: Asymptotic analysis of finite-source M/M/1 retrial queueing system with collisions and server subject to breakdowns and repairs. Ann. Oper. Res. **277**, 1–17 (2018)

18. Nazarov, A., Kvach, A., Yampolsky, V.: Asymptotic analysis of closed Markov retrial queuing system with collision. In: Dudin, A., Nazarov, A., Yakupov, R., Gortsev, A. (eds.) ITMM 2014. CCIS, vol. 487, pp. 334–341. Springer, Cham (2014). https://doi.org/10.1007/978-3-319-13671-4_38

19. Nazarov, A., Sztrik, J., Kvach, A.: Comparative analysis of methods of residual and elapsed service time in the study of the closed retrial queuing system M/GI/1//N with collision of the customers and unreliable server. In: Dudin, A., Nazarov, A., Kirpichnikov, A. (eds.) ITMM 2017. CCIS, vol. 800, pp. 97–110. Springer, Cham (2017). https://doi.org/10.1007/978-3-319-68069-9_8

20. Nazarov, A., Moiseeva S.P.: Methods of asymptotic analysis in queueing theory. NTL Publishing House of Tomsk University (2006). (In Russian)

21. Peng, Y., Liu, Z., Wu, J.: An M/G/1 retrial G-queue with preemptive resume priority and collisions subject to the server breakdowns and delayed repairs. J. Appl. Math. Comput. **44**(1–2), 187–213 (2014)

22. Roszik, J.: Homogeneous finite-source retrial queues with server and sources subject to breakdowns and repairs. Ann. Univ. Sci. Budap. Rolando Eötvös, Sect. Comput. **23**, 213–227 (2004)

23. Sztrik, J., Almási, B., Roszik, J.: Heterogeneous finite-source retrial queues with server subject to breakdowns and repairs. J. Math. Sci. **132**, 677–685 (2006)

24. Tóth, Á., Bérczes, T., Sztrik, J., Kvach, A.: Simulation of finite-source retrial queueing systems with collisions and non-reliable server. In: Vishnevskiy, V.M., Samouylov, K.E., Kozyrev, D.V. (eds.) DCCN 2017. CCIS, vol. 700, pp. 146–158. Springer, Cham (2017). https://doi.org/10.1007/978-3-319-66836-9_13

25. Wang, J., Zhao, L., Zhang, F.: Analysis of the finite source retrial queues with server breakdowns and repairs. J. Ind. Manag. Optim. **7**(3), 655–676 (2011)

26. Zhang, F., Wang, J.: Performance analysis of the retrial queues with finite number of sources and service interruptions. J. Korean Stat. Soc. **42**(1), 117–131 (2013)

Available Throughput of Transport Connection with Selective Repeat Mode in the Loaded Data Transmission Path

Pavel Mikheev, Anastasiya Pichugina, Pavel Pristupa, and Sergey Suschenko[✉]

National Research Tomsk State University, Lenina str., 36, 634050 Tomsk, Russia
ssp.inf.tsu@gmail.com

Abstract. An indicator model of transport connection with multiple subscribers competing for the throughput is proposed for selective failure mode. The indicator of competition is the queue of competitive data flows in transit nodes of transport connection with specified parameters. The analysis of available throughput in different conditions when competing is carried out.

Keywords: Transport protocol · Selective failure mode · Competition for resources · Throughput · Protocol parameters · Round-trip delay · Mathematical model · Markov chain

1 Introduction

The most important operational characteristic of a subscriber connection controlled by a computer network transport protocol is its throughput. This indicator is largely determined by the intensity of external flows relative to this connection, which has at least a part of common route with it. The main indicator of "external" load on the path in which the studied transport connection is laid is the size of the queues ahead of protocol data blocks of analyzed connection in transit nodes. Monitoring such an indicator allows us to evaluate the distribution of queue lengths in transit nodes from external network streams in regard to analyzed connection and to use when calculating the operational characteristics of the connection and the choice of protocol parameters for the communication time between a given pair of subscribers. Known models of asynchronous control procedures of a separate data link and transport protocol [1–7] do not allow us to take into account the load on shared network resources which is provided by the neighboring with other virtual connections, aggregated on different sections of the path in separate links of the route of a given subscriber connection, and present itself as "External" queues in transit nodes. The study of data transmission process in a loaded transport connection [8,9] was carried

The research supported by Russian Ministry of Education and Science (project no. 2.4218.2017/4.6).

A. Dudin et al. (Eds.): ITMM 2019, CCIS 1109, pp. 28–37, 2019.
https://doi.org/10.1007/978-3-030-33388-1_3

out with significant restrictions on the values of the protocol parameters and characteristics of data transmission path. The paper proposes a mathematical model of transport connection controlled by transport protocol in the selective failure mode, which takes into account, apart from the distortion factor in the forward and reverse data transmission paths and retransmission mechanisms, caused by distortions and the timeout of non-reception of response from the recipient of the information flow, and also non-zero queues lengths from "external" inter-subscriber connections for end-to-end timeout durations with interval and below restrictions.

2 Indicator Data Transmission Path Model

Let us consider the exchange between subscribers connected by a multi-link data path. Assume that the following assumptions are true: The nodes of the path are connected by duplex communication channels having the same speed in both directions. The length of the tract, expressed in the number of hop is equal to D_n. The return channel, on which confirmation is delivered to the sender about the validity of the reception of sequence of data segments, has a length D_o. The probabilities of segment distortion in the communication channel are specified for the forward $R_n(d)$, $d = \overline{1, D_n}$ and reverse—$R_o(d)$, $d = \overline{1, D_o}$ the transmission directions of each segment of the hop is given. Then the reliability of transmission of data segments along the path from the source to the addressee and back will be $F_n = \prod_{d=1}^{D_n}(1 - R_n(d))$; $F_o = \prod_{d=1}^{D_o}(1 - R_o(d))$. The processing time of segments in path nodes is the same. Interacting subscribers have an unlimited flow of segments for transmission, and the exchange is carried out by segments of the same length. The recipient's confirmation of the validity of received data is transferred in the segments of the counter flow. We believe that the retransmission of segments is organized in accordance with the selective rejection procedure [1]. We also assume that the loss of segments due to the absence of buffer memory at the nodes of the path does not occur. Probability function is given b_n, $n = \overline{0, N}$ that each segment from the flow of analyzed connection in transit nodes will meet a queue of size $n \leq N$, where N is the maximum queue size determined by the capacity of buffer pools of transit nodes. We will call cycle time t necessary for output of a segment to a line. The cycle is determined by the sum of the segment output time to the line, the signal distribution time in the communication channel and processing time of the segment by the receiving node. The timeout S, expressed in duration t, runs before the start of transmission of the first segment of sequence and is fixed for all segments within the window width. We assume that the size of the controlled protocol window is determined by the value of W, and $S > W$—sets the duration of the timeout for waiting for confirmation of validity of data delivery. It is obvious that the sum of lengths of forward and reverse data paths $D = D_n + D_o$ can be interpreted as the duration of the round-trip delay the unloaded path, expressed in cycles t. After next segment is transferred, protocol copies it to the queue of transmitted but not confirmed data and starts a timeout. As soon as the queue size becomes equal to the width of the window W, the control protocol suspends transmission while waiting for the acknowledgement or the expiration of timeout S for confirmation. Upon receiving the confirmation, the segments that reached the addressee without distortion are removed from the queue. When S timeout expires, the corresponding segment is retransmitted and timeout starts again. Then the time of confirmation by the sender of end-to-end acknowledgement is distributed according to the geometric law with the parameter F_o and the duration of the sampling cycle t. The operation of virtual connection controlled by transport protocol in a loaded multilink data transmission path with segment queues before sending data or confirmations can be described by a markovized process of the dynamics of a queue of transmitted but not confirmed segments in which the queue size ahead of the forward or reverse data flow of the test connection is additional variable of Markov process. In the state of Markov chain (i, n) the source sent a sequence of size $i - n$ segments, which in the process of transfer in one of the links met a queue with length of n segments. The coordinates $i = \overline{0, W + n}$, $n = \overline{0, N}$ of the states of Markov chain correspond to the number of segments transmitted but not confirmed by the recipient and the time from the beginning of the transmission of the sequence, while the values $i = \overline{W + n + 1, S - 1}$, $n = \overline{0, N}$ correspond to the time during which the sender is not active and is waiting for confirmation of a receipt of the validity of transmitted sequence from the W segments. We define by $P(i, n)$, $i = \overline{0, S - 1}$, $n = \overline{0, N}$, the probabilities of the states of the Markov chain. Then the sequence of transmitted, but not confirmed data segments of considered virtual connection with a zero-length queue grows to the state of a Markov chain with coordinates $(D - 1, 0)$ with probability b_0. Further increase in size of this sequence occurs with probability $b_0(1 - F_o)$. In states (i, n), $i = \overline{D - 1 + n, S - 1}$, $n = \overline{0, N}$, it is possible for the sender to receive acknowledgement and depending on the delivery results, the sender transmits new segments (with a positive acknowledgement), or repeatedly—distorted. Since transmitted sequence of segments of the virtual connection under study may encounter a queue of non-zero length at any moment of transferring process (on the path of the sequence to the addressee or when transferring confirmation to the sender of information flow), the transition from state $(i, 0)$, $i = \overline{0, S - 2}$ to state (i, n), $i = \overline{0, S - 2}$, $n = \overline{1, N}$ occurs with probability b_n.

3   State Probabilities of Markov Chain

Let us define π_{in}^{jm} the transition probabilities of Markov chain, where (i, n) are the coordinates of the initial one, and (j, m) are the altered states of the chain. Then the dynamics of the process of transmitting information flow in the selective failure mode in loaded data transmission path can be set with the following values of transition probabilities:

$$
\pi_{in}^{jm} = \begin{cases}
b_0, & j = i+1, m = 0; \ i = \overline{0, D-2}, n = 0; \\
b_0(1 - F_o), & j = i+1, m = 0; \ i = \overline{D-1, S-2}, n = 0; \\
b_m, & j = i, m = \overline{1, N}; \ i = \overline{0, S-2}, n = 0; \\
b_0 F_o, & j = D-1, m = 0; \ i = \overline{D-1, W-1}, n = 0; \\
b_0 F_o, & j = W+D-2-i, m = 0; \ i = \overline{W, W+D-2}, n = 0; \\
b_0 F_o, & j = 0, m = 0; \ i = \overline{W+D-1, S-2}, n = 0; \\
1, & j = 0, m = 0; \ i = S-1, n = \overline{0, N}; \\
1, & j = i+1, m = n; \ i = \overline{0, D-2+n}, n = \overline{1, N}; \\
1 - F_o, & j = i+1, m = n; \ i = \overline{D-1+n, S-2}, n = \overline{1, N}; \\
F_o, & j = D-1, m = 0; \ i = \overline{D-1+n, W-1+n}, n = \overline{1, N}; \\
F_o, & j = W+n+D-2-i, m = 0; \\
& \quad i = \overline{W+n, W+n+D-2}, n = \overline{1, N}; \\
F_o, & j = 0, m = 0; \ i = \overline{W+n+D-1, S-2}, n = \overline{1, N}.
\end{cases}
\tag{1}
$$

The variety of solutions of system of equilibrium equations for Markov chain state probabilities is determined by relations between the protocol parameters W, S, the total path length D and the maximum length of queues N. Since time-out length must exceed the window width, and be no shorter than the round-trip delay ($S \geq D$), exceeding the waiting time in queues from protocol data blocks of corresponding traffic prior to transmission in transit nodes a wide variety of solutions for different areas of change in the values of the Protocol parameters and queue lengths are distinguished. Analysis of the transmission process in analytical form for arbitrary values of protocol parameters in the conditions of competition for network resources is possible only under the assumption that the "external" queues have a non-zero length ($b_0 = 0$).

4   Analysis of Transmission Process with Lower Restrictions on Duration of Timeout

Consider the transfer process for protocol parameters related to the total path length and the maximum queue size of the form inequalities $W \geq D$, $S \geq D + W + N - 1$. The system of equilibrium equations is written as follows:

$$
P(0,0) = F_o \sum_{n=1}^{N} \sum_{i=D+W+n-2}^{S-2} P(i,n) + \sum_{n=0}^{N} P(S-1, n);
\tag{2}
$$

$$
P(i,0) = F_o \sum_{n=1}^{N} P(D+W+n-2-i, n), \ i = \overline{1, D-2};
\tag{3}
$$

$$P(D-1,0) = F_o \sum_{n=1}^{N} \sum_{i=D+n-1}^{W+n-1} P(i,0); \qquad (4)$$

$$P(0,n) = b_n P(0,0), \ n = \overline{1,N}; \qquad (5)$$

$$P(i,n) = b_n P(i,0) + P(i-1,0), \ i = \overline{1,D-1}, \ n = \overline{1,N}; \qquad (6)$$

$$P(i,n) = P(i-1,n), \ i = \overline{D, D+n-1}, \ n = \overline{1,N}; \qquad (7)$$

$$P(i,n) = (1 - F_o)P(i-1,n), \ i = \overline{D+n, S-1}, \ n = \overline{1,N}. \qquad (8)$$

Let us find a solution to this system of equations. According to Eq. (7), we obtain: $P(i,n) = P(D-1,n), \ i = \overline{D, D+n-1}, \ n = \overline{1,N}$, and from (8) we have: $P(i,n) = (1 - F_o)^{i-D-n+1} P(D-1,n), \ i = \overline{D+n, S-1}, \ n = \overline{1,N}$. Taking into account these relations from (3), (4) for $i = \overline{1, D-1}$ we find:

$$P(i,0) = F_o (1 - F_o)^W \sum_{m=1}^{N} P(D-1,m), \ i = \overline{1, D-2},$$

$$P(D-1,0) = \left(1 - (1 - F_o)^{W-D+1}\right) \sum_{m=1}^{N} P(D-1,m).$$

Substituting the relations found in (6) with (5) taken into account, we obtain

$$P(i,n) = b_n \left[P(0,0) + (1 - F_o)^{W-i-1} \left(1 - (1 - F_o)^i\right) \sum_{m=1}^{N} P(D-1,m) \right],$$

$$i = \overline{1, D-2}, \ n = \overline{1,N},$$

$$P(D-1,n) = b_n \left[P(0,0) + \left(1 - (1 - F_o)^{W-1}\right) \sum_{m=1}^{N} P(D-1,m) \right], \ n = \overline{1,N}.$$

Hence, we successively express for arbitrary $n = \overline{1,N}$ through $P(D-1,n)$ the probabilities of states $P(D-1,m) \ m = \overline{n+1, N}$:

$$P(D-1,n) = \frac{b_n}{1 - \left(1 - (1 - F_o)^{W-1}\right) \sum_{m=1}^{N} b_m} \left[P(0,0) + \right.$$

$$\left. + \left(1 - (1 - F_o)^{W-1}\right) \sum_{m=n+1}^{N} P(D-1,m) \right], \ n = \overline{1,N}. \quad (9)$$

When $n = N$ from here we come to: $P(D-1,N) = \frac{b_N P(0,0)}{1 - F_o^{W-1}}$. Substituting this relation into (9) for values of n from $N-1$ to 1, we recursively find functional expressions for the probabilities of states $P(D-1,n)$ through $P(0,0)$: $P(D-1,n) = \frac{b_n P(0,0)}{1 - F_o^{W-1}}, \ n = \overline{1,N}$. Hence, from previously found relations, we finally obtain the probability distribution of states of Markov chain

$$P(i,0) = F_o \frac{P(0,0)}{(1-F_o)^i}, \quad i = \overline{1, D-2};$$

$$P(D-1,0) = \frac{\left(1 - (1-F_o)^{W-D+1}\right) P(0,0)}{(1-F_o)^{W-1}};$$

$$P(i,n) = \frac{b_n P(0,0)}{(1-F_o)^i}, \quad i = \overline{0, D-2}, \ n = \overline{1, N};$$

$$P(i,n) = \frac{b_n P(0,0)}{(1-F_o)^{W-1}}, \quad i = \overline{D-1, D+n-1}, \ n = \overline{1, N};$$

$$P(i,n) = \frac{b_n (1-F_o)^{i-D-n+1} P(0,0)}{(1-F_o)^{W-1}}, \quad i = \overline{D+n-1, S-1}, \ n = \overline{1, N},$$

and from the normalization condition we find the probability of the initial state

$$P(0,0) = \frac{F_o(1-F_o)^{W-1}}{1 + F_o(1+\bar{N}) + (1-F_o)^{W-D+1} - (1-F_o)^W - \sum_{m=1}^{N} b_m(1-F_o)^{S-D+1-m}},$$

where $\bar{N} = \sum_{n=1}^{N} n b_n$.

Let us consider the solution found in a number of special cases. For deterministic return path ($F_o = 1$), the space of significant states (i,n) forms a plane of an isosceles along coordinates i and n triangle $i = \overline{D-1, D-1+n}, \ n = \overline{0, N}$:

$$P(D-1,0) = \frac{1}{2+\bar{N}};$$

$$P(i,0) = \frac{b_n}{2+\bar{N}}, \quad i = \overline{D-1, D-1+n}, \ n = \overline{1, N}.$$

With an unlimited width of the window ($w = \infty$) the states (i,n), $i = \overline{0, D-2}$, $n = \overline{0, N}$ are non-recurrent ($P(i,n) = 0$) and probabilities of the state of Markov chain take the form

$$P(D-1,0) = \frac{F_o}{1 + F_o(1+\bar{N})};$$

$$P(i,n) = \frac{b_n F_o}{1 + F_o(1+\bar{N})}, \quad i = \overline{D-1, D-1+n}, \ n = \overline{1, N};$$

$$P(i,n) = \frac{b_n(1-F_o)^{i-D-n+1} P(0,0)}{1 + F_o(1+\bar{N})}, \quad i \geq D-1+n, \ n = \overline{1, N}.$$

Let us consider the process of data transfer in conditions where the width of the window does not exceed the duration of the round-trip delay ($W \leq D$), and the size of the timeout is limited from below ($S \geq D + W + N - 1$). According to (1) the system of equilibrium equations given above will change as follows. Equations (2), (5) and (8) will remain unchanged, (3)—true for $i = \overline{1, W-1}$, Eq. (4) will take the form $P(D-1,0) = 0$, Eq. (6)—true for $i = \overline{1, W-1}$,

$n = \overline{1, N}$, Eq. (7)—for $i = \overline{W, D - 1 + n}$, $n = \overline{1, N}$. The solution of the system of equilibrium equations is as follows

$$P(i,0) = \frac{F_o P(0,0)}{(1 - F_o)^i}, \ i = \overline{1, W - 1};$$

$$P(i,n) = \frac{b_n P(0,0)}{(1 - F_o)^i}, \ i = \overline{0, W - 1}, \ n = \overline{1, N};$$

$$P(i,n) = \frac{b_n P(0,0)}{(1 - F_o)^{W-1}}, \ i = \overline{W - 1, D + n - 1}, \ n = \overline{1, N};$$

$$P(i,n) = \frac{b_n (1 - F_o)^{i-D-n+1} P(0,0)}{(1 - F_o)^{W-1}}, \ i = \overline{D + n - 1, S - 1}, \ n = \overline{1, N},$$

and from the normalization condition we obtain the probability of the initial state

$$P(0,0) = \frac{F_o(1 - F_o)^{W-1}}{2 + F_o(D - W + \bar{N}) + (1 - F_o)^W - (1 - F_o)^{S-D+1} - \sum\limits_{m=1}^{N} \frac{b_m}{(1-F_o)^m}}.$$

If $F_o = 1$ only states are significant

$$P(W - 1, 0) = \frac{F_o^2}{D - W + N + 2};$$

$$P(i,0) = \frac{b_n F_o^2}{D - W + N + 2}, \ i = \overline{W - 1, D - 1 + n}, \ n = \overline{1, N}.$$

The unlimited duration of timeout leads to probability of the initial state of the following form:

$$P(0,0) = \frac{F_o(1 - F_o)^{W-1}}{2 + F_o(D - W + \bar{N}) - (1 - F_o)^W}.$$

For start-stop protocol ($W = 1$) we obtain

$$P(0,0) = \frac{F_o}{1 + F_o(D + \bar{N}) - (1 - F_o)^{S-D+1} \sum\limits_{m=1}^{N} \frac{b_m}{(1-F_o)^m}}.$$

5 Available Throughput of the Transport Connection

The capacity of a transport connection under the conditions of competition of flows of different corresponding subscribers for the throughput of data transmission path is defined as the relation of the average amount of data transmitted between two consecutive acknowledgement to the average time acknowledgement [4,5]. Contribution to the speed of the virtual connection is given by those states of Markov chain for which it is possible to obtain acknowledgement. Normalized per unit throughput of virtual connection in loaded path is determined by the relation of the average number of data segments transmitted by the sender between two consecutive acknowledgement to the average time between acknowledgement expressed in the number of intervals of duration t: $Z(W, S) = \overline{V}/\overline{T}$. Since acknowledgements are transferred in each segment independently and arrive to the sender every cycle t, provided that they are not distorted in the path of length D from the recipient to the sender of the information flow the average time between acknowledgement is distributed according to the geometric law with the parameter F_o and will be: $\overline{T} = 1/F_o$. The average volume of data transmitted between acknowledgements taking into account the fact that each segment of the test connection with the probability b_n, $n = \overline{0, N}$ meets the size of the queue n and contributes to the amount of information transmitted inversely proportional to the value $n + 1$, is determined by generalizing the relation given in [4]

$$\overline{V} = \sum_{n=0}^{N} \frac{1}{n+1} \left[\sum_{l=2D-1+n}^{W+2D-2+n} \bar{l}P(l, n) + \sum_{l=W+2D-1+n}^{S-1} \overline{W}P(l, n) \right].$$

Values $\bar{l}$ and $\overline{W}$ are determined by the average number of segments that reached the addressee in selective procedure for repeating distorted segments:

$$\bar{l} = (l - 2D - n + 2)F_n, \quad \overline{W} = WF_n.$$

Then dependence of throughput of the virtual connection on the protocol parameters (W, S), characteristics of transmitting path (D, F_n, F_o) and load parameters $(b_n, n = \overline{1, N})$ will take the form:

$$Z(W, S) = F_n F_o \sum_{n=0}^{N} \frac{1}{n+1} \left[\sum_{l=2D-1+n}^{W+2D-2+n} (l - 2D + 2 - n)P(l, n) + W \sum_{l=W+2D-1+n}^{S-1} P(l, n) \right].$$

Hence, for an arbitrary width of the window when $S \geq D + W + N - 1$ we finally get

$$Z_c(W, S) = \begin{cases} F_n \dfrac{\sum\limits_{n=1}^{N} \frac{b_n}{n+1} \left[1 - (1-F_o)^W - WF_o(1-F_o)^{S-D-n+1} \right]}{2 + F_o(D-W+\tilde{N}) - (1-F_o)^W - \sum\limits_{n=1}^{N} b_n(1-F_o)^{S-D-n+1}}, & W < D; \\[6mm] F_n \dfrac{F_o^2 \left(1 - (1-F_o)^{W-D+1}\right) + \sum\limits_{n=1}^{N} \frac{b_n}{n+1} \left[1 - (1-F_o)^W - WF_o(1-F_o)^{S-D-n+1} \right]}{1 + F_o(\tilde{N}+1) + (1-F_o)^{W-D+1} - (1-F_o)^W - \sum\limits_{n=1}^{N} b_n(1-F_o)^{S-D-n+1}}, & W \geq D. \end{cases}$$

For interval limits on the duration of the timeout and the queue size of competitors $1 \leq N \leq D - 2$ the speed of transport connection in a competitive data transmission environment will be

$$Z_c(W,S) = F_n \left\{ F_o^2 \left(1 - (1 - F_o)^{W-D+1}\right) + \sum_{n=1}^{N} b_n - \sum_{n=1}^{S-D-W+1} \frac{b_n}{n+1}\left[(1 - F_o)^W \right. \right.$$

$$\left. + W F_o (1 - F_o)^{S-D-n+1}\right] - \sum_{n=S-D-W+2}^{N} \frac{b_n}{n+1}(1 - F_o)^{S-D-n+1}\left[1 + \right.$$

$$\left. \left. + F_o(S - D - n + 1)\right] \right\} \Big/ \left\{ 1 + F_o(1 + \bar{N}) + (1 - F_o)^{W-D+1} - (1 - F_o)^W - \right.$$

$$\left. - \sum_{n=1}^{N} b_n (1 - F_o)^{S-D-n+1} \right\}.$$

In the case of an absolutely reliable return channel ($F_o = 1$), available through-put of transport connection $W \le D$ is largely determined by the proximity of window width to the duration of round-trip delay

$$Z_c(W,S) = \frac{F_n}{2 + D - W + \bar{N}} \sum_{n=1}^{N} \frac{b_n}{n+1},$$

and for $W \ge D$—is invariant to D

$$Z_c(W,S) = \frac{F_n}{2 + \bar{N}} \left[1 + \sum_{n=1}^{N} \frac{b_n}{n+1} \right].$$

The unlimited duration of the timeout ($S \to \infty$) when $W < D$ leads to the dependence of the form

$$Z_c(W,\infty) = \frac{F_n\left(1 - (1 - F_o)^W\right) \sum_{n=1}^{N} \frac{b_n}{n+1}}{2 + F_o(D - W + \bar{N}) - (1 - F_o)^W},$$

and for unlimited increasing width of the window we obtain

$$Z_c(\infty,\infty) = \frac{F_n}{1 + F_o(\bar{N} + 1)} \left[F_o^2 + \sum_{n=1}^{N} \frac{b_n}{n+1} \right].$$

Numerical analysis shows that the available throughput for the transport con-nection $W \ge D$ is practically invariant to the duration of round-trip delay, sig-nificantly decreasing from the saturation range at $W = D$ and $F_o < 1$. In case $W < D$ available throughput is underloaded and the effective data transmission rate is significantly reduced. With increasing competition between subscribers for the throughput of the transmission path, the queue size increases, and the speed of information transfer decreases rapidly.

6   Conclusion

The analysis of competitor process of information flows of various inter-subscriber connections for the throughput on shared sections of the path has been carried out. An indicator model of transport connection, competing for the throughput of individual sections of the route, in the form of Markov chain with discrete time, describing the dynamics of queue of sent but not confirmed protocol data blocks, is proposed. The distribution of states of Markov chain under various operating conditions of transport connection is obtained. Analytical dependencies of transport connection speed are found for different ratios between parameters of transport protocol, the characteristics of network channels and load parameters. Numerical studies of available throughput of transport connection in selective re-transmission mode showed that the transmission rate between subscribers is determined by the reliability of data transmission, distribution of queue length of protocol units in transit nodes, and the ratio between duration of round-trip delay and the window width. The direction of further research is to single out the task of analyzing the available throughput of transport connections with interval restrictions on the size of the queues of competitive flows and duration of the end-to-end timeout of transport protocol. It is important to analyze the efficiency of application of forward error correction procedures at transport protocol level with exclusive and competitive use of network communication channels.

References

1. Boguslavskii, L.B.: Data Flow Control in Computer Networks. Energoatomizdat, Moscow (1984). (In Russian)
2. Gelenbe, E., Labetoulle, J., Pugolle, G.: Performance evaluation of the HDLC protocol. Comput. Netw. **2**, 409–415 (1978)
3. Borodikhin, E.A., Korotaev, I.A.: Analysis of functioning and optimization of HDLC protocol. Avtomatika i Vychislitel'naya Tekhnika (2), 47–51 (1993)
4. Sushchenko, S.P.: Analytical models of asynchronous procedures for data link control. Avtomatika i Vychislitel'naya Tekhnika (2), 32–40 (1988)
5. Kokshenev, V.V., Sushchenko, S.P.: Performance analysis of asynchronous procedure of data link control. Vichislitel'nie technologii **13**(5), 61–65 (2008)
6. Ewald, N.L., Kemp, A.H.: Analytical model of TCP NewReno through a CTMC. In: Bradley, J.T. (ed.) EPEW 2009. LNCS, vol. 5652, pp. 183–196. Springer, Heidelberg (2009). https://doi.org/10.1007/978-3-642-02924-0_15
7. Padhye, J., Firoiu, V., Towsley, D.F., Kurose, J.F.: Modeling TCP Reno performance: a simple model and its empirical validation. IEEE/ACM Trans. Netw. **8**(2), 133–145 (2000)
8. Kokshenev, V.V., Mikheev, P.A., Sushchenko, S.P.: Analysis of the selective mode of the transport protocol in loaded data path. Vestnik TSU. Seriya upravlenie, vichislitelnaya technika i informatika. No 3(24), 78–94 (2013)
9. Herrero, R.: Modeling and comparative analysis of Forward Error Correction in the context of multipath redundancy. Telecommun. Syst. Modell. Anal. Design Manag. **65**(4), 783–794 (2017)

Artificial Regeneration Based Regenerative Estimation of Multiserver System with Multiple Vacations Policy

Alexander Rumyantsev[1,2]([✉]) [iD] and Irina Peshkova[2] [iD]

[1] Institute of Applied Mathematical Research, Karelian Research Centre of RAS,
Petrozavodsk, Russia
ar0@krc.karelia.ru
[2] Petrozavodsk State University, Petrozavodsk, Russia

Abstract. A common approach to performance estimation of queueing system models is based on point estimators. Confidence intervals provide more insight on the system performance, and basically a regenerative structure of the studied process is required. However, classical regeneration epochs (arrivals into empty system) are too rare (or even might not exist) in complex queueing systems (such as high performance computing clusters), thus are inappropriate for practical usage. Instead, we propose the so-called artificial regeneration based on exponential splitting, which provide the desired estimators with reasonable accuracy. We illustrate the approach considering a multiserver model with energy efficiency management by activating the so-called sleep mode of servers which we hope is relevant to energy/performance tradeoff in cloud computing.

Keywords: Regenerative estimation · Confidence interval · Multiserver system · Performance · Energy efficiency · Artificial regeneration

1 Introduction

Regeneration of stochastic process is one of the basic and efficient methods of estimating stationary performance of queueing systems [3, 7–10, 12, 13]. The regenerating stochastic process (describing the systems behavior) from time to time enters a special state (restarts) at regeneration points and trajectory between such points can be separated into interdependent (or, in general, weakly dependent) and identically distributed groups. It allows to apply well-developed classical statistical methods based on the special form of the Central Limit Theorem (CLT) for confidence estimation of Quality of Service (QoS) parameters. At the same time the efficiency of the approach depends strongly on the accuracy of estimation which, in turn, depends on the frequency of the regeneration.

However, in many practical cases the system does not necessary experience classical regeneration, or the regeneration points are too rare to be practically

© Springer Nature Switzerland AG 2019
A. Dudin et al. (Eds.): ITMM 2019, CCIS 1109, pp. 38–50, 2019.
https://doi.org/10.1007/978-3-030-33388-1_4

useful. Then various methods may be applied, including more general constructions, such as wide sense regeneration [6], renovation [5] or accelerated regeneration [1]. Another method, closely related to the aforementioned, is the so-called artificial regeneration [2]. The artificial regeneration method is based on constructing a new, equivalent to the original, system by using the splitting property, which allows to change the internal structure of continuous-valued random variable in such a way to obtain a sequence of times at which the distribution has some specific properties, and these times occur with positive probability (possibly by enriching the probability space). If the splitting is based on exponential distribution, then the system at this particular points has memoryless property, and thus, regenerates in classical sense. In this work we propose to use artificial regeneration in the case when classical regeneration do not exist or is too rare to provide the desired accuracy in an acceptable simulation time. This approach can be effectively applied for estimation of QoS parameters such as stationary mean queue size or stationary mean waiting time, when e.g. service times have heavy-tailed distribution. In this paper we develop the method introduced in [2], and apply the artificial regeneration method to analysis of a multiserver system with energy efficiency management, focusing on practical aspects of method implementation.

This paper is organized as follows. In Sect. 2 we give some necessary background on regeneration and regenerative estimation based on the corresponding version of CLT. In Sect. 3 we define artificial regeneration points in general case. In Sect. 4 we demonstrate the method by considering a multiserver system with energy efficiency management by switching servers to power saving mode, and study the model. We perform some numerical experiments in Sect. 5. We give conclusions in Sect. 6.

2   Regeneration in Multiserver Systems

The main idea of regenerative approach is to separate the original process $\Theta = \{\Theta(t)\}_{t \geqslant 0}$ trajectory into (independent or weakly dependent) *regeneration cycles* $G_k = \{\Theta(t), \beta_k \leqslant t < \beta_{k+1}, k \geqslant 0\}$ (the parts of trajectory between *regeneration points* $\beta = \{\beta_n\}_{n \geqslant 1}$) with independent and identically distributed (iid.) cycle lengths $\alpha_k = \beta_{k+1} - \beta_k, k \geqslant 0$. It allows to study the process as a sequence of iid. cycles.

The process Θ is called *classical regenerative* if the distribution of $\{\Theta(\beta_k + t), t \geqslant 0, (\beta_k - \beta_i), i \geqslant k\}$ is independent on $k \geqslant 0$ and does not depend on the prehistory $\{\Theta(t), t < \beta_k; \beta_1, \ldots, \beta_k\}$. A frequently used definition of regeneration point for the process Θ describing a queueing system is related to such time epochs, when the number of customers in the system hits zero, i.e. $\beta_{n+1} = \inf\{k > \beta_n : \nu_k = 0\}$, where ν_n the number of customers in the system.

However, in multiserver systems classical regeneration is not guaranteed by stability of the system (e.g. each client enters the system before a service completion of another client), on the contrast to a single-server system. Thus, appropriate widening of the regeneration notion was required, and wide sense regeneration was introduced [6], allowing the (finite depth) dependence (the case $k = 1$

is denoted as *weak regeneration*) of regeneration cycles G_k, keeping the independence between regeneration cycles and regeneration epochs.

In general, it is hard to obtain weak regeneration epochs constructively. From practical point of view, the method of *renovation events* can be used for weak regeneration points construction. The method is based on detection of the so-called renovation periods and is well developed for stochastic recurrent sequences widely used in the analysis of queueing systems [5]. Note however, that such a construction depends on the complexity of the system, and detection of renovation events is difficult for complex systems. At that, an alternative constructive method is based on obtaining an artificially defined regeneration points by considering a stochastically equivalent process instead of original one [2]. In this paper we develop the artificial regeneration method and apply it to a multiserver queueing model.

Once the regeneration points are detected, an appropriate version of the CLT (classical or weakly-dependent version, w.r.t. the type of regeneration used) is applied to regeneration cycles in order to obtain performance estimates of a system.

2.1 Regenerative Estimation

To estimate some performance characteristic $\chi(\Theta)$ of the process Θ, e.g. $\chi(\Theta)$ is the stationary average customer delay in the system or energy consumption of the system per unit time. Let $Y_j = \sum_{i=\beta_j}^{\beta_{j+1}-1} \chi(\Theta_i)(t_{i+1} - t_i)$ – the sum of the values of $\chi(\Theta_i)$ on j-th cycle (w.r.t. intervals $t_{i+1} - t_i$ between events).

For classical case the statistical estimation of the process characteristic $\chi(Z)$ is reduced to the following: in the presence of independent and identically distributed observations (Y_j, α_j), $j \geq 0$ it is necessary to estimate [3]

$$r = \frac{\mathrm{E}Y_1}{\mathrm{E}\alpha_1},$$

where $\mathrm{E}Y$ – expectation of Y. Suppose that $\mathrm{E}(Y_1 + \alpha_1)^2 < \infty$, the using CLT we can obtain $(1 - 2\gamma)\%$ confidence interval for r [4]:

$$\left[\overline{r}_n \pm \frac{h_\gamma \sqrt{\overline{\mathrm{Var}}(n)}}{\sqrt{n}\overline{\alpha}_n} \right], \tag{1}$$

where $\overline{\alpha_n}$ ($\overline{Y_n}$) – sample mean for α_i (Y_i), n – the number of cycles, $\overline{\mathrm{Var}}(n)$ – sample estimate of $\mathrm{Var}(Y_1) - 2r\mathrm{cov}(Y_1, \alpha_1) + r^2\mathrm{Var}(\alpha_1)$, $h_\gamma = \Phi^{-1}((1 - \gamma)/2)$, $\Phi(x)$ – Laplace function.

Note that the width of constructed confidence intervals, and hence the efficiency of estimation, depends on the number of regeneration cycles n.

3   Artificial Regeneration

In this section we develop the method of construction of artificial regeneration points introduced in [2]. The method is based on replacement of the original process by a stochastically equivalent process with enlarged probability space, which, in turn, hits the desired state with positive probability, infinitely often. This allows, after obtaining the regeneration points, to apply regenerative estimation and construct confidence intervals to analyze performance of the original process. The method is based on the general construction of splitting, which is introduced below.

3.1 Exponential Splitting of a Continuous Random Variable

The density f of continuous random variable (r.v.) T can be split [14] if there exists some $0 < p < 1$ and density f_0, such that:

$$f \geqslant pf_0. \tag{2}$$

Then we define a new Bernoulli random variable I called *splitting indicator* s.t. $P(I = 1) = p$. Now we construct a split r.v. T' (equivalent the r.v. T) as follows:

$$T' = IT_0 + (1 - I)T_1, \tag{3}$$

where T_0 has density f_0, and T_1 has density

$$f_1 = \frac{f - pf_0}{1 - p}. \tag{4}$$

It is clear that the r.v. T' is equivalent to T (only noting that T' is defined on an extended probability space due to introduction of the splitting indicator I). If $I = 1$, then we say that the split r.v. T' is governed by f_0 (and governed by f_1 otherwise). We note that, due to the introduction of splitting indicator, the subsets in an extended probability space corresponding to f_0 and f_1 are disjoint. The r.v. T' is called *splitting representation* of T. The purpose of such a transformation is to let the r.v. T' have the desired (say, memoryless etc.) properties of the distribution f_0 with probability not less than p.

Now we consider *exponentially split* r.v. We say that (positive valued) r.v. T is exponentially split if there exist constants $\lambda > 0$, $\tau_0 \geqslant 0$ and $0 < p < 1$ such that

$$f(x) \geqslant p\lambda e^{-\lambda(x-\tau_0)}, \quad x \geqslant \tau_0. \tag{5}$$

That is, splitting representation T' of exponentially split r.v. T consists of T_0 being left truncated exponentially distributed r.v. with truncation point τ_0, rate λ, density

$$f_0(x) = \begin{cases} 0, & x \leqslant \tau_0, \\ \lambda e^{-\lambda(x-\tau_0)}, & x > \tau_0, \end{cases}$$

and T_1 having density

$$f_1(x) = \begin{cases} \dfrac{f(x)}{1-p}, & x \leqslant \tau_0, \\ \dfrac{f(x) - p\lambda e^{-\lambda(x-\tau_0)}}{1-p}, & x > \tau_0. \end{cases}$$

From simulation point of view splitting means that, instead of generation of r.v. T with density f, we generate a triple T_0, T_1, I and construct T' as in (3). The property (2) together with specially constructed (4) guarantee, that the simulated value T' is stochastically equivalent to T. Moreover, if $I = 1$, that is, T' is governed by truncated exponential distribution, then for any $t > \tau_0$ the remaining part $T' - t = T_0 - t$ may be replaced by (classical) exponentially distributed r.v. τ due to memoryless property of the remainder:

$$\mathrm{P}(T_0 - t > x + y | T_0 - t > x) = \frac{e^{-\lambda(x+y+t-\tau_0)}}{e^{-\lambda(x+t-\tau_0)}} = e^{-\lambda y}, \quad x, y > 0, t > \tau_0. \quad (6)$$

Following [2], consider the following example. Let $f(x)$ be the density of Gamma distribution:

$$f(x) = \frac{\alpha(\alpha x)^{\beta-1}}{\Gamma(\beta)} e^{-\alpha x}, x \geqslant 0,$$

then for any α and $\beta \geqslant 1$ it is easy to set parameters λ, τ_0, p of the splitting in such a way to satisfy minorization condition (5). Indeed, let

$$\tau_0 = \frac{1}{\alpha}, \ \lambda = \alpha, \ p = \frac{1}{e\Gamma(\beta)}. \quad (7)$$

Then it is easy to check that (5) holds for $x > \tau_0$, and thus $f(x)$ is exponentially split.

3.2 Artificial Regeneration of Stochastic Process with Discrete and Continuous Components

In many practical applications obtaining analytical results is hardly possible. In such a case simulation is used to perform a numerical study and derive (less general, however) results of practical impact. One of the powerful simulation methods is the discrete event simulation (DES). When performing a DES, it is assumed that the system state significantly changes only at some specific time epochs (related to the discrete events appearing in the system, e.g. customer arrival, service completion, server failure etc.), whereas the system state evolves in some predictable way in-between (e.g. the work in the system decreases linearly etc.). This property allows to evolve the system state only at the aforementioned specific event epochs, and derive system performance measures after the simulation based on the trace obtained. In many cases such a model may be obtained by means of the stochastic recursion, see e.g. [11].

A DES model of a queueing system naturally contains two types of components: the discrete-valued component is related to some countable quantities in the system (e.g. the number of customers in the queue, the server state etc.), and the continuous-valued component is related to some continuous resource (e.g. the remaining time before a customer arrival, the remaining work in the system/at server etc.) that are changing in time t. Motivated by that, we consider a stochastic process $\Theta = \{\mathbf{X}(t), \mathbf{T}(t)\}_{t \geqslant 0}$, with discrete (vector) component

$\mathbf{X} = (X_1, \ldots, X_n)$ and continuous component $\mathbf{T} = (T_1, \ldots, T_m) \geqslant \mathbf{0}$ (hereafter we denote vector-valued variables in bold and assume the dimension is clear from context, if not sub-indexed explicitly). Moreover, we assume that $\mathbf{T}$ decreases *linearly* with time, that is, $\mathbf{T}(t + \delta) = \mathbf{T}(t) - \delta\mathbf{1}$, for $\delta > 0$, and define an *event* as a time epoch t_i such that $T_{j^*}(t_i) = 0$ for some $j^* \in \{1, \ldots, m\}$ (in this case we say that the ith event occurring at t_i is of j^*th type). We assume that the discrete component $\mathbf{X}$ is changed only at event epochs, e.g. by some recurrent relation

$$\mathbf{X}(t_i+) = G_{j^*}\left(\mathbf{X}(t_i-)\right), \tag{8}$$

where G_{j^*} is the recursion related to type j^* event, or, in general, assume the change is governed by some Markovian probability

$$\mathrm{P}_{j^*}(x, x') = \mathrm{P}_{j^*}\{\mathbf{X}(t_i+) = x' | \mathbf{X}(t_i-) = x\}.$$

The component $\mathbf{X}(t)$ does not change if $\mathbf{T}(t) > \mathbf{0}$, that formally means

$$\mathbf{X}(t) = \mathbf{X}(t_i+), i = \max\{k : t_k < t\}.$$

We also assume that in the vector $\mathbf{T}$ the zeroed at t_i- component T_{j^*} is initialized at t_i+ from some distribution with density f_{j^*}, that is,

$$f_{j^*}(u, x, x') = \mathrm{P}(T_{j^*}(t_i+) \in du | \mathbf{X}(t_i-) = x, \mathbf{X}(t_i+) = x'). \tag{9}$$

Now we assume, that if $\mathbf{X}(t) = x^*$ for some specific x^*, then the process Θ allows to perform exponential splitting of the continuous component $\mathbf{T}$, that is for each continuous component T_j there exist constants $\tau_0(j), \lambda(j), p(j)$ such that (5) holds for the density $f_j(u, x, x^*)$, with corresponding $f_{0,j}(u) = \lambda_j e^{-\lambda_j(u - \tau_0(j))}$ and $f_{1,j}$ defined appropriately. At that, we introduce a split process $\Theta' = \{\mathbf{X}(t), \mathbf{T}'(t), \mathbf{Z}(t)\}_{t \geqslant 0}$ enhanced with auxiliary phase variable $\mathbf{Z} \in \{0, 1, 2\}^m$ denoting the phases of components of $\mathbf{T}$ defined componentwise in a recursive manner as follows

$$Z_{j^*}(t_i+) = \begin{cases} 1, & \text{if } \mathbf{X}(t_i+) = x^* \text{ and } I_{j^*}(t_i+) = 1, \\ 2, & \text{otherwise,} \end{cases} \tag{10}$$

$$Z_k(t_i+) = Z_k(t_i-), k \neq j^*, \tag{11}$$

where t_i is the epoch of type j^* event, and I_{j^*} is the splitting indicator corresponding to the component T'_{j^*} with success probability $p(j^*)$. However, starting from $Z_{j^*}(t_i+) = 1$ at time t_i, the component Z_{j^*} makes a step between the events:

$$Z_{j^*}(t) = \begin{cases} 1, t_i < t < t_i + \tau_0(j^*) \\ 0, t \geqslant t_i + \tau_0(j^*), \end{cases} \tag{12}$$

that is, the phase Z_{j^*} is changed from one to zero when not less than $\tau_0(j^*)$ time passes since the epoch t_i corresponding to the last event of j^*th type. In case $Z_j \neq 1$, the component Z_j does not change during inter-event time.

In fact, the phase $Z_k, k \neq j^*$ at time t_i is interpreted as follows:

- if $Z_k(t_i) = 0$, then the component $T'_k(t_i)$ is in the pure *exponential* phase, that is, the previous event of type k appeared at time $t < t_i - \tau_0(k)$ and the splitting indicator $I_k(t+) = 1$, so that more than $\tau_0(k)$ time has passed since that epoch, and hence the value of the component $T'_k(t_i+)$ may be sampled from classical exponential distribution with rate $\lambda(j)$ independently of the current value of $T'_k(t_i-)$;
- if $Z_k(t_i-) = 1$, then the splitting of T'_k was performed at some time epoch t of the previous event of type k, $I_k(t+) = 1$, but $t + \tau_0(k) > t_i$ so that still $T'_k(t_i)$ is in the *pre-exponential* phase (we adopt the terms from [2]) corresponding to the shift $\tau_0(k)$, this also means replacement of $T'_k(t_i+)$ by independent exponential is impossible yet;
- if $Z_k(t_i-) = 2$, then T'_k is in *non-exponential* phase, that is, either the splitting of T'_k was not performed at previous kth type event epoch t (if $\mathbf{X}(t+) \neq x^*$), or the split r.v. $T'_k(t+)$ is governed by $T'_{1,k}$, that is, $I_k(t+) = 0$.

We stress that exponential splitting of T'_j is performed only at time epoch t corresponding to event of type j, that is, at the time of initialization of (the new) $T'_j(t+)$ after zeroing, and only if $X_j(t+) = x^*_j$. If $X_j(t+) \neq x^*_j$, then T'_j is sampled from the original distribution (9) and we set $Z_j(t+) = 2$. The necessity of the pre-exponential phase is explained by the memoryless property (6) of a left truncated exponential distribution. However, to perform the simulation efficiently, we need to define the *pseudo-event* as the time epoch when step defined in (12) occurs. The pseudo-event corresponds to initialization of the (pure) exponential phase of the component T'_j and appears exactly at time $t + \tau_0(j)$, where t is the splitting time of T'_j. Note that pseudo-event does not correspond to transition in $\mathbf{X}$. Thus, we use the following (adopted from [2]) procedure of splitting of the component T'_j: first we initialize $T'_j(t+) = \tau_0(j)$, if at $t+$ the event j occurs, and set $Z_j(t+) = 1$. Then we wait for occurrence of pseudo-event of type j at some $t' > t$ (that is, the component $T'_j(t'-) = 0$, and $Z_j(t'-) = 1$), initialize $T'_j(t'+)$ by sampling from classical exponential distribution with rate $\lambda(j)$ and set $Z_j(t'+) = 0$. By this trick we eliminate the necessity to find the previous event of jth type, and thus, keep the system Markovian. (Note that in fact $t' = t + \tau_0(j)$).

It remains to define the artificial regeneration epochs as the (increasing) sequence of *pseudo-event* epochs $\{\beta_k\}_{k \geqslant 1}$ such that

$$\beta_{k+1} = \min\{t > \beta_k : \mathbf{X}(t+) = x^*, \mathbf{Z}(t+) = \mathbf{0}\}. \qquad (13)$$

3.3 Algorithmic Implementation

In this section we present an algorithmic implementation of the simulation of process Θ' and obtaining the regeneration epochs $\{\beta_k\}_{k \geqslant 1}$. The Algorithm 1 runs until N artificial regeneration epochs are obtained.

Algorithm 1. Artificial Regeneration

1: $\mathbf{Z} \leftarrow \mathbf{0}$	▷ Exponential phases for all components
2: $\mathbf{X} \leftarrow x^*$	▷ Start from artificial regeneration
3: $\mathbf{T} \leftarrow (\exp(\lambda_1), \ldots, \exp(\lambda_m))$	▷ Initialize continuous component
4: $t \leftarrow 0$	▷ Initialize modeling time
5: $k \leftarrow 1$	▷ Initialize regeneration counter
6: $\beta_k \leftarrow 0$	▷ Set first regeneration epoch at zero

7: **while** $k < N$ **do**
8:      $j^* \leftarrow \arg\min_{i \in \{1,\ldots,m\}} T_i$
9:      $t \leftarrow t + T_{j^*}$ ▷ Increase modeling time
10:      **for all** $j \neq j^*$ **do**
11:          **if** $Z_j > 0$ **then**
12:              $T_j \leftarrow T_j - T_{j^*}$
13:          **else**   ▷ Re-initialize all exponential phase components
14:              $T_j \leftarrow \exp(\lambda(j))$
15:          **end if**
16:      **end for**
17:      **if** $Z_{j^*} == 1$ **then**   ▷ Pseudo-event occurrence
18:          $Z_{j^*} \leftarrow 0$
19:          $T_{j^*} \leftarrow \exp(\lambda(j^*))$
20:          **if** $\mathbf{Z} == \mathbf{0}$ & $\mathbf{X} == x^*$ **then**   ▷ Artificial regeneration epoch
21:              $k \leftarrow k + 1$
22:              $\beta_k \leftarrow t$
23:          **end if**
24:      **else**   ▷ Event occurrence
25:          $\mathbf{X} \leftarrow G_{j^*}(\mathbf{X})$ ▷ Make a step in discrete component
26:          $Z_{j^*} \leftarrow 2$ ▷ For convenience
27:          **if** $X_{j^*} == x^*_{j^*}$ **then**
28:              $I_{j^*} \leftarrow \text{Bernoulli}(p(j^*))$
29:              **if** $I_{j^*} == 1$ **then**   ▷ Splitting occurs
30:                  $T_{j^*} \leftarrow \tau_0(j^*)$
31:                  $Z_{j^*} \leftarrow 1$ ▷ Pre-exponential phase starts
32:              **else**
33:                  $T_{j^*} \leftarrow f_{1,j}$ ▷ Initialize from density $f_{1,j}$
34:              **end if**
35:          **else**
36:              $T_{j^*} \leftarrow f_j$ ▷ Initialize from density f_j
37:          **end if**
38:      **end if**
39: **end while**

To simplify comprehension, we also illustrate the main loop of the algorithm, see Fig. 1.

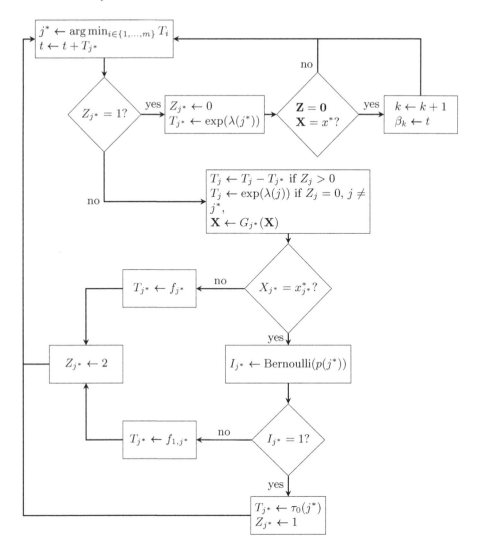

Fig. 1. Flowchart of the main loop of artificial regeneration algorithm

4   Artificial Regeneration in Multiserver System

In this section we apply the method of artificial regeneration to performance estimation of a multiserver system with energy efficiency control. We consider the following model. Let an m-server system receive a renewal input of customers waiting in a single queue if all servers are busy, and proceeding from queue to service in the order of arrival (First Come First Served queueing discipline). If at the ith service completion epoch the queue is empty, then the emptying server goes to the so-called sleep mode (with low energy consumption) for a (generally distributed) random time C_i having density $c(\cdot)$. The server during sleep mode period remains uninterrupted until the end of such a period, and no customer service is possible at this particular server. The interarrival time T_i between ith and $i + 1$th clients is generally distributed with density $a(\cdot)$, and service time of ith client S_i is generally distributed with density $b(\cdot)$ (note that for simplicity we assume identical servers, but heterogeneity of servers may be incorporated straightforwardly). Now we define the discrete and continuous components of the simulated process Θ. For convenience, we start indexing with zero. Let $\tau_0(t)$ be the queue size at time t, and $X_i(t)$ be the mode of ith server, $i = 1, \ldots, m$ defined as follows: $X_1(t) = 1$ if the server is active, and $X_1(t) = 0$ if the server is in the sleep mode. Now we define continuous component $\mathbf{T}(t)$ Let $T_0(t)$ be the time before the next arrival epoch, and $T_i(t)$ be the remaining times of current activity (service/sleep) of ith server, $i = 1, \ldots, m$. We stress that an arriving customer never starts receiving service immediately, but instead goes to the queue and waits for the nearest sleep mode completion epoch.

Now we define the transition functions $G_i, i = 0, \ldots, m$. For convenience denote e_i the vector with ith component equal to one, and other components equal zero, $i = 0, \ldots, m$. An arrival invokes the 0 type event, and since the arriving customer goes directly to the queue,

$$G_0(\mathbf{X}) = \mathbf{X} + e_0.$$

At the activity completion epoch of server $i = 1, \ldots, m$, the server starts sleep mode period if the queue is empty, and starts service period otherwise, hence,

$$G_i(\mathbf{X}) = \mathbf{X}(1 - e_i)\mathrm{I}\{\tau_0 = 0\} + (\mathbf{X}(1 - e_i) + e_i - e_0)\mathrm{I}\{\tau_0 > 0\}.$$

Now we define the transition functions for the continuous components. Since interarrival time always has density a, then for any x, x'

$$f_0(u, x, x') = f_0(u) = a(u).$$

The time before next activity completion of server $i = 1, \ldots, m$ depends on the current state of the queue, and thus

$$f_i(u, x, x') = b(u)\mathrm{I}\{\tau_0 > 0\} + c(u)\mathrm{I}\{\tau_0 = 0\}$$

where I is used as the comparison function for nonrandom arguments.

Now to define the regeneration epoch, we set $x^* = \{k, 1, \ldots, 1\}$ for some $k \geqslant 0$ and recall that artificial regeneration occurs at such pseudo-event epoch t that $\mathbf{X}(t+) = x^*$ (i.e. all servers are busy and k customers are waiting in the queue), and all the exponentially split components $T_0, \ldots, T_m$ are in exponential phases, that is, $\mathbf{Z}(t+) = \mathbf{0}$.

5 Numerical Illustration

To perform a practical illustration of the method, we consider an $m = 3$-server system with identical distributions of interarrival, service and sleep times

$$a(x) = b(x) = c(x) = \frac{\alpha(\alpha x)^{\beta-1}}{\Gamma(\beta)}e^{-\alpha x}, x \geqslant 0,$$

where we take $\alpha = 1.5, \beta = 1.5$. Now we set the parameters of truncated exponential distribution given in (7). As a performance measure, we consider the number of active (working) servers, that is, we estimate the following theoretical quantity

$$\chi(\Theta) = \lim_{t \to \infty} \frac{1}{t} \int_0^t \sum_{i=1}^m X_i(u)\, du.$$

Note that this performance measure is directly related to cost function in terms of energy efficiency (proportional to stationary average energy consumption per unit time). We run the DES of the system following Algorithm 1 until $N = 100$ regeneration epochs are obtained. Assume $0 = i_1, i_2, \ldots, i_{100}$ are the event epochs of regeneration, i.e.

$$\beta_j = t_{i_j}, j = 1, \ldots, 100.$$

Then we accumulate the per-cycle performance estimate as the following sum

$$Y_j = \sum_{k=i_j}^{i_{j+1}-1} \sum_{v=1}^m X_v(t_k)(t_{k+1} - t_k),$$

and perform confidence estimation as described in Sect. 2. Next, we simulate 5 trajectories of the original process Θ and do simple average estimator of the desired quantity as follows

$$\frac{1}{t_j} \sum_{k=1}^{j-1} \sum_{v=1}^m X_v(t_k)(t_{k+1} - t_k).$$

We plot the confidence intervals as well as the trajectories on Fig. 2. It may be seen that most of the trajectories fit the confidence corridor nicely, and the interval gets more narrow with increasing modeling time. However, as it may be seen from the plot, the simple average estimate may be highly displaced due to slow convergence to theoretical value.

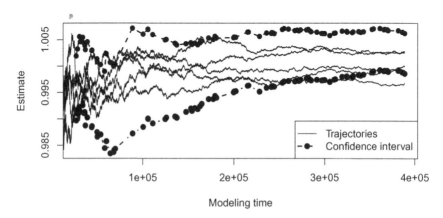

Fig. 2. Confidence interval based on artificial regeneration points for 3-server system with Gamma-distributed interarrival, service and sleep times.

6   Conclusion and Discussion

The method of artificial regeneration is a constructive method of obtaining regeneration points for regenerative estimation of performance of sophisticated queueing systems. In this paper we developed the variant of artificial regeneration method originally introduced in [2] and clarified algorithmic implementation from practical view. We applied the method to an important multiserver model with energy efficiency management, and demonstrated the results by numerical investigation. We hope that the method will allow to analyze many sophisticated models, such as a model of a high performance cluster (which is an extension of the multiserver model), but leave this for future research.

Acknowledgements. This research is partially supported by RFBR, projects 18-07-00147, 18-07-00156, 18-37-00094, 19-07-00303. Authors thank the ICWSNUCA Conference in Hyderabad, India, the place for early discussions of the method, that helped to improve the paper. Authors thank Prof. Evsey Morozov for helpful discussions.

References

1. Andradóttir, S., Calvin, J.M., Glynn, P.W.: Accelerated regeneration for Markov chain simulations. Probab. Eng. Inf. Sci. **9**(04), 497–523 (1995). https://doi.org/10.1017/S0269964800004022
2. Andronov, A.: Artificial regeneration points for stochastic simulation of complex systems. In: 10th European Simulation Symposium on Simulation Technology: Science and Art, ESS 1998, pp. 34–40. SCS, Delft (1998)
3. Asmussen, S.: Applied Probability and Queues. Springer, New York (2003)
4. Crane, M.A., Lemoine, A.J.: An Introduction to the Regenerative Method for Simulation Analys. Lecture Notes in Control and Information Sciences, vol. 4. Springer, Heidelberg (1977). https://doi.org/10.1007/BFb0007339
5. Foss, S.G., Kalashnikov, V.V.: Regeneration and renovation in queues. Queueing Syst. **8**(1), 211–223 (1991). https://doi.org/10.1007/BF02412251
6. Glynn, P.: Wide-sense regeneration for Harris recurrent Markov processes: an open problem. Queueing Syst. **68**(3–4), 305–311 (2011). https://doi.org/10.1007/s11134-011-9238-x
7. Glynn, P.W.: Some topics in regenerative steady-state simulation. Acta Appl. Math. **34**, 225–236 (1994)
8. Glynn, P.W., Iglehart, D.L.: Simulation methods for queues: an overview. Queueing Syst. **3**(3), 221–255 (1988). https://doi.org/10.1007/BF01161216
9. Glynn, P.W., Iglehart, D.L.: Conditions for the applicability of the regenerative method. Manag. Sci. **39**(9), 1108–1111 (1993). http://www.jstor.org/stable/2632816
10. Henderson, S.G., Glynn, P.W.: Regenerative steady-state simulation of discrete-event systems. ACM Trans. Model. Comput. Simul. **11**(4), 313–345 (Oct 2001). http://portal.acm.org/citation.cfm?doid=508366.508367. https://doi.org/10.1145/508366.508367
11. Rumyantsev, A., Zueva, P., Kalinina, K., Golovin, A.: Evaluating a single-server queue with asynchronous speed scaling. In: German, R., Hielscher, K.-S., Krieger, U.R. (eds.) MMB 2018. LNCS, vol. 10740, pp. 157–172. Springer, Cham (2018). https://doi.org/10.1007/978-3-319-74947-1_11

12. Sigman, K., Wolff, R.W.: A review of regenerative processes. SIAM Rev. **35**(2), 269–288 (1993)
13. Sigman, K.: Queues as Harris recurrent Markov chains. Queueing Syst. **3**(2), 179–198 (1988). https://doi.org/10.1007/BF01189048
14. Thorrison, H.: Coupling, Stationarity, and Regeneration. Springer, New York (2000)

On Combining Stochastic and Deterministic Models

Narayanan C. Viswanath[(⊠)] [iD]

Government Engineering College, Thrissur, Kerala, India
narayanan_viswanath@yahoo.com

Abstract. Stochastic models which are applied to model queues where the traffic intensity is greater than one and to health care models which involves huge sized populations such as human cells, viruses etc., faces the challenge of very huge number of states and hence demands huge computational cost. We studied the upper level stochastic effects in some of these models. We found that the upper level stochastic effects can't be omitted in many situations; but can be remarkably similar to the upper level effects in a differential equation model, closely related to the stochastic model. Given a huge dimensioned stochastic model, we replaced the upper level stochastic effects with the upper level dynamics of a suitably defined differential equation model, to study certain system performance measures. This resulted in a reduction in the computational cost involved with the use of the stochastic model with a large number of states.

Keywords: Stochastic models · Deterministic models · Expected time · Computational cost

1 Introduction

Stochastic modeling is a widely used tool in many fields of science. Queueing, Inventory, Reliability Engineering are some such areas. A stochastic model can consider several aspects of real world phenomena as its states, and collect information in terms of state probabilities. This information could be collected at an arbitrary time or in the steady state. However, in many situations this involves huge computational costs [1]. This may be a reason, why these tools are less popular in fields like health care.

Differential equation models have remained as first choice of mathematical modeling in the health care field [2–8]. Simplicity in formation, small dimension and ability to collect useful information are some advantages enjoyed by the deterministic models. However compared to deterministic models, stochastic models have the advantage that they can incorporate an abundance of information in the modeling of real world scenarios. As an example, in the place of five or six differential equations in a deterministic model [3], a Markov chain model [9, 10] may involve millions of differential or linear equations. This show that incorporation of more details often demands huge computational power for the analysis of a stochastic model. Scarcity of analytical results except in some special cases [9] can be seen to be another reason why differential equation models, despite being less effective in capturing lower level changes of variables [9], are preferred over stochastic models in the healthcare field. With the

© Springer Nature Switzerland AG 2019
A. Dudin et al. (Eds.): ITMM 2019, CCIS 1109, pp. 51–59, 2019.
https://doi.org/10.1007/978-3-030-33388-1_5

emergence of modern computational tools to handle the huge dimension, stochastic models especially Markov models are getting acceptance [9–18] in the health care field.

The huge dimension problem can be seen to exist in fields where Markov modeling is popular. For example consider an M/M/1 queueing model in which the arrival rate is greater than the service rate. It is well known that the steady state does not exist for such a model. However a network engineer, who may have to handle such a queueing system, may wish to know how much time it will take for the queue size to grow beyond certain limit. A similar problem exists for a birth death process, when the birth rate is greater than the death rate. Theoretically, one can prove that the above time has a phase-type distribution with suitable parameters. However, calculating its expected value may involve solving a linear system of millions of equations.

The above problem may demand several computational hours of an ordinary desktop computer. One may solve it faster using a workstation or a super computer. However such fast computing could not yet achieved using the hardware in a mobile station.

To speed up the computation, researchers have applied different techniques [9, 10, 14]. The HIV latently infected cell model in [14] assumes that stochastic effects are important only for small values of the random variable representing the number of actively infected cells. They thus proceed with a fast deterministic numerical integration technique for higher values of the variable. However, such an assumption may seem to contradict findings in [13] that stochastic effects in viral evolution can't be neglected even when the HIV population sizes are large.

In the current paper, we studied some Markov models and the corresponding deterministic models, which could be useful in many fields such as queueing and healthcare, to check whether it is possible to discard the upper level stochastic effects. We also studied whether the upper level stochastic effects are similar to that in a corresponding deterministic model. We demonstrate that capturing the upper level dynamics through deterministic models and lower level dynamics through stochastic models, hence combining the results from two types of models of the same problem could be a very useful technique.

This paper is arranged as follows. In Sect. 2, we discuss the method. In Sect. 3, we discuss some of its applications to different queueing/healthcare models. Section 4 concludes the discussion.

2   Methods

2.1   Importance of Upper Level Stochastic Effects

Let $\{N(t)\}$ be a pure death process, with death rate μ and $N(0) = n$. When $N(t) = i$, the cumulative death rate is $i\mu$. The state space of $\Xi = \{N(t)|t \geq 0\}$ is $\{0, 1, 2, ..., n\}$. The transition rate from state i to $i - 1$ is $i\mu$. Ξ starts from the state n and eventually reaches the state 0. Let $E(i)$ denote the expected time taken by Ξ to reach the state 0 from state i. Since $N(t)$ can only decrease, the average time $N(t)$ remains in the state i is $\frac{1}{i\mu}$. This implies the equations

$$E(i) = \frac{1}{i\mu} + E(i-1); 2 \leq i \leq n,$$

$$E(1) = \frac{1}{\mu}$$

These equations in turn give,

$$E(n) = \frac{1}{\mu} S(n),$$

where

$$S(n) = \sum\nolimits_{i=1}^{n} \frac{1}{i}$$

Distribution of the time for $N(t)$ to reach the state 0 from n can be found in [19].

Now to consider a real world situation, let us assume that $N(t)$ denote the number of HIV latently infected cells in the body of an HIV patient [20]. Then $E(n)$ denotes the expected time for the eradication of the latent cell reservoir. A half-life $\tau_{1/2} = 44$ months of latently infected cells [21] implies a death rate $\mu = \frac{\ln(2)}{\tau_{1/2}} = 5.2 \times 10^{-4} \, d^{-1}$. Now, if the initial population of latent cells is $n = 10^5$, it follows that $E(10^5) = 63.7$ years and if $n = 10^6$, $E(10^6) = 75.8$ years. The difference between $E(10^5)$ and $E(10^6)$ suggest that upper level stochastic effects does make a difference. We find that $S(10^6) - S(10^5)$ is just 2.3 and hence the difference between $E(10^5)$ and $E(10^6)$ was contributed by the death rate μ, which is small.

If we had assumed $N(t)$ denote the number of viruses in the body of the HIV patient, their death rate could have been much higher say $\mu = 9.1 \, d^{-1}$ [22]. In this case $E(n)$ denotes the expected time for the initial viral load n to reach the level 0. It follows that there is not much difference between $E(10^5)$, which is 1.3 days and $E(10^6)$, which is 1.6 days. This shows that upper level stochastic effects did not contribute much in this example.

These examples suggests that for population variables with faster downward movement rate than upward movement rate, upper level stochastic effects may become less important and it does matter in the opposite case. Motivated by the above examples we hope to present a method, like the one in [14] which applied a deterministic numerical integration, for capturing upper level stochastic effects when it really matters.

2.2 A Relation Between Stochastic and Deterministic Models

Going back to the Markov chain Ξ, define $P_i(t)$ as the probability that $N(t) = i$. Thus $H(t) = \sum_{i=1}^{n} iP_i(t)$ is the expected value of $N(t)$. $P_i(t)$ satisfies the set of differential equations

$$P'_i(t) = -i\mu P_i(t) + (i+1)\mu P_{i+1}(t); 0 \le i \le n-1$$
$$P'_n(t) = -n\mu P_n(t).$$

Differentiating $H(t)$, we get $H'(t) = \sum_{i=1}^{n} iP'_i(t)$. Substituting for $P'_i(t)$ from the above equations, we get

$$H'(t) = -\sum_{i=1}^{n-1} i^2 \mu P_i(t) + \sum_{i=1}^{n-1} i(i+1)\mu P_{i+1}(t) - n^2 \mu P_n(t)$$

that is

$$H'(t) = -\sum_{i=1}^{n-1} i^2 \mu P_i(t) + \sum_{i=1}^{n-1} (i+1)^2 \mu P_{i+1}(t) - \sum_{i=1}^{n-1} (i+1)\mu P_{i+1}(t) - n^2 \mu P_n(t)$$

that is

$$H'(t) = -\mu P_1(t) - \sum_{i=1}^{n-1} (i+1)\mu P_{i+1}(t)$$

that is

$$H'(t) = -\mu H(t).$$

On solving the above differential equation with the initial condition $H(0) = n$, we get $H(t) = ne^{-\mu t}$. This shows that the average value in the stochastic model satisfies a deterministic model, which establishes a relation between the two models.

2.3   Combining Stochastic and Deterministic Models

We take $H'(t) = -\mu H(t)$ as the deterministic model corresponding to the Markov model $\{N(t)\}$ given in Sect. 2.1.

The expression $H(t) = ne^{-\mu t}$ gives $t = -\frac{1}{\mu} log\left(\frac{H(t)}{n}\right)$.

Let us define $\hat{t}_{n,m}$ as the time taken by $N(t)$ in the deterministic model (by this, we mean the movement of $H(t)$) to reach the level m after starting from the initial level n. Then $\hat{t}_{n,m} = -\frac{1}{\mu} log\left(\frac{m}{n}\right)$.

Similarly, let $t_{n,m}$ be the average time taken by $N(t)$ in the stochastic model Ξ to reach the state m after starting from the initial state n. Then $t_{n,m} = E(n) - E(m)$.

With $n = 10^5$ and $\mu = 5.2 \times 10^{-4}$ d^{-1}, we find from Table 1 that $t_{n,m}$ and $\hat{t}_{n,m}$ are very close and that $\hat{t}_{n,m}$ can be taken as an approximate value of $t_{n,m}$.

Now combining the results from the stochastic and deterministic models, we write the approximate value $AE(m)$ of the expected time $E(n)$ as $AE(m) = \hat{t}_{n,m} + E(m)$. Notice that $\hat{t}_{n,m}$ has been calculated from the deterministic model and $E(m)$ from the

Table 1. Comparison of the average time $t_{n,m}$ (in years), to reach the state m from the state n in the Markov model Ξ (pure death process), with $\hat{t}_{n,m}$, that in the deterministic model. Parameters: $n = 10^5$ and $\mu = 5.2 \times 10^{-4} \, d^{-1}$

m	1	10	50	100	1000	10^4
$t_{n,m}$	58.43	48.27	39.99	36.37	24.26	12.13
$\hat{t}_{n,m}$	60.66	48.53	40.05	36.39	24.26	12.13

stochastic model. To check how good the approximation is, we present the following results.

$$AE(1) = \hat{t}_{n,1} + E(1) = 60.66 + 5.27 = 65.93 \text{ yrs}$$

$$AE(100) = \hat{t}_{n,100} + E(100) = 36.39 + 27.33 = 63.72 \text{ yrs}$$

$$AE(10^4) = \hat{t}_{n,10^4} + E(10^4) = 12.13 + 51.57 = 63.7 \text{ yrs}$$

We find that $AE(100)$ and $AE(10^4)$ gives good approximations for the actual value of E (n), which is 63.7. Notice that for finding $E(10^5)$ using the stochastic model alone, we had to consider 10^5 states of the Markov chain. This reduces to just 100 states using AE (100). Thus, there is a substantial reduction in the computational power if we combine the deterministic and stochastic models. This method depends on the fact that $t_{n,m}$ and $\hat{t}_{n,m}$ are remarkably close when m is large; that is the upper level stochastic effects are similar to that in a corresponding differential equation model.

3 Examples of Application of the Method

3.1 A Birth-Death Model

Let $\Psi = \{X(t)\}$ denote a birth-death process with birth rate λ and death rate μ. Assume that $\lambda > \mu$, so that the population size $X(t)$ becomes infinity as time tends to infinity. Let $S(M)$ denote the time taken by $X(t)$ to reach the state M starting from state 0. It follows that $S(M)$ has a phase-type distribution with representation (β, T), where β is a 1 x $(M + 1)$ row vector of the form

$$\beta = (1, \ 0, \ 0, \ \ldots, \ 0)$$

and T is an $M + 1$ square matrix whose $(i, j)^{th}$ entry, T_{ij} is given by

$$T_{ij} = \begin{cases} -\lambda, & j = i = 1 \\ -(i-1)(\lambda + \mu), & 2 \le j = i \le M+1 \\ \lambda, & i = 1, j = 2 \\ (i-1)\lambda, & 2 \le i \le M, j = i+1 \\ (i-1)\mu, & 2 \le i \le M+1, j = i-1 \end{cases}$$

The expected value of $S(M)$ is then given by

$$\mu_{S(M)} = -\beta T^{-1} e.$$

Now we shall derive the differential equation satisfied by the expected value $E(X(t))$. For this note that

$$X(t+\Delta t) - X(t) = \begin{cases} +1, & \text{with probability } \lambda X(t)\Delta t \\ -1, & \text{with probability } \mu X(t)\Delta t \\ 0, & \text{with probability } 1 - (\lambda+\mu)X(t)\Delta t \end{cases}$$

and hence

$$E(X(t+\Delta t) - X(t)) = (\lambda - \mu)X(t)\Delta t.$$

The above equation leads to the differential equation

$$\frac{d}{dt}E(X(t)) = (\lambda - \mu)E(X(t)).$$

Solving, we get

$$E(X(t)) = X(0)e^{(\lambda-\mu)t},$$

which in turn gives

$$t = \frac{1}{(\lambda-\mu)}\log\left(\frac{E(X(t))}{X(0)}\right).$$

Let $u_{L,M}$, be the time required for $E(X(t))$ to move from L to M, where $L < M$. We have

$$u_{L,M} = \frac{1}{(\lambda-\mu)}\log\left(\frac{M}{L}\right).$$

As in Sect. 2.3, we define an approximate value $AES(L)$ for the time $\mu_{S(M)}$ as

$$AES(L) = \mu_{S(L)} + u_{L,M}.$$

Table 2 presents a numerical example of the above procedure with $M = 5 \times 10^5$, $\lambda = 3.7$ and $\mu = 2.0$. The actual average time $\mu_{S(M)}$ is 8.647. The table shows that AES (1000) gives a very good approximation of $\mu_{S(M)}$.

Table 2. Approximate value of the actual average time $\mu_{S(M)} = 8.647$, in the birth-death model using $AES(L)$. Parameters: $M = 5 \times 10^5$, $\lambda = 3.7$ and $\mu = 2.0$.

L	10	50	100	500	1000	10^4
$\mu_{S(L)}$	2.156	3.208	3.627	4.581	4.990	6.346
$u_{L,M}$	6.364	5.418	5.010	4.063	3.656	2.301
$AES(L)$	8.521	8.626	8.637	8.645	8.646	8.647

3.2   A PH/M/1 Queue

Next we consider a PH/M/1 queue. The inter-arrival time follow a phase-type distribution with representation (α, H) of order k and service time is exponentially distributed with parameter μ. We take the average arrival rate as $\lambda = \pi H^0$, where $H^0 = -He$ and π is the steady state distribution of the generator matrix $H + H^0\alpha$. As in Sect. 3.1, we assume that $\lambda > \mu$, so that the queue size tends to infinity as time tends to infinity. Let $Y(t)$ denote the number of customers in the system including the one getting service (if any). $\Phi = \{Y(t) | t \geq 0\}$ forms a Markov chain.

Let $S(M)$ has the same definition as in Sect. 3. It again follows that $S(M)$ has a phase-type distribution with representation (β, T), where β is a $1 \times (M + 1)k$ row vector of the form

$$\beta = (\alpha, 0, 0, \ldots, 0)$$

and T is an $(M + 1)k$ square matrix given by

$$T = \begin{bmatrix} H & H^0\alpha & & & \\ \mu I & H - \mu I & H^0\alpha & & \\ & \ddots & \ddots & \ddots & \\ & & \mu I & H - \mu I & H^0\alpha \\ & & & \mu I & H - \mu I \end{bmatrix}.$$

The expected value of $S(M)$ is then given by

$$\mu_{S(M)} = -\beta T^{-1} e.$$

The differential equation satisfied by $E(Y(t))$ is taken as

$$\frac{d}{dt}E(Y(t)) = (\lambda - \mu)E(Y(t)).$$

Let $u_{L,M}$, $\mu_{S(L)}$ and $AES(L)$ have the same definition and expression as in Sect. 3.1. Table 3 presents a numerical example of the discussed procedure with parameters as: $M = 1 \times 10^5$, $\alpha = (0.4, 0.6)$, $H = \begin{bmatrix} -12 & 1 \\ 2 & -15 \end{bmatrix}$, $\lambda = \pi H^0 = 12.03$ and $\mu = 2.0$. It shows that the actual average time $\mu_{S(M)} = 9973.126$ is obtained with $L = 10$.

Table 3. Approximate value of the actual average time $\mu_{S(M)} = 9973.126$, in the PH/M/1 queueing model using $AES(L)$. Parameters: $M = 1 \times 10^5$, $\alpha = (0.4, 0.6)$, $H = \begin{bmatrix} -12 & 1 \\ 2 & -15 \end{bmatrix}$, $\lambda = \pi H^0 = 12.03$ and $\mu = 2.0$.

L	10	50	100	500	10000
$\mu_{S(L)}$	1.077	5.066	10.053	49.945	997.384
$u_{L,M}$	9972.048	9968.059	9963.073	9923.180	8975.741
$AES(L)$	9973.126	9973.126	9973.126	9973.126	9973.126

4 Conclusions

Very huge state space has always been a challenge for Markov models in queues and several other fields. Examples are number of customers in a queueing model with bursty traffic and virus cell population in a health care model. In many problems, a deterministic model involving the same variables could be formulated. We have demonstrated that the upper level stochastic effects in a Markov model may be very much identical to the upper level dynamics in a suitably formulated deterministic model. We thus combined the upper level results from a deterministic model with the lower level results from a Markov model involving same variables to reduce the computational cost associated with the Markov model. Since Markov models are highly useful in collecting important information in real world scenarios like a disease progression or data packet transmission, the method discussed in the paper might improve their applicability further.

References

1. Mehmood, R., Lu, J.A.: Computational Markovian analysis of large systems. J. Manuf. Technol. Manag. **22**(6), 804–817 (2011)
2. Nowak, M.A., Anderson, R.M., Mc Lean, A.R., Wolfs, T.F.W., Goudsmit, J., May, R.M.: Antigenic diversity thresholds and the development of AIDS. Science **254**, 963–969 (1991)
3. Rong, L., Perelson, A.S.: Modeling HIV persistence, the latent reservoir, and viral blips. J. Theor. Biol. **260**, 308–331 (2009)
4. Nosyk, B., Min, J.E., Lima, V.D., Hogg, R.S., Montaner, J.S.G.: Cost-effectiveness of population-level expansion of highly active antiretroviral treatment for HIV in British Columbia, Canada: a modelling study. Lancet HIV **2**, 393–400 (2015)
5. Wang, X., Tang, S., Song, X., Rong, L.: Mathematical analysis of an HIV latent infection model including both virus-to-cell infection and cell-to-cell transmission. J. Biol. Dyn. (2016). https://doi.org/10.1080/17513758.2016.1242784
6. Astacio, J., Briere, D., Guillén, M., Martínez, J., Rodríguez, F., Valenzuela-Campos, N.: Mathematical models to study the outbreaks of Ebola. Technical report, Biometrics Unit Technical Report BU-1365-M, Cornell University (1996). https://ecommons.cornell.edu/handle/1813/31962/BU-1365-M.pdf
7. Moolgavkar, S.H., Venzon, D.J.: Two-event model for carcinogenesis: incidence curves for childhood and adult tumors. Math. Biosci. **47**, 55–77 (1979)

8. Altrock, P.M., Liu, L.L., Michor, F.: The mathematics of cancer: integrating quantitative models. Nat. Rev. Cancer **15**, 730–745 (2015)
9. Conway, J.M., Coombs, D.A.: stochastic model of latently infected cell reactivation and viral blip generation in treated HIV patients. PLoS Comput. Biol. **7**(4), e1002033 (2011)
10. Viswanath, N.C.: Calculating the expected time to eradicate HIV-1 using a Markov chain. IEEE/ACM Trans. Comput. Biol. Bioinf. **15**(1), 60–67 (2018)
11. Aalen, O.O., Farewell, V.T., De Angelis, D., Day, N.E., Gill, O.N.: A Markov model for HIV disease progression including the effect of HIV diagnosis and treatment: application to AIDS prediction in England and Wales. Stat. Med. **16**(19), 2191–2210 (1997)
12. Granich, R.M., Gilks, C.F., Dye, C., De Cock, K.M., Williams, B.G.: Universal voluntary HIV testing with immediate antiretroviral therapy as a strategy for elimination of HIV transmission: a mathematical model. Lancet **373**, 48–57 (2009)
13. Read, E.L., Tovo-Dwyer, A.A., Chakraborty, A.K.: Stochastic effects are important in intrahost HIV evolution even when viral loads are high. Proc. Natl. Acad. Sci. U.S.A. **10**, 19727–19732 (2012)
14. Hill, A.L., Rosenbloom, D.I.S., Fu, F., Nowak, M.A., Siliciano, R.F.: Predicting the outcomes of treatment to eradicate the latent reservoir for HIV-1. Proc. Natl. Acad. Sci. U.S. A. **111**, 13475–13480 (2014)
15. Konrad, B.P., Taylor, D., Conway, J.M., Ogilvie, G.S., Coombs, D.: On the duration of the period between exposure to HIV and detectable infection. Epidemics (2017). https://doi.org/10.1016/j.epidem.2017.03.002
16. Toth, D.J.A., Gundlapalli, A.V., Khader, K., et al.: Estimates of outbreak risk from new introductions of Ebola with immediate and delayed transmission control. Emerg. Infect. Dis. **21**(8), 1402–1408 (2015). https://doi.org/10.3201/eid2108.150170
17. Bellan, S.E., Pulliam, J.R.C., Pearson, C.A.B., et al.: Statistical power and validity of Ebola vaccine trials in Sierra Leone: a simulation study of trial design and analysis. Lancet Inf. Dis. **15**, 703–710 (2015)
18. Sun, X., Bao, J., Shao, Y.: Mathematical modeling of therapy-induced cancer drug resistance: connecting cancer mechanisms to population survival rates. Sci. Rep. (2016). https://doi.org/10.1038/srep22498
19. Solomon, D.L.: Time to extinction of a pure death process. Biometrics Unit Mimeo Series BU-360-M. https://ecommons.cornell.edu/bitstream/handle/1813/32450/BU-360-M.pdf
20. Blankson, J.N., Persaud, D., Siliciano, R.F.: The challenge of viral reservoirs in HIV-1 infection. Ann. Rev. Med. **53**, 557–593 (2002)
21. Finzi, D., Blankson, J., Siliciano, J.D., et al.: Latent infection of CD4+ T cells provides a mechanism for lifelong persistence of HIV-1, even in patients on effective combination therapy. Nat. Med. **5**(5), 512–517 (1999)
22. Ramratnam, B., Bonhoeffer, S., Binley, J., et al.: Rapid production and clearance of HIV-1 and hepatitis C virus assessed by large volume plasma apheresis. Lancet **354**, 1782–1785 (1999)

Estimation of the Probability Density Parameters of the Interval Duration Between Events in Correlated Semi-synchronous Event Flow of the Second Order by the Method of Moments

Lyudmila Nezhelskaya and Diana Tumashkina[✉]

National Research Tomsk State University, Tomsk, Russia
ludne@mail.ru, diana1323@mail.ru

Abstract. We consider a correlated semi-synchronous event flow of the second order with two states; it is one of the mathematical models for an incoming stream of claims (events) in modern digital integral servicing networks, telecommunication systems and satellite communication networks. We solve the problem of estimating the probability density parameters of the values of the interval duration between the moments of the events occurrence by the method of moments for general and special cases of setting the flow parameters. The results of statistical experiments performed on a flow simulation model are given.

Keywords: Correlated semi-synchronous event flow of the second order · Probability density · Estimation of the parameters · Method of moments

1 Introduction

Mathematical models of queueing theory are widely used to describe real economic, physical, and other processes. In modern times, thanks to the fast development of information technologies, another important fields of queueing theory applications are the design and creation of digital integrated service networks (DISN). Since in practice, the parameters defining the event flow change randomly in time, the doubly stochastic event flows are adequate mathematical models of information flows of messages operating in the DISN [1–8]. These flows are characterized by double randomness: the moments when events occur are random and the intensity of the flow is a random process. This leads to the study of doubly stochastic event flows.

Depending on how the transition from state to state occurs, these event flows can be divided into three types: (1) synchronous flows, the transition from state

© Springer Nature Switzerland AG 2019
A. Dudin et al. (Eds.): ITMM 2019, CCIS 1109, pp. 60–72, 2019.
https://doi.org/10.1007/978-3-030-33388-1_6

to state in which depends directly on the occurrence of the event; (2) asynchronous flows, the transition from state to state in which does not depend on whether an event has occurred or not; (3) semi-synchronous flows, for which the definition of the first type is true for one state and the second type for the second state. A semi-synchronous event flow of the second order is the object of studying in this work. The main problems in the studying of doubly stochastic event flows are problems that are realized by observing the moments when events occur: (1) estimating the states of an event flow [9,10]; (2) estimating flow parameters [11–14].

The problem of optimal states estimation for the considered event flow under its complete observability was solved in [10] and with partial observability in [15].

In this paper, we find the explicit form of the probability density of the values of the interval duration between the moments of the events occurrence for general and special cases of setting the parameters of a correlated semi-synchronous event flow of the second order. The problem of estimating density parameters is solved by the method of moments for each considered case.

2 Problem Setting

We consider the stationary operation mode of a semi-synchronous doubly stochastic event flow of the second order (hereinafter flow), the accompanying random process of which is a piecewise constant process $\lambda(t)$ with two states S_1 and S_2. Hereinafter, the ith state of the process is understood as the state S_i, $i = 1, 2$.

The duration of the interval between the flow events at the first state is determined by the random variable $\eta = min(\xi^{(1)}, \xi^{(2)})$, where random variable $\xi^{(1)}$ has distribution function $F_1^{(1)}(t) = 1 - e^{-\lambda_1 t}$, random variable $\xi^{(2)}$ has distribution function $F_1^{(2)}(t) = 1 - e^{-\alpha_1 t}$; $\xi^{(1)}$ and $\xi^{(2)}$ are independent random variables.

At the moment of the flow event occurrence, the process $\lambda(t)$ transits from the state S_1 to S_j either with probability $P_1^{(1)}(\lambda_j|\lambda_1)$, or with probability $P_1^{(2)}(\lambda_j|\lambda_1)$, depending on what value the random variable η has taken, $j = 1, 2$. Here $\sum_{j=1}^{2} P_1^{(k)}(\lambda_j|\lambda_1) = 1$, $k = 1, 2$. The duration of the interval between the flow events at the first state is random variable with distribution function $F(t) = 1 - e^{-(\lambda_1+\alpha_1)t}$.

The time during which the process $\lambda(t)$ remains at the second state is random variable with distribution function $F_2(t) = 1 - e^{-\alpha_2 t}$. During the time when the process $\lambda(t)$ is in the second state, there is a Poisson event flow with parameter λ_2.

Hereinafter, it is assumed that the state S_i (ith state) of the process $\lambda(t)$ takes place if $\lambda(t) = \lambda_i$, $i = 1, 2$; $\lambda_1 > \lambda_2 \geq 0$.

The infinitesimal characteristics matrices for the process $\lambda(t)$ are as follows

$$\mathbf{D}_0 = \left\| \begin{matrix} -(\lambda_1 + \alpha_1) & 0 \\ \alpha_1 & -(\lambda_2 + \alpha_2) \end{matrix} \right\|,$$

$$\mathbf{D}_1 = \left\| \begin{matrix} \lambda_1 P_1^{(1)}(\lambda_1|\lambda_1) + \alpha_1 P_1^{(2)}(\lambda_1|\lambda_1) & \lambda_1 P_1^{(1)}(\lambda_2|\lambda_1) + \alpha_1 P_1^{(2)}(\lambda_2|\lambda_1) \\ 0 & \lambda_2 \end{matrix} \right\| .$$

Elements of the matrix $\mathbf{D}_1$ are the intensities of the process transitions from state to state with an event occurrence. Nondiagonal elements of the matrix $\mathbf{D}_0$ are the intensities of transitions from state to state without an event. In turn, the diagonal elements of the matrix $\mathbf{D}_0$ are the intensities of the process exit from its states taken with the opposite sign.

An example of one of the realizations of the process $\lambda(t)$ and the event flow are shown on Fig. 1, where $t_1, t_2, \ldots$ denote the moments when events occur in the flow.

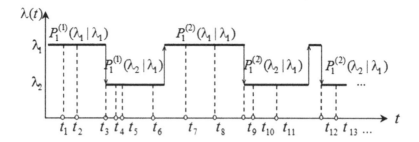

Fig. 1. Semi-synchronous event flow of the second order

Since the process $\lambda(t)$ is unobservable in principle, and we can only observe time moments $t_1, t_2, \ldots$ when events occur in the flow, then $\lambda(t)$ is a hidden Markov process or an unobservable accompanying Markov process. The sequence $\{\lambda(t_k)\}$ at the time moments $t_1, t_2, \ldots, t_k, \ldots$ of events occurrence is an embedded Markov chain.

We denote by $\tau_k = t_{k+1} - t_k$, $k = 1, 2, \ldots$, the value of interval duration between neighboring events, and by $p(\tau)$ the probability density of the value of interval duration between neighboring events in the observed flow. Since we consider the stationary operation mode of the observed flow then $p(\tau_k) = p(\tau)$ for all $k = 1, 2, \ldots$, $\tau \geq 0$. Then we can let the moment of event occurrence t_k equal to zero without loss of generality, i.e. the moment of the event occurrence is $\tau = 0$.

3   Derivation of Probability Density $p(\tau)$

We introduce the conditional probability $p_{ij}(\tau)$ that there are no events on the interval $(0, \tau)$ and that the process value $\lambda(\tau) = \lambda_j$ at the time moment τ, provided that the process value $\lambda(0) = \lambda_i$, $i, j = 1, 2$ at the time moment $\tau = 0$ [16].

3.1   Probability Density for the General Case of Setting Flow Parameters

Lemma 1. *The conditional probabilities* $p_{ij}(\tau)$, $i,j = 1,2$, *in a correlated semi-synchronous event flow of the second order are given by the following*

$$p_{11}(\tau) = e^{-(\lambda_1+\alpha_1)\tau}, \quad p_{12}(\tau) = 0, \quad p_{22}(\tau) = e^{-(\lambda_2+\alpha_2)\tau},$$

$$p_{21}(\tau) = \frac{\alpha_2}{(\lambda_1+\alpha_1)-(\lambda_2+\alpha_2)}[e^{-(\lambda_2+\alpha_2)\tau} - e^{-(\lambda_1+\alpha_1)\tau}], \quad \tau \geq 0, \qquad (1)$$

where $(\lambda_1+\alpha_1)-(\lambda_2+\alpha_2) \neq 0$.

Proof. The proof is carried out by solving differential equations for $p_{ij}(\tau)$, $i,j = 1,2$.

Lemma 2. *The probability densities* $\tilde{p}_{ij}(\tau)$, $i,j = 1,2$, *in a correlated semi-synchronous event flow of the second order are given by the following formulas*

$$\tilde{p}_{1j}(\tau) = [\lambda_1 P_1^{(1)}(\lambda_j|\lambda_1) + \alpha_1 P_1^{(2)}(\lambda_j|\lambda_1)]e^{-(\lambda_1+\alpha_1)\tau}, \quad j = 1,2,$$

$$\tilde{p}_{21}(\tau) = \frac{\alpha_2[\lambda_1 P_1^{(1)}(\lambda_1|\lambda_1) + \alpha_1 P_1^{(2)}(\lambda_1|\lambda_1)]}{(\lambda_1+\alpha_1)-(\lambda_2+\alpha_2)}[e^{-(\lambda_2+\alpha_2)\tau} - e^{-(\lambda_1+\alpha_1)\tau}], \qquad (2)$$

$$\tilde{p}_{22}(\tau) = \frac{\alpha_2[\lambda_1 P_1^{(1)}(\lambda_2|\lambda_1) + \alpha_1 P_1^{(2)}(\lambda_2|\lambda_1)]}{(\lambda_1+\alpha_1)-(\lambda_2+\alpha_2)}[e^{-(\lambda_2+\alpha_2)\tau} - e^{-(\lambda_1+\alpha_1)\tau}] + $$
$$ + \lambda_2 e^{-(\lambda_2+\alpha_2)\tau},$$

where $\tau \geq 0$, $(\lambda_1+\alpha_1)-(\lambda_2+\alpha_2) \neq 0$.

Proof. We introduce the joint probability that without flow events occurrence at the interval $(0,\tau)$ the process $\lambda(\tau)$ transited from the first state to the first at this interval, then the event occurred at the half-interval $[\tau, \tau + \Delta\tau)$ with probability $1 - e^{-\lambda_1\Delta\tau}$ and, at the moment of the flow event occurrence, the process $\lambda(\tau)$ transited from the first state to the first $(S_1 \rightarrow S_1)$ with probability $P_1^{(1)}(\lambda_1|\lambda_1)$ or, on the half-interval $[\tau, \tau + \Delta\tau)$ a flow event occurred with probability $1 - e^{-\alpha_1\Delta\tau}$ and at the moment of the flow event occurrence the process $\lambda(\tau)$ transited from the first state to the first $(S_1 \rightarrow S_1)$ with probability $P_1^{(2)}(\lambda_1|\lambda_1)$. This joint probability is as follows $p_{11}(\tau)[\lambda_1 P_1^{(1)}(\lambda_1|\lambda_1) + \alpha_1 P_1^{(2)}(\lambda_1|\lambda_1)]\Delta\tau + o(\Delta\tau)$. We note that the joint probability under consideration can be represented as $p_{11}(\tau)[\lambda_1 P_1^{(1)}(\lambda_1|\lambda_1) + \alpha_1 P_1^{(2)}(\lambda_1|\lambda_1)]\Delta\tau + o(\Delta\tau) = \tilde{p}_{11}(\tau)\Delta\tau + o(\Delta\tau)$, where $\tilde{p}_{11}(\tau)$ is the probability density corresponding to the joint probability. Writing the last equality in the form $\tilde{p}_{11}(\tau) + o(\Delta\tau)/\Delta\tau = [\lambda_1 P_1^{(1)}(\lambda_1|\lambda_1) + \alpha_1 P_1^{(2)}(\lambda_1|\lambda_1)]p_{11}(\tau) + o(\Delta\tau)/\Delta\tau$ and tending $\Delta\tau$ to zero, we find $\tilde{p}_{11}(\tau) = [\lambda_1 P_1^{(1)}(\lambda_1|\lambda_1) + \alpha_1 P_1^{(2)}(\lambda_1|\lambda_1)]p_{11}(\tau)$. We define the remaining joint probabilities in the same way. Thus, the probability density $p_{ij}(\tau)$ that the process $\lambda(\tau)$ transits from the state S_i to S_j, $i,j = 1,2$, at the interval $(0,\tau)$, without the flow event occurrence at this interval and with the event occurrence at the moment τ, is written in the form

$$\tilde{p}_{1j}(\tau) = [\lambda_1 P_1^{(1)}(\lambda_j|\lambda_1) + \alpha_1 P_1^{(2)}(\lambda_j|\lambda_1)]p_{11}(\tau), \ j = 1,2,$$

$$\tilde{p}_{21}(\tau) = [\lambda_1 P_1^{(1)}(\lambda_1|\lambda_1) + \alpha_1 P_1^{(2)}(\lambda_1|\lambda_1)]p_{21}(\tau),$$

$$\tilde{p}_{22}(\tau) = [\lambda_1 P_1^{(1)}(\lambda_2|\lambda_1) + \alpha_1 P_1^{(2)}(\lambda_2|\lambda_1)]p_{21}(\tau) + \lambda_2 p_{22}(\tau). \tag{3}$$

Substituting (1) into (3), we obtain (2).

Lemma 3. *The transition probabilities* p_{ij}, $i,j = 1,2$, *in a correlated semi-synchronous event flow of the second order are given by the following*

$$p_{1j} = (\lambda_1 P_1^{(1)}(\lambda_j|\lambda_1) + \alpha_1 P_1^{(2)}(\lambda_j|\lambda_1))/(\lambda_1 + \alpha_1), \ j = 1,2,$$

$$p_{21} = (\alpha_2[\lambda_1 P_1^{(1)}(\lambda_1|\lambda_1) + \alpha_1 P_1^{(2)}(\lambda_1|\lambda_1)])/[(\lambda_1 + \alpha_1)(\lambda_2 + \alpha_2)],$$

$$p_{22} = (\alpha_2[\lambda_1 P_1^{(1)}(\lambda_2|\lambda_1) + \alpha_1 P_1^{(2)}(\lambda_2|\lambda_1)])/[(\lambda_1 + \alpha_1)(\lambda_2 + \alpha_2)] + \lambda_2/(\lambda_2 + \alpha_2). \tag{4}$$

Proof. We define p_{ij}, $i,j = 1,2$, the probability of the process $\lambda(\tau)$ transition from the state S_i to S_j for a time which will pass from the moment $\tau = 0$ until the next flow event occurrence. Since τ is an arbitrary time moment, then these transition probabilities are defined as

$$p_{ij} = \int_0^{\infty} \tilde{p}_{ij}(\tau)d\tau, \ i,j = 1,2. \tag{5}$$

Thus, substituting (2) into (5), we obtain (4).

Let us consider the conditional stationary probability $\pi_i(0)$ that the process $\lambda(\tau)$ is in the state S_i at the time $\tau = 0$, provided that a flow event has occurred at the time $\tau = 0$, $i,j = 1,2$, $\pi_1(0) + \pi_2(0) = 1$.

Lemma 4. *The conditional stationary probabilities* $\pi_i(0)$, $i = 1,2$, *in a correlated semi-synchronous event flow of the second order are given by the following formulas*

$$\pi_1(0) = \frac{\alpha_2[\lambda_1 P_1^{(1)}(\lambda_1|\lambda_1) + \alpha_1 P_1^{(2)}(\lambda_1|\lambda_1)]}{\lambda_2[\lambda_1 P_1^{(1)}(\lambda_2|\lambda_1) + \alpha_1 P_1^{(2)}(\lambda_2|\lambda_1)] + (\lambda_1 + \alpha_1)\alpha_2},$$

$$\pi_2(0) = \frac{(\lambda_2 + \alpha_2)[\lambda_1 P_1^{(1)}(\lambda_2|\lambda_1) + \alpha_1 P_1^{(2)}(\lambda_2|\lambda_1)]}{\lambda_2[\lambda_1 P_1^{(1)}(\lambda_2|\lambda_1) + \alpha_1 P_1^{(2)}(\lambda_2|\lambda_1)] + (\lambda_1 + \alpha_1)\alpha_2}, \ \pi_1(0) + \pi_2(0) = 1. \tag{6}$$

Proof. Since the sequence of the moments $t_1, t_2, ..., t_k, ...$ forms an embedded Markov chain $\{\lambda(t_k)\}$ then the following equations are valid for probabilities $\pi_i(0)$

$$\pi_1(0) = p_{11}\pi_1(0) + p_{21}\pi_2(0), \quad \pi_2(0) = p_{12}\pi_1(0) + p_{22}\pi_2(0), \tag{7}$$

where the probabilities p_{ij}, $i, j = 1, 2$, are defined by (4).

From (7), taking into account that $\pi_1(0) + \pi_2(0) = 1$, we find

$$\pi_1(0) = p_{21}/(p_{12} + p_{21}), \quad \pi_2(0) = p_{12}/(p_{12} + p_{21}). \tag{8}$$

Substituting (4) into (8), we obtain (6).

Lemmas 2 and 4 yield the following theorem.

Theorem 1. *The probability density of the value of interval duration between neighboring events in a correlated semi-synchronous event flow of the second order is given by the following*

$$p(\tau) = \gamma z_1 e^{-z_1 \tau} + (1 - \gamma) z_2 e^{-z_2 \tau}, \ \tau \geq 0, \tag{9}$$

$$\gamma = \frac{\pi_1(0)(\lambda_1 + \alpha_1 - \lambda_2) - \alpha_2}{(\lambda_1 + \alpha_1) - (\lambda_2 + \alpha_2)}, \ z_1 = \lambda_1 + \alpha_1, \ z_2 = \lambda_2 + \alpha_2,$$

where $(\lambda_1 + \alpha_1) - (\lambda_2 + \alpha_2) \neq 0$, *and the probability* $\pi_1(0)$ *is defined in (6).*

Proof. Due to the fact that the process $\lambda(t)$ has a Markov property, if its evolution is considered starting from the time moment t_k, $k = 1, 2, ...$, of the flow event occurrence, the probability density $p(\tau)$ of the value of interval duration between neighboring events in the flow under consideration is determined as

$$p(\tau) = \sum_{i=1}^{2} \pi_i(0) \sum_{j=1}^{2} \tilde{p}_{ij}(\tau), \ \tau \geq 0. \tag{10}$$

Substituting the expressions (2) and (6) into (10), after the necessary transformations, we obtain (9).

3.2 Probability Density for the Special Case of Setting Flow Parameters

Let us consider the case when the coefficient $(\lambda_1 + \alpha_1) - (\lambda_2 + \alpha_2) = 0$ in (9).

Lemma 5. *The conditional probabilities* $p_{ij}(\tau)$, $i, j = 1, 2$, *in a correlated semi-synchronous event flow of the second order in the case when* $(\lambda_1 + \alpha_1) - (\lambda_2 + \alpha_2) = 0$ *are given by the following*

$$p_{11}(\tau) = e^{-(\lambda_1 + \alpha_1)\tau}, \ p_{12}(\tau) = 0, \ p_{22}(\tau) = e^{-(\lambda_1 + \alpha_1)\tau},$$

$$p_{21}(\tau) = (\lambda_1 + \alpha_1 - \lambda_2)\tau e^{-(\lambda_1 + \alpha_1)\tau}, \ \tau \geq 0. \tag{11}$$

Proof. The proof is carried out by solving differential equations for $p_{ij}(\tau)$, $i, j = 1, 2$.

Then, based on Lemmas 4 and 5, we formulate the following theorem.

Theorem 2. *The probability density of the value of interval duration between neighboring events in a correlated semi-synchronous event flow of the second order in the case when* $(\lambda_1 + \alpha_1) - (\lambda_2 + \alpha_2) = 0$ *is given by the following*

$$p(\tau) = [(\lambda_1 + \alpha_1) - \pi_2(0)(\lambda_1 + \alpha_1 - \lambda_2)(1 - (\lambda_1 + \alpha_1)\tau)]e^{-(\lambda_1+\alpha_1)\tau}, \ \tau \geq 0, \ (12)$$

where the probability $\pi_2(0)$ *is defined in (6).*

Proof. Substituting (11) into (3), we find the densities $\tilde{p}_{ij}(\tau)$; substituting (3) into (5) and (5) into (8), we obtain the probabilities $\pi_i(0)$. Substituting $\tilde{p}_{ij}(\tau)$ and $\pi_i(0)$, $i, j = 1, 2$, into (10), as a result of the necessary transformations we obtain (12).

4   Estimation of the Distribution Parameters by the Method of Moments

Let us consider the statistics $C_l = \frac{1}{n}\sum_{k=1}^{n} \tau_k^l$, where $\tau_k = t_{k+1} - t_k$.

4.1   General Case of Setting Flow Parameters

Let us have a sample $\tau_1, \tau_2, ..., \tau_n$ from the distribution $p(\tau|z_1, z_2, \gamma)$ depending on three unknown parameters z_1, z_2, γ. Let $M(\tau^l) = \int_0^\infty \tau^l p(\tau|z_1, z_2, \gamma)d\tau$ be the initial theoretical moment of the lth order which is a function of the unknown parameters. Then it is close to the corresponding selective moment $\bar{\tau}^l$ which is the $C_l = \frac{1}{n}\sum_{k=1}^{n} \tau_k^l$ statistics. For the first three initial moments we write the moment equations

$$M(\tau^l) = C_l, \ l = 1, 2, 3. \tag{13}$$

Given the kind of density (9), we get $M(\tau^l) = l!\gamma/z_1^l + l!(1-\gamma)/z_2^l$, $l = 1, 2, 3$. Then the system (13) takes the following form

$$z_1 z_2 C_1 - z_2\gamma - z_1(1 - \gamma) = 0, \ (z_1 + z_2)C_1 - z_1 z_2 C_2/2 = 1, \tag{14}$$

$$(z_1 + z_2)C_2 - z_1 z_2 C_3/3 = 2C_1.$$

Solving the system of equations (14), we find parameter estimates of $p(\tau)$

$$\hat{z}_{1,2} = \frac{1}{2}\left(-\frac{2(C_3 - 3C_1 C_2)}{3C_2^2 - 2C_1 C_3} \pm \sqrt{\left(\frac{2(C_3 - 3C_1 C_2)}{3C_2^2 - 2C_1 C_3}\right)^2 + 4\frac{6(C_2 - 2C_1^2)}{3C_2^2 - 2C_1 C_3}}\right). \tag{15}$$

Since $z_1 = \lambda_1 + \alpha_1$, $z_2 = \lambda_2 + \alpha_2$, where $\lambda_1 > \lambda_2 \geq 0$, and the relationship between parameters α_1 and α_2 is unknown, then the relationship between parameters z_1 and z_2 is unknown too. Thus, in order to determine which root of equation (15) can be chosen as the parameter estimate $\hat{z}_1$ and which parameter as the estimate $\hat{z}_2$, additional information about the flow is needed.

From the first equation of system (14), we obtain the estimate of the parameter γ

$$\hat{\gamma} = \hat{z}_1(1 - C_1\hat{z}_2)/(\hat{z}_1 - \hat{z}_2), \quad \hat{z}_2 \neq \hat{z}_1. \tag{16}$$

Let us introduce the density $p(\tau)$ in the following form

$$p(\tau) = \gamma(\lambda_1 + \alpha_1)e^{-(\lambda_1+\alpha_1)\tau} + (1 - \gamma)(\lambda_2 + \alpha_2)e^{-(\lambda_2+\alpha_2)\tau}, \quad \tau \geq 0.$$

And let us solve the problem of estimating parameters λ_1, λ_2, α_1, α_2 of the density function by the method of moments with a known value of the parameter γ. We note that the parameters under estimation coincide with the flow parameters.

Let us have a sample τ_1, τ_2, ..., τ_n from the distribution $p(\tau|\lambda_1, \lambda_2, \alpha_1, \alpha_2)$ depending on four unknown parameters. So for estimating λ_1, λ_2, α_1, α_2 it is necessary to have four moment equations, i.e. $M(\tau^l) = C_l$, $l = 1, 2, 3, 4$. As a result of the necessary transformations, the system of moments equations takes the form

$$(\lambda_1 + \alpha_1)(\lambda_2 + \alpha_2)C_1 - (\lambda_2 + \alpha_2)\gamma - (\lambda_1 + \alpha_1)(1 - \gamma) = 0,$$

$$((\lambda_1 + \alpha_1) + (\lambda_2 + \alpha_2))C_1 - 1/2(\lambda_1 + \alpha_1)(\lambda_2 + \alpha_2)C_2 = 1,$$

$$((\lambda_1 + \alpha_1) + (\lambda_2 + \alpha_2))C_2 - 1/3(\lambda_1 + \alpha_1)(\lambda_2 + \alpha_2)C_3 = 2C_1, \tag{17}$$

$$((\lambda_1 + \alpha_1) + (\lambda_2 + \alpha_2))C_3 - 1/4(\lambda_1 + \alpha_1)(\lambda_2 + \alpha_2)C_4 = 3C_2.$$

Theorem 3. *The system (17) for the unknown parameters λ_1, λ_2, α_1, α_2 of a semi-synchronous event flow of the second order is incompatible.*

Proof. We reduce the system (17) to a linearly inhomogeneous system by letting

$$x_1 = \lambda_2+\alpha_2-\lambda_1-\alpha_1, x_2 = \lambda_1+\alpha_1, x_3 = \lambda_2+\alpha_2+\lambda_1+\alpha_1, x_4 = (\lambda_1+\alpha_1)(\lambda_2+\alpha_2).$$

The resulting system of linear inhomogeneous equations for four unknowns x_i, $i = 1, 2, 3, 4$, is the following

$$\gamma x_1 + x_2 - C_1 x_4 = 0, \ C_1 x_3 - C_2 x_4/2 = 1, \ C_2 x_3 - C_3 x_4/3 = 2C_1, \tag{18}$$

$$C_3 x_3 - C_4 x_4/4 = 3C_2.$$

The system (18) has no solutions, i.e. the system is incompatible and it is impossible to estimate the flow parameters λ_1, λ_2, α_1, α_2 having only information about the density $p(\tau)$.

4.2   Special Case of Setting Flow Parameters

Let us introduce the density (12) in the following form

$$p(\tau) = [z - a(1 - z\tau)]e^{-z\tau}, \; z = \lambda_1 + \alpha_1, \; a = \pi_2(0)\alpha_2, \; \tau \geq 0. \quad (19)$$

Let us have a sample $\tau_1, \tau_2, ..., \tau_n$ from the distribution $p(\tau|z, a)$ depending on two unknown parameters z, a. Let $M(\tau^l) = \int_0^\infty \tau^l p(\tau|z, a)d\tau$ be the initial theoretical moment of the lth order which is a function of the unknown parameters. Considering the form of the density (19), we obtain $M(\tau^l) = l!/z^i [1 + a/z]$, $l = 1, 2$.

Then the system of moments equations takes the following form

$$[z + a]/z^2 = C_1, \; [z + 2a]/z^3 = C_2. \quad (20)$$

Solving the system (20), we find

$$\hat{z}^{(1)} = \frac{1}{C_2}\left(2C_1 - \sqrt{4C_1^2 - 2C_2}\right)/C_2, \; \hat{z}^{(2)} = \frac{1}{C_2}\left(2C_1 + \sqrt{4C_1^2 - 2C_2}\right). \quad (21)$$

In this case, the following conditions must be satisfied

$$\hat{z}^{(1)}\hat{z}^{(2)} = 2/C_2 > 0, \; \hat{z}^{(1)} + \hat{z}^{(2)} = 4C_1/C_2 > 0, \; 4C_1^2 - 2C_2 \geq 0.$$

Substituting the estimates $\hat{z}^{(1)}$ and $\hat{z}^{(1)}$ into the first equation of (20), we obtain

$$\hat{a}^{(1)} = C_1(\hat{z}^{(1)})^2 - \hat{z}^{(1)}, \; \hat{a}^{(2)} = C_1(\hat{z}^{(2)})^2 - \hat{z}^{(2)}. \quad (22)$$

The question naturally arises of which pair $\{\hat{z}^{(1)}, \hat{a}^{(1)}\}$ and $\{\hat{z}^{(2)}, \hat{a}^{(2)}\}$ to choose as a solution of the problem. Taking into account the explicit form of the parameters $z = \lambda_1 + \alpha_1$, $a = \pi_2(0)\alpha_2$ and the condition $(\lambda_1 + \alpha_1) - (\lambda_2 + \alpha_2) = 0$, it is easy to show that $\hat{z}$ and $\hat{a}$ must satisfy the conditions $\hat{a} > 0$ and $\hat{z} - \hat{a} > 0$. During the analytical test of the obtained pairs for the fulfillment of these conditions, the following conclusions were made. If $C_2/2 \leq C_1^2 < 2C_2/3$ then both pairs are equal and any of them can be chosen as an estimate; if $C_1^2 \geq 2C_2/3$ then only $\{\hat{z}^{(1)}, \hat{a}^{(1)}\}$ can be chosen as the solution of the estimation problem.

Remark. It can be shown that for the special case of setting flow parameters $(\lambda_1 + \alpha_1) - (\lambda_2 + \alpha_2) = 0$, a system of three or more moment equations is incompatible.

5   Results of Numerical Calculations

In order to establish the quality of estimation, we developed the algorithm for calculating parameter estimates of the density. The algorithm consists of two stages. The simulation of the semi-synchronous event flow of the second order is performed directly at the first stage of the implementation algorithm. Parameter estimates are calculated at the second stage of the algorithm using the formulas obtained.

Let us consider general case of setting the parameters of the flow. For this case we compute the estimates $\hat{z}_1$, $\hat{z}_2$, γ by the formulas (15), (16) and we obtain sample averages $\hat{M}(\hat{\theta}) = \frac{1}{N}\sum_{k=1}^{N}\hat{\theta}^{(k)}$ and offset estimates $\delta(\theta) = |\hat{M}(\hat{\theta}) - \theta|$, where $\theta \in \{z_1, z_2, \gamma\}$, $\hat{\theta} \in \{\hat{z}_1, \hat{z}_2, \hat{\gamma}\}$.

In the first statistical experiment, we consider the dependence of $\hat{M}(\hat{\theta})$, $\delta(\theta)$ from the values $\lambda_1 = 2, 3, 4, 5, 6$ for simulation time $T = 100$ units of time, $N = 100$, probabilities $P_1^{(1)}(\lambda_1|\lambda_1) = P_1^{(2)}(\lambda_2|\lambda_1) = 0,4$, $P_1^{(1)}(\lambda_2|\lambda_1) = P_1^{(2)}(\lambda_1|\lambda_1) = 0,6$ and parameters $\lambda_2 = 0,8$, $\alpha_1 = 2$, $\alpha_2 = 0,8$. The experiment results are given in Table 1, where the last three rows are the real values of the parameters under evaluation.

Table 1. The results of the first statistical experiment for z_1, z_2, γ

λ_1	2	3	4	5	6
$\hat{M}(\hat{z}_1)$	4,3298	5,2167	6,1552	7,0959	8,0483
$\delta(z_1)$	0,3298	0,2167	0,1552	0,0959	0,0483
$\hat{M}(\hat{z}_2)$	3,1902	3,0743	3,0377	2,9545	2,9151
$\delta(z_2)$	0,3902	0,2743	0,2377	0,1545	0,1151
$\hat{M}(\hat{\gamma})$	0,3621	0,3393	0,3291	0,2909	0,2717
$\delta(\gamma)$	0,2510	0,1846	0,1513	0,0989	0,0701
z_1	4	5	6	7	8
z_2	2,8	2,8	2,8	2,8	2,8
γ	0,1111	0,1547	0,1778	0,1920	0,2016

Analyzing the numerical results given in Table 1, we can make the following conclusions. There is a displacement in the obtained estimates by an amount $\delta(\theta)$ relative to the initial values of the estimated parameters. The offset $\delta(\theta)$ decreases with the increase of the parameter λ_1. This is due to the fact that the frequency of transitions from the first state to the second of the process $\lambda(t)$ increases with the increase of the parameter λ_1, which has a positive effect on the conditions of states distinguishability.

Let us consider special case of setting the parameters of the flow. For this case we compute the estimates $\hat{z}$, $\hat{a}$, γ by the formulas (21), (22) and we obtain sample averages $\hat{M}(\hat{\theta})$ and offset estimates $\delta(\theta)$, where $\theta \in \{z, a\}$, $\hat{\theta} \in \{\hat{z}, \hat{a}\}$.

In the second statistical experiment, we consider the dependence of $\hat{M}(\hat{\theta})$, $\delta(\theta)$ from the simulation time values T_m for fixed $N = 100$, for the probabilities $P_1^{(1)}(\lambda_1|\lambda_1) = P_1^{(2)}(\lambda_1|\lambda_1) = 0,65$, $P_1^{(1)}(\lambda_2|\lambda_1) = P_1^{(2)}(\lambda_2|\lambda_1) = 0,35$ and flow parameters $\lambda_1 = 4$, $\lambda_2 = 1,5$, $\alpha_1 = 0,5$, $\alpha_2 = 3$. The results of the experiment are given in Table 2, where the last two rows are the real values of the parameters under evaluation.

Analysis of the numerical results given in Table 2 shows that the offset $\delta(\theta)$, $\theta \in \{z, a\}$, decreases with the increase of the T_m, which is quite normal. In other

Table 2. The results of the second statistical experiment for z, a

T_m	50	100	150	200	...	900	950	1000
$\hat{M}(\hat{z})$	4,8894	4,7864	4,7664	4,7463	...	4,5398	4,5397	4,5388
$\delta(z)$	0,3894	0,2864	0,2664	0,2463	...	0,0398	0,0397	0,0388
$\hat{M}(\hat{a})$	1,6402	1,5211	1,4984	1,4698	...	1,3777	1,3775	1,3769
$\delta(a)$	0,2998	0,1807	0,1580	0,1294	...	0,0373	0,0371	0,0365
z	4,5	4,5	4,5	4,5	...	4,5	4,5	4,5
a	1,3404	1,3404	1,3404	1,3404	...	1,3404	1,3404	1,3404

words, the quality of estimating the density parameters is the better (in the sense of reducing the offset estimates), the larger the simulation time T_m.

In the third statistical experiment, we consider the dependence of $\hat{M}(\hat{\theta})$, $\delta(\theta)$ from the parameter $\lambda_1 = 2, 3, 4, 5, 6$ for simulation time $T = 100$ units of time, $N = 100$, probabilities $P_1^{(1)}(\lambda_1|\lambda_1) = P_1^{(2)}(\lambda_2|\lambda_1) = 0,4$, $P_1^{(1)}(\lambda_2|\lambda_1) = P_1^{(2)}(\lambda_1|\lambda_1) = 0,6$ and flow parameters $\lambda_2 = 1$, $\alpha_1 = 0,8$. And the parameter α_2 is calculated by the formula $\alpha_2 = \lambda_1 + \alpha_1 - \lambda_2$. The results of this experiment are given in Table 3, where the last two rows are the real values of the parameters under evaluation.

Table 3. The results of the third statistical experiment for z, a

λ_1	2	3	4	5	6
α_2	1,8	2,8	3,8	4,8	5,8
$\hat{M}(\hat{z})$	3,1823	3,9902	4,9281	5,8682	6,8279
$\delta(z)$	0,3823	0,1902	0,1281	0,0682	0,0279
$\hat{M}(\hat{a})$	1,5620	1,9810	2,4835	3,0254	3,5955
$\delta(a)$	0,3942	0,2133	0,1165	0,0592	0,0299
z	2,8	3,8	4,8	5,8	6,8
a	1,1678	1,7677	2,3670	2,9662	3,5656

Analyzing the numerical results obtained in Table 3, we can make the following conclusions. With the increase of the parameter value λ_1 and, respectively, with the increase of the parameter α_2, the offset value $\delta(\theta)$ decreases, since the states of the process $\lambda(t)$ become more distinguishable ($\lambda_1 > \lambda_2$).

6   Conclusion

In this paper, the semi-synchronous event flow of the second order was considered, and the explicit form of the probability density of the values of the interval duration between the moments of the events occurrence was obtained for general and special cases of setting the flow parameters. The estimates of probability density parameters were found by the method of moments. The expressions for parameter estimates are obtained explicitly, which allows for calculations without the use of numerical methods. The algorithm for calculating estimates of density parameters is implemented in C# with Visual Studio 2013. In order to establish the quality of estimation, statistical experiments were performed, the numerical results of which do not contradict the physical interpretation.

References

1. Basharin, G.P., Kokotushkin, V.A., Naumov, V.A.: On the equivalent substitutions method for computing fragments of communication networks. Izv. Akad. Nauk USSR. Tekhn. Kibern. **6**, 92–99 (1979). (in Russian)
2. Neuts, M.F.: A versatile Markov point process. J. Appl. Probab. **16**, 764–779 (1979)
3. Cox, D.R.: The analysis of non-Markovian stochastic processes by the inclusion of supplementary variables. In: Proceedings of the Cambridge Philosophical Society, vol. 51, no. 3, pp. 433–441 (1955)
4. Lucantoni, D.M.: New results on the single server queue with a bath Markovian arrival process. Commun. Stat. Stoch. Models **7**, 1–46 (1991)
5. Dudin, A.N., Klimenok, V.I.: Queueing Systems with Correlated Flows. Belarus Gos. Univ., Minsk (2000). (in Russian)
6. Basharin, G.P., Gaidamaka, Y.V., Samouylov, K.E.: Mathematical theory of teletraffic and its application to the analysis of multiservice communication of next generation networks. Autom. Control Comput. Sci. **47**(2), 62–69 (2013)
7. Klimenok, V., Dudin, A., Vishnevsky, V.: Tandem queueing system with correlated input and cross-traffic. In: Kwiecień, A., Gaj, P., Stera, P. (eds.) CN 2013. CCIS, vol. 370, pp. 416–425. Springer, Heidelberg (2013). https://doi.org/10.1007/978-3-642-38865-1_42
8. Vishnevsky, V.M., Semenova, O.V.: Polling Systems: Theory and Applications for Broadband Wireless Networks. Academic Publishing, London (2012)
9. Nezhelskaya, L.: Optimal state estimation in modulated MAP event flows with unextendable dead time. In: Dudin, A., Nazarov, A., Yakupov, R., Gortsev, A. (eds.) ITMM 2014. CCIS, vol. 487, pp. 342–350. Springer, Cham (2014). https://doi.org/10.1007/978-3-319-13671-4_39
10. Nezhelskaya, L., Tumashkina, D.: Optimal state estimation of semi-synchronous event flow of the second order under its complete observability. In: Dudin, A., Nazarov, A., Moiseev, A. (eds.) ITMM/WRQ -2018. CCIS, vol. 912, pp. 93–105. Springer, Cham (2018). https://doi.org/10.1007/978-3-319-97595-5_8
11. Kalyagin, A.A., Nezhelskaya, L.A.: Comparison of MP- and MM-estimates of the duration of the dead time in generalized semi-synchronous event flow. Tomsk State Univ. J. Control Comput. Sci. **3**(32), 23–32 (2015). (in Russian)
12. Nezhel'skaya, L.: Probability density function for modulated MAP event flows with unextendable dead time. In: Dudin, A., Nazarov, A., Yakupov, R. (eds.) ITMM 2015. CCIS, vol. 564, pp. 141–151. Springer, Cham (2015). https://doi.org/10.1007/978-3-319-25861-4_12
13. Gortsev, A.M., Nezhel'skaya, L.A.: Estimate of parameters of synchronously alternating Poisson stream of events by the moment method. Telecommun. Radio Eng. **50**(1), 56–63 (1996)

14. Gortsev, A.M., Klimov, I.S.: Estimation of the parameters of an alternating Poisson stream of events. Telecommun. Radio Eng. **48**(10), 40–45 (1993)
15. Nezhelskaya, L.A., Tumashkina, D.A.: Optimal state estimation of semi-synchronous event flow of the second order with non-extending dead time. Tomsk State Univ. J. Control Comput. Sci. **46**, 73–82 (2019). (in Russian)
16. Nezhelskaya, L.A.: The joint probability density of the duration of the intervals of modulated MAP-flow of events and the conditions of the flow recurrence. Tomsk State Univ. J. Control Comput. Sci. **1**(30), 57–67 (2015). (in Russian)

Special Retrial Queues
with State-Dependent Input Rate

Ivan Atencia[1], Eugene Lebedev[2], Vadym Ponomarov[2(✉)], and Hanna Livinska[2]

[1] Department of Applied Mathematics, Higher Polytechnic School,
University of Malaga, 29071 Malaga, Spain
`iatencia@ctima.uma.es`
[2] Applied Statistics Department, Taras Shevchenko National University of Kyiv,
Volodymyrska Str., 64, Kyiv 01601, Ukraine
`leb@unicyb.kiev.ua`, `vponomarov@gmail.com`, `livinskaav@gmail.com`

Abstract. This paper studies Markovian models of controlled systems with retrials, where the rate of the flow of retrials does not depend on the amount of their sources. The input flow of customers is controlled according to a multithreshold strategy. The conditions for the existence of a stationary regime are defined for this type of system, and clear formulae of vector-matrix type for stationary probabilities are obtained. The problem of multicriteria optimization of the system's profit are also considered.

Keywords: Queueing · Repeated calls · Constant retrial rate · Stationary regime · Optimal control

1 Introduction

Queueing systems with retrials are a specific class of stochastic models that allow consideration of an important peculiarity of the service process. The customer that upon its arrival finds all the servers busy will join a source of retrial customers for trying for service later on. This type of models is widely used in computer and communication networks, call centers, planes landing management in airports, etc.

The mathematical analysis of systems with retrials has its peculiarities. The consideration of retrials leads to a multidimensional service process, typical for stochastic networks, and, as a consequence, it complicates the theory. The main models of systems with retrials, their problems and some results can be found in [7,9].

This paper studies a specific class of Markovian systems with retrials that have two peculiarities. Firstly, it is admitted that the rate of the input flow depends on the number of retrials at the present moment of time. The situation when this dependence is represented by a piecewise constant function or by a threshold strategy is studied in details. Threshold strategies can solve the

© Springer Nature Switzerland AG 2019
A. Dudin et al. (Eds.): ITMM 2019, CCIS 1109, pp. 73–85, 2019.
https://doi.org/10.1007/978-3-030-33388-1_7

problems of optimal choice of the model parameters. The solutions for this type of problems are given at the end of this paper.

Secondly, unlike in the conventional models of systems with retrials it is assumed that the rate of retrials does not depend on the number of their sources. In papers [3,4] systems with a constant intensity of retrials have been used to model multi-access protocols (CSMA/CD, ALOHA). Other uses in the information flows modeling in Internet can be found in [2].

2   Mathematical Model of the Service Process

In order to define the service process in the retrial system, let us introduce a two-dimensional Markov chain with continuous time $X(t) = (X_1(t), X_2(t))'$, $X_1(t) \in \{0, 1, ..., c\}$, $X_2(t) \in \{0, 1, ...\}$, defined by its local parameters $\alpha_{(i,j)(i',j')}$, $(i,j),(i',j') \in S(X) = \{0, 1, ..., c\} \times \{0, 1, ...\}$ in the following way.

1. For $0 \leq i \leq c - 1$

$$
\alpha_{(i,j)(i',j')} = \begin{cases} \lambda_j, & when & (i',j') = (i+1,j); \\ i\nu, & when & (i',j') = (i-1,j); \\ \mu, & when & (i',j') = (i+1,j-1); \\ -(\lambda_j + i\nu + \mu), & when & (i',j') = (i,j); \\ 0, & otherwise. \end{cases}
$$

2. For $i = c$

$$
\alpha_{(i,j)(i',j')} = \begin{cases} \lambda_j, & when & (i',j') = (c,j+1); \\ c\nu, & when & (i',j') = (c-1,j); \\ -(\lambda_j + c\nu), & when & (i',j') = (c,j); \\ 0, & otherwise. \end{cases}
$$

The two-dimensional Markov chain, whose local parameters matrix has the described above structure, can be interpreted as a service process in a retrial system that consists of c identical servers. The service rate on each server is $\nu > 0$, the rate of retrials $\mu > 0$, and $\lambda_j > 0$ is the rate of the input flow with the condition that there are j sources of retrials in the system. The first component $X_1(t) \in \{0, 1, ..., c\}$ indicates the number of busy servers at the instant $t \geq 0$, and the second one $X_2(t) \in \{0, 1, ...\}$ is the number of retrial sources. Further, the process $X(t) = (X_1(t), X_2(t))'$ is the main subject of our investigation.

Let us write the states of $X(t)$ as $S(X) = \{(0,0), ..., (c,0), (0,1), ..., (c,1), ...\}$. Then the infinitesimal matrix $\mathcal{Q}$ of the Markov chain $X(t)$ can be represented in a matrix-block form:

$$
\mathcal{Q} = \begin{pmatrix} \mathcal{Q}^{(0,0)} & \mathcal{Q}^{(0,+1)} & & \\ \mathcal{Q}^{(1,-1)} & \mathcal{Q}^{(1,0)} & \mathcal{Q}^{(1,+1)} & \\ \ddots & \ddots & \ddots & \\ & \mathcal{Q}^{(j,-1)} & \mathcal{Q}^{(j,0)} & \mathcal{Q}^{(j,+1)} \\ & & \ddots & \ddots & \ddots \end{pmatrix},
$$

where

$$\mathcal{Q}^{(j,-1)} = \|q_{nm}^{(j,-1)}\|_{n,m=0}^{c} = \begin{pmatrix} 0 & \mu & & & \\ & 0 & \mu & & \\ & & \ddots & \ddots & \ddots & \\ & & & 0 & \mu \\ & & & & 0 \end{pmatrix},$$

$$\mathcal{Q}^{(j,0)} = \|q_{nm}^{(j,0)}\|_{n,m=0}^{c} = \begin{pmatrix} s_{0j} & \lambda_j & & & \\ \nu & s_{1j} & \lambda_j & & \\ & 2\nu & s_{2j} & \lambda_j & \\ & & \ddots & \ddots & \ddots \\ & & & (c-1)\nu & s_{c-1j} & \lambda_j \\ & & & & c\nu & s_{cj} \end{pmatrix},$$

$$s_{ij} = -[\lambda_j + i\nu + (1 - \delta_{ic})(1 - \delta_{j0})\mu],$$

where δ_{ij} is a Kronecker delta and $\mathcal{Q}^{(j,+1)} = \|q_{nm}^{(j,+1)}\|_{n,m=0}^{c} = diag(0, \ldots, 0, \lambda_j)$ is a diagonal matrix.

It follows that $X(t)$ is a quasi-birth-and-death process (QBD process), whose blocks, generally speaking, depend on the level number (see for example [11], p. 189). If $\lambda_j = \lambda$, then we have a normal Poisson process of rate λ at the system input, and the matrices $\mathcal{Q}^{(j,\pm1)} = \mathcal{Q}^{(\pm1)} = \|q_{nm}^{(\pm1)}\|_0^c$, $\mathcal{Q}^{(j,0)} = \mathcal{Q}^{(0)} = \|q_{nm}^{(0)}\|_0^c$ become independent from $j \geq 1$. Thus, $X(t)$ becomes a level independent QBD process.

The ergodicity conditions, with an easy interpretation, are well known for these processes.

Let the infinitesimal matrix $\tilde{\mathcal{Q}} = \|\tilde{q}_{nm}\|_0^c = \mathcal{Q}^{(-1)} + \mathcal{Q}^{(0)} + \mathcal{Q}^{(+1)}$ be irreducible and $\rho = (\rho_0, \rho_1, \ldots, \rho_c)'$ be the unique solution of

$$\rho' \tilde{\mathcal{Q}} = 0'_{c+1}, \qquad \rho' 1_{c+1} = 1$$

where $0_{c+1}, 1_{c+1}$ are the $(c + 1)$-dimensional vectors formed by zeros and ones, respectively.

Then the irreducible QBD process is ergodic when

$$\rho' \mathcal{Q}^{(+1)} 1_{c+1} < \rho' \mathcal{Q}^{(-1)} 1_{c+1}. \tag{1}$$

This result persists if the process $X(t)$ is close to the level independent QBD process.

Lemma 1. *Let* $X(t) = (X_1(t), X_2(t))'$ *be an irreducible level dependent QBD process, that takes values in* $S(X) = \{0, 1, ..., c\} \times \{0, 1, ...\}$, *for which the following conditions are satisfied*

$$\lim_{j \to \infty} \mathcal{Q}^{(j,\pm1)} = \mathcal{Q}^{(\pm1)}, \qquad \lim_{j \to \infty} \mathcal{Q}^{(j,0)} = \mathcal{Q}^{(0)},$$

where the matrix convergence is understood in componentwise meaning. If matrix $\tilde{\mathcal{Q}}$ *is irreducible and condition (1) is satisfied, then* $X(t)$ *is ergodic.*

Proof. For proving the lemma, we use the method of test functions (Lyapunov functions). To create such a function, let us introduce some terms:

$$q_n^{(\pm 1)} = \sum_{m=0}^{c} q_{nm}^{(\pm 1)}, \quad d_n = q_n^{(-1)} - q_n^{(+1)}, \quad n = 0, 1, \dots, c,$$

and

$$(\delta_0, \dots, \delta_{c-1})' = (Q^*)^{-1}[(d_0, \dots, d_{c-1})' - \varepsilon 1_c'], \tag{2}$$

where $\varepsilon = \sum_{n=0}^{c} p_n d_n > 0$, $Q^* = \|q_{\tilde{n}m}\|_0^{c-1}$ is the matrix of size $c \times c$, that is obtained from matrix $\tilde{Q}$ by deleting the last row and the last column; 1_c is a vector of size c, formed by ones.

Following paper [6], we use $\varphi(s) = \delta_n + j$ as a Lyapunov function $\varphi(s)$, $s = (n, j) \in S(X)$ ($\delta_n = 0$, when $n = c$). Let us verify the validity of the conditions of the ergodicity criterion from [7], p. 97.

Let

$$q_n^{(j, \pm 1)} = \sum_{m=0}^{c} q_{nm}^{(j, \pm 1)}, \quad d_n^{(j)} = q_n^{(j, -1)} - q_n^{(j, +1)},$$

$$\tilde{Q}^{(j)} = \|\tilde{q}_{nm}^{(j)}\|_{n,m=0}^{c} = Q^{(j, -1)} + Q^{(j, 0)} + Q^{(j, +1)},$$

$$\tilde{Q}^{(j)*} = \|\tilde{q}_{nm}^{(j)}\|_{n,m=0}^{c-1}, \quad y_s = y_{(n,j)} = \sum_{p \in S(X)} \alpha_{sp}(\varphi(p) - \varphi(s)), \quad s = (n, j) \in S(X).$$

Using (2), we obtain

$$(y_{(0,j)}, \dots, y_{(c-1,j)})' = -\varepsilon 1_c' + [(d_0, \dots, d_{c-1})' - (d_0^{(j)}, \dots, d_{c-1}^{(j)})] +$$
$$+ [Q^{(j)*} - Q^*](Q^*)^{-1}[(d_0, \dots, d_{c-1})' - \varepsilon 1_c'].$$

For $n = c$ we have

$$y_{(c,j)} = -\varepsilon + (\tilde{q}_{c0}^{(j)} - \tilde{q}_{c0}, \dots, \tilde{q}_{cc-1}^{(j)} - \tilde{q}_{cc-1})(\delta_0, \dots, \delta_{c-1})' - (d_c^{(j)} - d_c).$$

Based on the conditions of the lemma, it is possible to find such a number $j_0 = j_0(\varepsilon)$, that for all $n = 0, 1, \dots, c$ and $j \geq j_0$

$$y_{(n,j)} \leq -\varepsilon/2.$$

Therefore, process $X(t)$ is ergodic. The lemma is proved.

As for the retrial process with a constant intensity, the conditions of Lemma 1 are satisfied when:

$$\lambda_j > 0, \quad j = 0, 1, \dots; \quad \lim_{j \to \infty} \lambda_j = \lambda; \tag{3}$$

and

$$\frac{\lambda(\lambda + \mu)^c}{c! \mu \sum_{i=0}^{c-1} \frac{(\lambda + \mu)^i}{i! \nu^{i-c}}} < 1. \tag{4}$$

Consequently, Lemma 1 allows to obtain the following result.

Corollary 1. *If conditions (3) and (4) are satisfied for the retrial process with a constant rate $X(t)$, then the process is ergodic.*

Further, we suppose for the rate of the input flow that its dependence on the number of sources of retrials is defined by the threshold strategy. It means that, for an integer $l \geq 2$, thresholds $0 = H_0 < H_1 < \ldots < H_{l-1} < H_l = \infty$, $H' = (H_1, \ldots, H_{l-1})$, are fixed. If at an instant $t \geq 0$ the number of retrial sources is $X_2(t) \in [H_{i-1}, H_i)$, then the system is working in the ith regime and the rate of the input flow is $\lambda^{(i)}$. Thus, λ_j is a piecewise constant function: $\lambda_j = \lambda^{(i)}$, $j \in [H_{i-1}, H_i)$, $i = 1, \ldots, l$.

If $\lambda^{(i)} > 0$, $i = 1, \ldots, l$, and for $\lambda = \lambda^{(l)}$ condition (4) is satisfied, then for the process $X(t)$ there exists a stationary regime and our next goal is to find its stationary probabilities π_{ij}, $(i, j) \in S(X)$.

3 Investigation of the Process in the Stationary Regime

In order to find the stationary probabilities π_{ij}, let us use the theorem about the equality of probability flows across the border of the closed area in a stationary regime (see, for example, [12], Chapter II). For every $j = 0, 1, \ldots$ we create a partition of the phase space $S(X) = S_j^{(1)}(X) \cup \bar{S}_j^{(1)}(X)$, $S_j^{(1)}(X) = \{(p, q) \in S(X) : q \leq j\}$. Equaling the flows of probabilities through the border of the area $S_j^{(1)}(X)$, we obtain:

$$\lambda_j \pi_{cj} = \mu \sum_{i=0}^{c-1} \pi_{ij+1}, \quad j = 0, 1, \ldots \tag{5}$$

Now, for $i = 0, 1, \ldots, c - 1$, $j = 0, 1, \ldots$ we create a partition of the phase space $S(X) = S_{ij}^{(2)}(X) \cup \bar{S}_{ij}^{(2)}(X)$, $S_{ij}^{(2)}(X) = \{(i, j)\}$. Equaling the flows of probabilities through the border of the area $S_{ij}^{(2)}(X)$, we obtain the following system of equations:

$$[\lambda_j + (1 - \delta_{j0})\mu]\pi_{0j} = \nu\pi_{1j}, \quad j = 0, 1, \ldots \tag{6}$$

$$[\lambda_j + (1 - \delta_{j0})\mu + i\nu]\pi_{ij} = \lambda_j \pi_{i-1j} + (i + 1)\nu\pi_{i-1j} + \mu\pi_{i-1j+1},$$
$$i = 1, \ldots, c - 1, \quad j = 0, 1, \ldots \tag{7}$$

Let us introduce notations for the matrices that depend on the system parameters:

- $A_j = \| a_{ik}(j) \|_{i,k=1}^{c}$, $j = 0, 1, \ldots$, are matrices with elements $a_{ii-1}(j) = \mu$, $i = 1, 2, \ldots, c - 1$,

$$a_{ck}(j) = \begin{cases} \dfrac{c\mu\nu}{\lambda_j}, & k \neq c - 1, \\[2mm] \dfrac{\mu[\lambda_j + c\nu]}{\lambda_j}, & k = c - 1, \end{cases}$$

and the rest of the elements equal to 0;

- $B_j =\| b_{ik}(j) \|_{i,k=1}^c$, $j = 0, 1, \ldots$, are three-diagonal matrices with elements $b_{ii-1}(j) = -\lambda_j$, $c_i = 2, \ldots, c$, $j = 0, 1, \ldots$, $b_{ii}(j) = \lambda_j + \mu + (i-1)\nu$, $i = 1, \ldots, c$, $j = 0, 1, \ldots$, $b_{ii+1}(j) = -i\nu$, $i = 1, \ldots, c-1$, $j = 0, 1, \ldots$;
- $C =\| c_{ik}(j) \|_{i,k=1}^c$, where $c_{11} = 1$, $c_{1k} = 0$, $k = 2, \ldots, c$, $c_{ik} = b_{i-1k}(H_{l-1})$, $i = 2, \ldots, c$, $k = 1, \ldots, c$.

Let us give the following statement whose proof is based on Hadamard criterion ([11], p. 419).

Lemma 2. *Matrices B_j and C are nonsingular.*

To find out a stationary distribution of the service process $X(t)$, let us study a similar process with a limited state space. This process describes the functioning of the stochastic system, similar to the original one, but that has a limit on the maximum length of the queue: new customers are lost when all the servers are busy and there are N sources of retrials in the system. Formally this kind of behavior is described by the Markov chain $X(t, N) = (X_1(t, N), X_2(t, N))'$, $X_1(t, N) \in 0, 1, \ldots, c$, $X_2(t, N) \in 0, 1, \ldots, N$. Its infinitesimal rates $a_{(i,j)(i',j')}^{(N)}$, $(i, j), (i', j') \in S(X, N) = \{0, 1, \ldots, c\} \times \{0, 1, \ldots, N\}$ coincide with the corresponding rates $a_{(i,j)(i',j')}$ of the chain $X(t)$ at all points except of the boundary case $i = c$, $j = N$,

$$
a_{(c,N)(i',j')}^{(n)} = \begin{cases} c\nu, & when \quad (i', j') = (c-1, N); \\ -c\nu, & when \quad (i', j') = (c, N); \\ 0, & otherwise \end{cases}
$$

Since the phase space $S(X, N)$ of the process $X(t, N)$ is finite, then there exists a stationary regime for $X(t, N)$, and we denote its stationary probabilities by $\pi_{ij}(N)$, $(i, j) \in S(X, N)$. Further, by the parameter N we realize the limit transition $N \to \infty$. Therefore, we assume that $N > H_{l-1}$.

For the vector of stationary probabilities $\pi_j(N) = (\pi_{0j}(N), \ldots, \pi_{c-1j}(N))'$, the following lemma holds.

Lemma 3. *The stationary probabilities of process $X(t, N)$ are linked by the following correlation:*

$$
\pi_j(N) = \Delta_j(N)\pi_{00}(N), \quad j = 0, 1, \ldots, N,
$$

where

$$
\Delta_0(N) = (B_0 - \mu E)^{-1} A_0 \Delta_1(N),
$$

$$
\Delta_j(N) = \frac{\left(\prod_{i=j}^{N-1} B_i^{-1} A_i \right) C^{-1} e_1}{e_1'(B_0 - \mu E)^{-1} B_0 \left(\prod_{i=j}^{N-1} B_i^{-1} A_i \right) C^{-1} e_1}, \quad j = 1, 2, \ldots,
$$

$$
e_1 = (\delta_{11}, \delta_{12}, \ldots, \delta_{1c})'.
$$

Proof. Probabilities $\pi_{ij}(N)$, $(i,j) \in S(X,N)$ of a truncated system satisfy the system of Eqs. (5)–(7) of the original system. Only a limit case $j = N$ requires a special consideration, for which the following correlations are met:

$$[\lambda_N + \mu]\pi_{0N}(N) = \nu\pi_{1N}(N),$$

$$[\lambda_N + \mu + i\nu]\pi_{iN}(N) = \lambda_N\pi_{i-1N}(N) + (i+1)\nu\pi_{i+1N}(N), \quad i = 1,\ldots,c-2.$$

Let us add the following identity into the last system of equations: $\pi_{0N}(N) = \pi_{0N}(N)$ and rewrite it in a vector-matrix form

$$C\pi_N(N) = e_1 \cdot \pi_{0N}(N).$$

From this we obtain

$$\pi_N(N) = C^{-1}e_1 \cdot \pi_{0N}(N). \tag{8}$$

The system of Eqs. (5)–(7) of the truncated system can be rewritten in a vector-matrix form:

$$A_j\pi_{j+1}(N) = B_j\pi_j(N), \quad j = 1,\ldots,N-1,$$

$$A_0\pi_1(N) = (B_0 - \mu E)^{-1}\pi_0(N).$$

Considering (8), we obtain

$$\pi_j(N) = \left(\prod_{i=j}^{N-1} B_i^{-1}A_i\right) C^{-1}e_1\pi_{0N}(N), \quad j = 1,\ldots,N, \tag{9}$$

$$\pi_0(N) = (B_0 - \mu E)^{-1}B_0\left(\prod_{i=0}^{N-1} B_i^{-1}A_i\right) C^{-1}e_1\pi_{0N}(N).$$

The last equation yields probability $\pi_{0N}(N)$:

$$\pi_{0N}(N) = \left\{e_1'(B_0 - \mu E)^{-1}B_0\left(\prod_{i=0}^{N-1} B_i^{-1}A_i\right) C^{-1}e_1\right\}^{-1} \cdot \pi_{00}(N).$$

Substituting the last equation into (9), we obtain the confirmation of the lemma.

The lemma is proved.

Let us investigate the limit behavior of vector $\Delta_j(N)$, for $N \to \infty$. The following result was obtained.

Lemma 4. *The limits of vectors $\Delta_j(N)$, $j = 0,1,\ldots$, for $N \to \infty$ are defined by the following correlations:*

$$\Delta_0 = (B_0 - \mu E)^{-1}A_0\Delta_1,$$

$$\Delta_j(N) = \frac{\left(B_{H_{k-1}}^{-1} A_{H_{k-1}}\right)^{H_{k-j}} \left(\prod_{i=k}^{l-2} \left(B_{H_i}^{-1} A_{H_i}\right)^{H_{i+1}-H_i}\right) u\nu' C^{-1} e_1}{e_1'(B_0 - \mu E)^{-1} B_0 \left(\prod_{i=0}^{l-2} \left(B_{H_i}^{-1} A_{H_i}\right)^{H_{i+1}-H_i}\right) u\nu' C^{-1} e_1},$$

$$j = H_{k-1}, \ldots, H_k - 1, \quad k = 1, \ldots, l-1,$$

$$\Delta_j(N) =$$
$$= \frac{u\nu' C^{-1} e_1}{e_1'(B_0 - \mu E)^{-1} B_0 \left(\prod_{i=0}^{l-2} \left(B_{H_i}^{-1} A_{H_i}\right)^{H_{i+1}-H_i}\right) \left(B_{H_{k-1}}^{-1} A_{H_{k-1}}\right)^{j-H_{l-1}} u\nu' C^{-1} e_1},$$

$$j = H_{l-1}, \ldots,$$

where $u = (u_1, u_2, \ldots, u_c)' > 0$, $\nu = (\nu_1, \nu_2, \ldots, \nu_c)' > 0$ are the right and the left eigenvectors of matrix $B_{H_{l-1}}^{-1} A_{H_{l-1}}$, which correspond to the Perron root r.

Proof. As the rate of input flow λ_j is a piecewise constant function, vector $\Delta_j(N)$ can be represented in the following way:

$$\Delta_j(N) = \frac{\left(\prod_{i=j}^{N-1} B_i^{-1} A_i\right) C^{-1} e_1}{e_1'(B_0 - \mu E)^{-1} B_0 \left(\prod_{i=0}^{N-1} B_i^{-1} A_i\right) C^{-1} e_1} =$$

$$= \frac{\left(\prod_{i=j}^{H_K-1} B_{H_{k-1}}^{-1} A_{H_{k-1}}\right) \cdot \ldots \cdot \left(\prod_{i=H_{l-1}}^{N-1} B_{H_{l-1}}^{-1} A_{H_{l-1}}\right) C^{-1} e_1}{e_1'(B_0 - \mu E)^{-1} B_0 \left(\prod_{i=H_0}^{H_1-1} B_{H_0}^{-1} A_{H_0}\right) \cdot \ldots \cdot \left(\prod_{i=H_{l-1}}^{N-1} B_{H_{l-1}}^{-1} A_{H_{l-1}}\right) C^{-1} e_1} =$$

$$= \frac{\left(B_{H_{k-1}}^{-1} A_{H_{k-1}}\right)^{H_k-j} \cdot \ldots \cdot \left(B_{H_{l-1}}^{-1} A_{H_{l-1}}\right)^{N-H_{l-j}} C^{-1} e_1}{e_1'(B_0 - \mu E)^{-1} B_0 \left(B_{H_0}^{-1} A_{H_0}\right)^{H_1-H_0} \cdot \ldots \cdot \left(B_{H_{l-1}}^{-1} A_{H_{l-1}}\right)^{N-H_{l-j}} C^{-1} e_1},$$

$$j = H_{k-1}, \ldots, H_k - 1, \quad k = 1, \ldots, l-1, \quad (10)$$

$$\Delta_j(N) = \frac{\left(B_{H_{l-1}}^{-1} A_{H_{l-1}}\right)^{N-j} C^{-1} e_1}{e_1'(B_0 - \mu E)^{-1} B_0 \left(B_{H_0}^{-1} A_{H_0}\right)^{H_1-H_0} \cdot \ldots \cdot \left(B_{H_{l-1}}^{-1} A_{H_{l-1}}\right)^{N-H_{l-j}} C^{-1} e_1}$$

$$= \frac{\left(B_{H_{l-1}}^{-1} A_{H_{l-1}}\right)^{N-j} C^{-1} e_1}{e_1'(B_0 - \mu E)^{-1} B_0 \left(B_{H_0}^{-1} A_{H_0}\right)^{H_1-H_0} \ldots \left(B_{H_{l-1}}^{-1} A_{H_{l-1}}\right)^{j-H_{l-j}} \left(B_{H_{l-1}}^{-1} A_{H_{l-1}}\right)^{N-j} C^{-1} e_1},$$

$$j = H_{l-1}, \ldots.$$

Matrix $B_{H_{l-1}}^{-1} A_{H_{l-1}} > 0$, therefore conditions of the theorem 8.2.8 [11] are met for it, and its ith power can be written as:

$$\left(B_{H_{l-1}}^{-1} A_{H_{l-1}}\right)^i = r^i u\nu' + o(r_1^i),$$

where $r_1 < r$, r is the Perron root of matrix $B_{H_{l-1}}^{-1} A_{H_{l-1}}$, u and ν are the right and the left eigenvectors, that correspond to r, $u'\nu = 1$.

Let us substitute this expression into the system of Eq. (10):

$$\Delta_j(N) =$$

$$\frac{\left(B_{H_{k-1}}^{-1} A_{H_{k-1}}\right)^{H_k-j} \left(\prod_{i=k}^{l-2}\left(B_{H_i}^{-1} A_{H_i}\right)^{H_{i+1}-H_i}\right)\left(r^{N-H_{i-1}} u\nu' + o\left(r_1^{N-H_{l-1}}\right)\right)C^{-1}e_1}{e_1'(B_0 - \mu E)^{-1} B_0\left(\prod_{i=0}^{l-2}\left(B_{H_i}^{-1} A_{H_i}\right)^{H_{i+1}-H_i}\right)\left(r^{N-H_{i-1}} u\nu' + o\left(r_1^{N-H_{l-1}}\right)\right)C^{-1}e_1},$$

$$j = H_{k-1},\ldots,H_k - 1, \quad k = 1,\ldots,l-1, \tag{11}$$

$$\Delta_j(N) =$$

$$= \frac{\left(r^{N-j} u\nu' + o\left(r_1^{N-j}\right)\right)C^{-1}e_1}{e_1'(B_0 - \mu E)^{-1} B_0\left(\prod_{i=0}^{l-2}\left(B_{H_i}^{-1} A_{H_i}\right)^{H_{i+1}-H_i}\right)\left(B_{H_{l-1}}^{-1} A_{H_{l-1}}\right)^{j-H_{l-1}}} \times$$

$$\times \frac{1}{\left(r^{N-j} u\nu' + o\left(r_1^{N-j}\right)\right)C^{-1}e_1}$$

$$j = H_{l-1},\ldots. \tag{12}$$

Now let us divide the numerator and the denominator of (11) by $r_1^{N-H_{l-1}}$, and the numerator and the denominator of (12) by r^{N-j} and proceed to the limit for $N \to \infty$:

$$\Delta_j = \lim_{N\to\infty} \Delta_j(N) =$$

$$= \frac{\left(B_{H_{k-1}}^{-1} A_{H_{k-1}}\right)^{H_k-j} \left(\prod_{i=k}^{l-2}\left(B_{H_i}^{-1} A_{H_i}\right)^{H_{i+1}-H_i}\right) u\nu'C^{-1}e_1}{e_1'(B_0 - \mu E)^{-1} B_0\left(\prod_{i=0}^{l-2}\left(B_{H_i}^{-1} A_{H_i}\right)^{H_{i+1}-H_i}\right) u\nu'C^{-1}e_1},$$

$$j = H_{k-1},\ldots,H_k - 1, \quad k = 1,\ldots,l-1,$$

$$\Delta_j = \lim_{N\to\infty} \Delta_j(N) =$$

$$= \frac{u\nu'C^{-1}e_1}{e_1'(B_0 - \mu E)^{-1} B_0\left(\prod_{i=0}^{l-2}\left(B_{H_i}^{-1} A_{H_i}\right)^{H_{i+1}-H_i}\right)\left(B_{H_{l-1}}^{-1} A_{H_{l-1}}\right)^{j-H_{l-1}} u\nu'C^{-1}e_1},$$

$$j = H_{k-1},\ldots.$$

Since the matrix $u\nu'$ is included in the numerator and the denominator for Δ_j, then the normalization condition $u\nu' = 1$ for eigenvectors u and v can be eliminated.

The lemma is proved.

Let us formulate and prove one more auxiliary result.

Lemma 5. *Let the condition of Lemma 1 be satisfied. Then*

$$\sum_{j=0}^{\infty} \frac{\lambda_{j-1} + \mu}{\lambda_{j-1}} 1_c' \Delta_j < \infty. \tag{13}$$

Proof. If the condition of Lemma 1 is satisfied, then the process $X(t)$ is ergodic, which means that there exists a probability $\pi_{00} > 0$. Using the results of the stochastic ordering of probability distributions for the migration processes from [7], we find

$$\lim_{N \to \infty} \pi_{00}(N) = \pi_{00} > 0. \tag{14}$$

From the normalization condition for the stationary distributions of the process $X(t, N)$, taking into consideration the results of Lemma 3, we obtain:

$$\pi_{00}(N) = \left\{ 1_c' \Delta_0(N) + \sum_{j=1}^{N} \frac{\lambda_{j-1} + \mu}{\lambda_{j-1}} 1_c' \Delta_j(N) \right\}^{-1}.$$

We prove the lemma using the method of contradiction. We will suppose that (13) does not converge. This means that for any large $L > 0$, there exists such a number $M = M(L)$ that

$$\sum_{j=0}^{M} \frac{\lambda_{j-1} + \mu}{\lambda_{j-1}} 1_c' \Delta_j > L.$$

It is not difficult to check that

$$\pi_{00}^{-1} = \lim_{N \to \infty} \pi_{00}^{-1}(N) = \lim_{N \to \infty} \left\{ 1_c' \Delta_0(N) + \sum_{j=1}^{N} \frac{\lambda_{j-1} + \mu}{\lambda_{j-1}} 1_c' \Delta_j(N) \right\} \geq$$

$$\geq \lim_{N \to \infty} \left\{ 1_c' \Delta_0(N) + \sum_{j=1}^{N} \frac{\lambda_{j-1} + \mu}{\lambda_{j-1}} 1_c' \Delta_j(N) \right\} =$$

$$1_c' \Delta_0(N) + \sum_{j=1}^{N} \frac{\lambda_{j-1} + \mu}{\lambda_{j-1}} 1_c' \Delta_j(N) > 1_c' \Delta_0(N) + L$$

Thus, $\pi_{00} = 0$, which contradicts (14).
The lemma is proved.

For $N \to \infty$ stationary distributions $\pi_{ij}(N)$ approximates the corresponding probabilities of the original system. On the basis of previously proved Lemmas we obtain the main result, which contains explicit vector-matrix formulae for the stationary probabilities of the system written via its parameters.

Theorem 1. *If the condition of Lemma 1 is met for the process $X(t)$, then:*

$$\pi_j = \lim_{N \to \infty} \pi_j(N) = \Delta_j \cdot \pi_{00} \quad j = 0, 1, \ldots,$$

$$\pi_{cj} = \frac{\mu}{\lambda_j} 1_c' \Delta_{j+1} \cdot \pi_{00} \quad j = 0, 1, \ldots, \quad (15)$$

where $\pi_{00} = \left\{ 1_c' \Delta_0 + \sum_{j=1}^{N} \frac{\lambda_{j-1}+\mu}{\lambda_{j-1}} 1_c' \Delta_j \right\}^{-1}$.

The theorem is a direct consequence of Lemmas 3 and 4. The probabilities $\pi_{cj}, j = 0, 1, \ldots$ are expressed by Eq. (5), and the probability π_{00} is found from the normalization condition.

Since $|B_j^{-1} A_j| = |B_j^{-1}| \cdot |A_j|$, then the characteristic equation $|B_j^{-1} A_j - zE| = 0$ has a root $z = 0$. Consequently, for $M/M/c/\infty$-systems for $c = 1, 2, \ldots, 5$, we can always express the Perron root and stationary probabilities explicitly in terms of the system parameters. To calculate the Perron roots and corresponding eigenvectors for $c > 5$ (and for $c \leq 5$) we can use efficient computational algorithms from [11] (Sec. 8.5).

4   Application in Optimization

As an example of using obtained results, let us consider the problem of optimizing the profit from the system's functioning.

Let $f_k(t, H)$ be the profit from the system's functioning in the kth regime, $k = 1, \ldots, l$; $f_{l+1}(t, H)$ be the number of customers that were refused for service and became sources of retrials; $f_{l+2}(t, H)$ be the number of switches in rate of the input flow. If the conditions of Lemma 1 are met, there exist limits $\lim_{t \to \infty} t^{-1} f_k(t, H)$, $k = 1, \ldots, l + 2$. Let us denote them by $f_k(H)$, $k = 1, \ldots, l + 2$.

The objective of maximizing the profit from the system's functioning is to find the values of the thresholds H_k, $k = 1, \ldots, l - 1$, which are the solution of the multi-criteria problem:

$$f_k(H) \to max, \quad k = 1, \ldots, l,$$

$$f_{l+1}(H) \to min, \quad f_{l+2}(H) \to min,$$

$$H_k \in \{0, 1, \ldots\}, \quad k = 1, \ldots, l - 1,$$

$$H_i < H_j, \quad i < j.$$

Similar optimization problems for other models of retrial queues were considered, for example, in [5,8,10]. If we take into account the economic nature of the problem, the most logical method to solve it in practice is the method of linear convolution of criteria. It consists in the fact that the solution of the original multi-objective problem is found by solving the one-criterion problem:

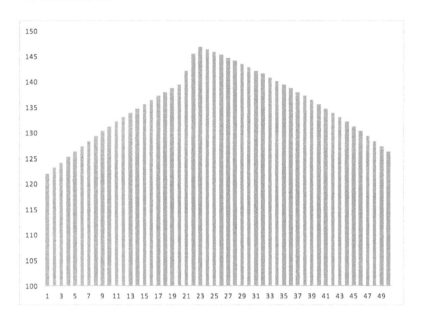

Fig. 1. Dependence of the objective function on the threshold value.

$$\sum_{i=1}^{l} C_i f_i(H) - C_{l+1} f_{l+1}(H) - C_{l+2} f_{l+2}(H) \to max,$$

$$H_i \in \{0, 1, \ldots\}, \quad k = 1, \ldots, l - 1,$$

$$H_i < H_j, \quad i < j.$$

where C_i, $i = 1, \ldots, l$, is the profit from the servicing of a customer when the system works in the ith regime; C_{l+1} is a fine for the service rejection; C_{l+2} is a fine for switching the service rate.

The limit functionals $f_k(H)$, $k = 1, \ldots, l + 2$, can be written using the stationary distribution of the system:

$$f_k(H) = \nu \sum_{i=1}^{c} \sum_{j=H_{k-1}}^{H_{k-l}} i \pi_{ij} \quad k = 1, \ldots, l,$$

$$f_{l+1}(H) = \sum_{j=0}^{\infty} \lambda_j \pi_{cj} = \sum_{i=1}^{l} \lambda^{(i)} \sum_{j=H_{i-1}}^{H_{i-l}} \pi_{cj},$$

$$f_{l+2}(H) = \sum_{i=1}^{l} \lambda^{(i)} \pi_{cH_i - 1} + \mu \sum_{i=1}^{l} \sum_{k=0}^{c-l} \pi_{kH},$$

which together with the result of Theorem 1 gives the algorithm of solving of the described optimization problem.

To illustrate the obtained results, let us consider an example of solving the problem of choosing the optimal threshold control strategy for the system with two functioning regimes and the following parameters: $\lambda^{(1)} = 13$, $\lambda^{(2)} = 8$, $\nu = 8.1$, $\mu = 1$. The above-described multi-criteria optimization problem will be solved by the method of linear convolution of criteria with coefficients $C_1 = 10, C_2 = 300, C_3 = 30, C_4 = 100$.

Figure 1 shows a graph of dependence of the objective function on the threshold value. As seen in the graph, the highest value of the criteria 146,87 is reached at $H = 23$.

5   Conclusions

The paper studies the stationary regime for systems with retrials of constant intensity, whose input flow of customers is controlled by multithreshold strategies. Explicit formulae of vector-matrix type for stationary probabilities via the system parameters are obtained. The obtained representation of stationary probabilities enables further systems analysis and calculation of their performance characteristics, as well as solving the optimization problems. As an example, a multicriteria problem of optimization of profit from the system's functioning was set and solved.

References

1. Artalejo, J.R., Gomez-Corral, A.: Retrial Queueing Systems: A Computational Approach. Springer, Heidelberg (2008). https://doi.org/10.1007/978-3-540-78725-9
2. Avrachenkov, K., Yechiali, U.: Retrial networks with finite buffers and their application to internet data traffic. Probab. Eng. Inf. Sci. **22**, 519–536 (2008)
3. Choi, B.D., Shin, Y.W., Ahn, W.C.: Retrial queues with collision arising from unsoltted CSMA/CD protocol. Queueing Syst. **11**, 335–356 (1992)
4. Choi, B.D., Park, K.K., Pearce, C.E.M.: An M/M/1 Retrial queue with control policy and general retrial times. Queueing Syst. **14**, 275–292 (1993)
5. Dudin, A.N., Klimenok, V.I.: Optimization of dynamic control of input load in node of informational-computing network. Autom. Technol. **3**, 25–31 (1991)
6. Falin, G.: Heavy traffic analysis of a random walk on a lattice semi-trip. Stoch. Models **11**(3), 395–409 (1995)
7. Falin, G.I., Templeton, J.G.C.: Retrial Queues. Chapman and Hall, London (1997)
8. Lebedev, E.A., Usar, I.Y.: Retrial queues with variable rate of input flow. Cybern. Syst. Anal. **3**, 151–159 (2013)
9. Phung-Duc, T.: Retrial queueing models: a survey on theory and applications. ArXiv abs arxiv:1906.09560, n. p. (2019)
10. Ponomarov, V., Lebedev, E.: Finite source retrial queues with state-dependent service rate. Commun. Comput. Inf. Sci. **356**, 140–146 (2013)
11. Roger, A.H., Charles, R.J.: Matrix Analysis. Cambridge University Press, Cambridge (1986)
12. Walrand, J.: An Introduction to Queueing Networks. Prentice Hall, Upper Saddle River (1988)

Redundant Queueing System with Hysteresis Backup Server Connection Strategy

Valentina Klimenok$^{1(\boxtimes)}$, Alexander Dudin^1, and Vladimir Vishnevsky^2

^{1} Department of Applied Mathematics and Computer Science,
Belarusian State University, 220030 Minsk, Belarus
{klimenok,dudin}@bsu.by
^{2} Institute of Control Sciences of Russian Academy of Sciences and Closed
Corporation "Information and Networking Technologies", Moscow, Russia
vishn@inbox.ru

Abstract. We consider an unreliable queueing system which can be used for modeling a broadband hybrid communication channel consisting FSO (Free Space Optics) channel and radio wave channel. The radio channel is backup and is connected when the optical channel is not available. In order to save energy, a hysteresis strategy for connecting a backup channel is used. This strategy is given by two thresholds: j_1 and $j_2, j_1 \leq j_2$. If the main channel fails, the backup channel connects to information transmission if the number i of customers in the system is such that $i > j_2$. If during backup channel operation the number of customers in the system becomes such that $i \leq j_1$, then the backup channel is disabled and the current customer continues to be serviced at the main server. If at some point in time the number i of customers in the system becomes such that $j_1 < i \leq j_2$, then the system works in the mode in which it worked until that time. We describe the system operation by two-dimensional Markov chain, calculate the steady state distribution and the main performance characteristics of the system. We introduce the cost function and present the example of numerical optimisation consisting in choosing the threshold values minimizing the cost function.

Keywords: Unreliable queueing system · Backup server · Hysteresis strategy · Steady state distribution · Performance characteristics · Cost function · Optimisation problem

1 Introduction

At the present day, the FSO-Free Space Optics technology has become widespread. The main advantages of atmospheric optical communication channel are high bandwidth, quality and confidentiality of communication. Therefore, laser systems are often used for a variety of applications that require high quality of data transmission, including financial, medical and military organizations.

© Springer Nature Switzerland AG 2019
A. Dudin et al. (Eds.): ITMM 2019, CCIS 1109, pp. 86–97, 2019.
https://doi.org/10.1007/978-3-030-33388-1_8

Along with the main advantages of wireless optical systems, their main disadvantages are also known: dependence of availability of the communication channel on weather conditions; the need to provide direct visibility between the emitter and receiver; limited range of communication. Unfavorable weather conditions such as rain, snow, fog (as well as sand dust, urban smog and various types of aerosols) can significantly reduce visibility and thus reduce the effective range of laser atmospheric communication lines. Therefore, in order to achieve required values of the reliability of the FSO communication channel, it is necessary to resort to the use of hybrid solutions based on laser and radio technology. The main mode of operation of the combined system is a laser mode. When the signal-to-noise ratio deteriorates, transmission of information is carried out in wide-band radio mode. After restoring the quality of the atmospheric optical channel transmission of information is carried out by this channel. When using the laser channel, the radio channel stops its work.

Due to the increased interest in hybrid systems in recent years, there have been many works in which their characteristics were studied, see e.g. [1–7]. However, in the majority of research, the authors use simulation and do not provide a comprehensive study of all the characteristics often limiting themselves to assessing the reliability of the hybrid channel.

Among the works devoted to the construction of mathematical models and their analysis, note [8–13]. The paper [9] is devoted to a hybrid communication system with so called hot standby, where the backup radio channel transmits data along with the FSO channel, but at low speed. In the paper [10], the hybrid communication system with cold redundancy is considered, where the radio-wave channel is assumed to be absolutely reliable and backs up FSO channel only in cases when the latter interrupts its functioning because of the unfavorable weather conditions. The paper [8] deals with a hybrid communication system where the millimeter-wave radio channel is used as a backup one. To model this system, the authors consider two-channel queueing system with unreliable heterogeneous servers which fail alternately. In further works [11–13], more complicated models of unreliable single-server queues are considered. They generalize models of [8–10] to the case of more complicated processes describing the operation of the hybrid communication systems.

It is assumed in all papers cited above that backup server connects to the service of customers immediately at the time of main server failure and turns off when the main server is being restored. It is clear that such a strategy of connecting a backup server may be ineffective due to the switching cost in situations where the switchings often occur. In the present paper, we consider queueing system suitable to model a hybrid communication channel with cold redundancy under more general, in comparison with the papers cited above, assumptions about strategy of switching between the main and backup server. This strategy (called as hysteresis strategy) is defined by two thresholds: j_1 and $j_2, j_1 \leq j_2$. If the main channel fails, the backup channel connects to information transmission if the number of customers i in the system is such that $i > j_2$. If during backup channel operation the number of customers in the system becomes such that $i \leq j_1$, then the backup channel is disabled and the current customer continues to be serviced at the main server. If at some point in time the number i of customers in the system becomes such that $j_1 < i \leq j_2$, then the system works in the mode in which it worked until that time. In the paper, we describe the system operation by two-dimensional Markov chain, calculate the steady state distribution and the main performance characteristics of the system. We introduce the cost function and present the example of numerical optimisation consisting in choosing the threshold values minimizing the cost function.

2   Mathematical Model

We consider a queuing system with an infinite waiting room and two servers, one of which (the main server) is unreliable and the other one (standby, backup server) is absolutely reliable. The latter one is in the so-called "cold" reserve. Interpretation: the unreliable server is the FSO (laser) channel, and the reliable one is a wireless radio channel IEEE 802.11n. Customers arrive at the system in the stationary Poisson flow with the rate λ. Service times on the main server and on the backup server are exponentially distributed with parameters μ_1 and μ_2, respectively.

Under the influence of weather conditions, the laser channel, i.e., the main server, can fail and immediately begins to be repaired. Breakdowns arrive at this server in a stationary Poisson flow with the intensity h. Repair time is exponentially distributed with the parameter τ. During the recovery, information is transmitted through the backup server, the transmission rate in which is lower than the transmission rate in the main server (we emphasize that the backup server may be involved in the servicing only during the repair time of the main server). In order to save energy, a hysteresis strategy for connecting the backup server is used, given by two thresholds j_1 and j_2 such as $j_1 \leq j_2$. With this strategy, the backup server and, accordingly, the whole system can work in two modes. If the main server fails, the backup server connected to the service of customer if the number i of customers in the system is such that $i > j_2 \geq 0$. In this case we will say that the system works in the 2nd mode. If during the backup server operation in the 2nd mode, the number of customers in the system becomes such that $i \leq j_1$, then the backup server is disabled (note that in such a scenario, the backup server can be switched on and off several times during repair time of the main sever). In this case we will say that the system works in the 1st mode. If at some point in time the number i of customers in the system becomes such that $j_1 < i \leq j_2$, then the system works in the mode in which it has worked until that moment. We will also assume that in the time intervals when the main server is fault free, the system works in the 0th mode.

After the restoration of the main server the backup server shuts down until the next failure of the main server. In such a case, the customer in service continues to be serviced on the main server. If at an arrival time the main sever is busy or under repair, then the arrival customer goes at the queue in the buffer of an infinite capacity. After the change of the server that provides service to the current customer, service continues from the moment of the change.

3   Process of the System States

The process of the system operation can be described by a regular irreducible continuous time Markov chain $\{i_t, n_t\}, t \geq 0$, where, at the time instant t :

- i_t is the number of customers in the system, $i_t \geq 0$;
- $n_t = 0$, if the main server is working, i.e., the system is working in mode 0;

$n_t = 1$, if the main server is under repair, the backup server is not yet on, i.e., the system is working in mode 1; $n_t = 2$, if the main server is under repair, the backup server is working, i.e., the system is working in mode 2.

Let us enumerate the states of the chain in the lexicographical order of its components. Denote by $Q_{i,l}$ the matrix of chain transition rates from the states corresponding to the value i of the first (countable) component to the states corresponding to the value l of this component, $i, l \geq 0$.

Lemma 1. *In the case $j_1 \neq j_2$ the infinitesimal generator Q of a Markov chain $\{i_t, n_t\}, t \geq 0$, has the block three-diagonal structure*

$$
Q = \begin{pmatrix}
Q_{0,0} & Q_{0,1} & O & O & O \cdots O & O & O & O & O & O \cdots \\
Q_{1,0} & Q_{1,1} & Q_{1,2} & O & O \cdots O & O & O & O & O & O \cdots \\
O & Q_{2,1} & Q_{2,2} & Q_{2,3} & O \cdots O & O & O & O & O & O \cdots \\
\vdots & \vdots & \vdots & \vdots & \vdots \ddots \vdots & \vdots & \vdots & \vdots & \vdots & \vdots \cdots \\
O & O & O & O & O \cdots O & Q_{j_2,j_2-1} & Q_{j_2,j_2} & Q_{j_2,j_2+1} & O & O \cdots \\
O & O & O & O & O \cdots O & O & Q_{j_2+1,j_2} & Q_1 & Q_2 & O \cdots \\
O & O & O & O & O \cdots O & O & O & Q_0 & Q_1 & Q_2 \cdots \\
O & O & O & O & O \cdots O & O & O & O & Q_0 & Q_1 \cdots \\
\vdots & \vdots & \vdots & \vdots & \vdots \ddots \vdots & \vdots & \vdots & \vdots & \vdots & \vdots \ddots
\end{pmatrix},
$$

where

$$
Q_{0,0} = \begin{pmatrix} -(\lambda + h) & h \\ \tau & -(\lambda + \tau) \end{pmatrix}, \quad Q_{0,1} = \lambda I_2,
$$

$$
Q_{i,i-1} = \begin{pmatrix} \mu_1 & 0 \\ 0 & 0 \end{pmatrix}, \quad Q_{i,i} = \begin{pmatrix} -(\lambda + h + \mu_1) & h \\ \tau & -(\lambda + \tau) \end{pmatrix}, \quad i = \overline{1, j_1},
$$

$$
Q_{i,i+1} = \lambda I_2, \ i = \overline{1, j_1 - 1}, \quad Q_{j_1,j_1+1} = \begin{pmatrix} \lambda & 0 & 0 \\ 0 & \lambda & 0 \end{pmatrix}, \quad Q_{j_1+1,j_1} = \begin{pmatrix} \mu_1 & 0 \\ 0 & 0 \\ 0 & \mu_2 \end{pmatrix},
$$

$$
Q_{j_1+1,j_1+2} = \begin{pmatrix} \lambda & 0 \\ 0 & \lambda \\ 0 & \lambda \end{pmatrix}, \ j_2 = j_1 + 1, \quad Q_{j_1+1,j_1+2} = \lambda I_3, \ j_2 > j_1 + 1,
$$

$$
Q_{j_1+1,j_1+1} = \begin{pmatrix} -(\lambda + h + \mu_1) & h & 0 \\ \tau & -(\lambda + \tau) & 0 \\ 0 & 0 & -(\lambda + \mu_2) \end{pmatrix}, \quad Q_{i,i-1} = \begin{pmatrix} \mu_1 & 0 & 0 \\ 0 & 0 & 0 \\ 0 & 0 & \mu_2 \end{pmatrix},
$$

$$Q_{i,i} = \begin{pmatrix} -(\lambda + h + \mu_1) & h & 0 \\ \tau & -(\lambda + \tau) & 0 \\ 0 & 0 & -(\lambda + \mu_2) \end{pmatrix}, \quad Q_{i,i+1} = \lambda I_3, \; i = \overline{j_1 + 2, j_2 - 1},$$

$$Q_{j_2,j_2-1} = \begin{pmatrix} \mu_1 & 0 & 0 \\ 0 & 0 & 0 \\ 0 & 0 & \mu_2 \end{pmatrix}, \quad Q_{\bar{j}_2,j_2+1} = \begin{pmatrix} \lambda & 0 \\ 0 & \lambda \\ 0 & \lambda \end{pmatrix},$$

$$Q_{j_2,j_2} = \begin{pmatrix} -(\lambda + h + \mu_1) & h & 0 \\ \tau & -(\lambda + \tau) & 0 \\ 0 & 0 & -(\lambda + \mu_2) \end{pmatrix}, \quad Q_{j_2+1,j_2} = \begin{pmatrix} \mu_1 & 0 & 0 \\ 0 & 0 & \mu_2 \end{pmatrix},$$

$$Q_{j_2+1,j_2+1} = \begin{pmatrix} -(\lambda + h + \mu_1) & h \\ \tau & -(\lambda + \tau + \mu_2) \end{pmatrix}, \quad Q_{j_2+1,j_2+2} = \lambda I_2;$$

$$Q_0 = \begin{pmatrix} \mu_1 & 0 \\ 0 & \mu_2 \end{pmatrix}, \quad Q_1 = \begin{pmatrix} -(\lambda + h + \mu_1) & h \\ \tau & -(\lambda + \tau + \mu_2) \end{pmatrix}, \quad Q_2 = \lambda I_2.$$

The proof of the lemma is carried out by analyzing the behavior of a chain over an infinitely small time interval.

Corollary 1. *The Markov chain $\{i_t, n_t\}, t \geq 0$, belongs to the class of continuous time quasi-Toeplitz Markov chains, see [14].*

Proof. The generator Q has block three-diagonal structure and, for $i > j_2 + 1$, its blocks $Q_{i,l}$ depend on the value i, l only via their difference $l - i$, precisely, $Q_{i,l} = Q_{l-i+1}$. Then, according the definition given in [14], the chain under consideration belongs to the class of quasi-Toeplitz Markov chains.

Corollary 2. *In the case $j_1 = j_2 = j$ the infinitesimal generator Q of a Markov chain $\{i_t, n_t\}, t \geq 0$, has the following block three-diagonal structure*

$$Q = \begin{pmatrix} Q_{00} & Q_{0,1} & O & \cdots & O & O & O & O & O & \cdots \\ Q_{1,0} & Q_{1,1} & Q_{1,2} & \cdots & O & O & O & O & O & \cdots \\ \vdots & \vdots & \vdots & \vdots & \vdots & \vdots & \vdots & \vdots & \vdots \\ O & O & O & \cdots & Q_{j-1,j-2} & Q_{j-1,j-1} & Q_{j-1,j} & O & O & \cdots \\ O & O & O & \cdots & O & Q_{j,j-1} & Q_{j,j} & Q_{j,j+1} & O & \cdots \\ O & O & O & \cdots & O & O & Q_0 & Q_1 & Q_2 & \cdots \\ O & O & O & \cdots & O & O & O & Q_0 & Q_1 & \cdots \\ \vdots & \vdots & \vdots & \vdots & \vdots & \vdots & \vdots & \vdots & \vdots & \ddots \end{pmatrix},$$

where

$$Q_{00} = \begin{pmatrix} -(h + \lambda) & h \\ \tau & -(\tau + \lambda) \end{pmatrix}, \quad Q_{0,1} = \lambda I_2,$$

$$Q_{i,i-1} = \begin{pmatrix} \mu_1 & 0 \\ 0 & 0 \end{pmatrix}, \quad Q_{i,i} = \begin{pmatrix} -(\mu_1 + h + \lambda) & h \\ \tau & -(\tau + \lambda) \end{pmatrix}, \quad Q_{i,i+1} = \lambda I_2, \; i = \overline{1, j},$$

$$Q_0 = \begin{pmatrix} \mu_1 & 0 \\ 0 & \mu_2 \end{pmatrix} \quad Q_1 = \begin{pmatrix} -(\mu_1 + h + \lambda) & h \\ \tau & -(\tau + \lambda + \mu_2) \end{pmatrix}, \quad Q_2 = \lambda I_2, \; i > j.$$

4   Stationary Distribution

For the system under consideration, the condition for the existence of the system stationary distribution coincides with the ergodicity condition for the chain $\{i_t, n_t\}, t \geq 0$, which is given by the following theorem.

Theorem 1. *The Markov chain $\{i_t, n_t\}, t \geq 0$, is ergodic if and only if*

$$\lambda < \frac{\tau \mu_1 + h \mu_2}{\tau + h}. \tag{1}$$

Proof. According to [14], the necessary and sufficient condition for the ergodicity of the quasi-Toeplitz Markov chain under consideration is expressed by the inequality

$$\mathbf{x} Q_2 \mathbf{e} < \mathbf{x} Q_0 \mathbf{e}, \tag{2}$$

where the vector $\mathbf{x} = (x_1, x_2)$ is the unique solution of the system

$$\mathbf{x}(Q_0 + Q_1 + Q_2) = \mathbf{0}, \quad \mathbf{x}\mathbf{e} = 1. \tag{3}$$

Solving the system (3), we obtain:

$$x_1 = \frac{\tau}{\tau + h}, \quad x_2 = \frac{h}{\tau + h}.$$

Substituting this solution into (2), we obtain the desired inequality (1).

In what follows, we assume that ergodicity condition (1) is fulfilled . Introduce the steady state probabilities of the chain:

$$p(i, n) = \lim_{t \to \infty} P\{i_t = i, n_t = n\}, i \geq 0, \, n = 0, 1,$$

and the vectors of these probabilities

$$\mathbf{p}_i = (p(i, 0), p(i, 1)), \, i = \overline{0, j_1}, \; \mathbf{p}_i = (p(i, 0), p(i, 1), p(i, 2)), \, i = \overline{j_1 + 1, j_2},$$

$$\mathbf{p}_i = (p(i, 0), p(i, 2)), \, i \geq j_2 + 1.$$

To calculate the vectors $\mathbf{p}_i$, we used a special case of the algorithm developed in [14] for quasi-Toeplitz Markov chain, adapted for the case of the chain under consideration.

Using the steady state probability vectors $\boldsymbol{p}_i, i \geq 0$, we can calculate various performance measures of the system. In the case when the stationary distribution $\boldsymbol{p}_i, i \geq 0$, is heavy tailed, the following result will be useful.

Lemma 2. *The generating function $\mathbf{P}(z) = \sum\limits_{i=j_2+1}^{\infty} \mathbf{p}_i z^i, |z| \leq 1$, satifies the following functional equation:*

$$\mathbf{P}(z)Q(z) = \mathbf{p}_{j_2+1} z^{j_2+1} Q_0 - z\mathbf{p}_0 (Q_{0,0} + Q_{0,1}z) - \sum_{i=1}^{j_2} \mathbf{p}_i z^i Q^{(i)}(z), \tag{4}$$

where

$$Q(z) = Q_0 + Q_1 z + Q_2 z^2, \; Q^{(i)}(z) = Q_{i,i-1} + Q_{i,i} z + Q_{i,i+1} z^2.$$

We will use formula (4) to calculate the values of the function $\mathbf{P}(z)$ and its derivatives at the point $z = 1$ without calculating infinite sums. The calculated values allow to find the moments of the number of customers in the system and a number of other characteristics of the system. Note that it is not possible to calculate directly the value of $\mathbf{P}(z)$ and its derivatives at the point $z = 1$ from Eq. (4), since the matrix $Q(1)$ is singular. This difficulty can be overcome by using the recursion formulas given below in Corollary 3.

Denote as $g^{(m)}(z)$ the mth derivative of $g(z)$, $m \geq 1$, and $g^{(0)}(z) = g(z)$. Denote as $\mathcal{B}(z)$ the right hand side of Eq. (4).

Corollary 3. *The derivatives of the vector generating function $\mathbf{P}(z)$, $|z| \leq 1$, at the point $z = 1$ are calculated recurrently as solutions of the following systems of linear algebraic equations:*

$$
\begin{cases}
\frac{d^m \mathbf{P}(z)}{dz^m}\big|_{z=1} Q(1) = \mathcal{B}^{(m)}(1) - \sum_{l=0}^{m-1} C_m^l \frac{d^l \mathbf{P}(z)}{dz^l}\big|_{z=1} Q^{(m-l)}(1), \\
\frac{d^m \mathbf{P}(z)}{dz^m}\big|_{z=1} Q'(1)\mathbf{e} = \frac{1}{m+1}[\mathcal{B}^{(m+1)}(1) - \sum_{l=0}^{m-1} C_{m+1}^l \frac{d^l \mathbf{P}(z)}{dz^l}\big|_{z=1} Q^{(m+1-l)}(1)]\mathbf{e}, \\
\qquad\qquad\qquad m \geq 0,
\end{cases}
$$

The proof of the corollary is implemented by analogy with the proof presented in [15].

5   Performance Measures. Optimization Problem

The list of performance measures of the system calculated using the results obtained are presented below.

1. Throughput of the system (maximum rate of the input flow which can be passed through the system)

$$
\varrho = \frac{\tau\mu_1 + h\mu_2}{\tau + h}.
$$

2. Probability that there are i customers in the system $p_i = \mathbf{p}_i\mathbf{e}, i \geq 0$.
3. Mean number of customers in the system

$$
L = \sum_{i=1}^{j_2} ip_i + \frac{d\mathbf{P}(z)}{dz}\big|_{z=1}\mathbf{e}.
$$

4. Variance of the number of customers in the system

$$
V = \sum_{i=1}^{j_2} i^2 p_i + \frac{d^2\mathbf{P}(z)}{dz^2}\big|_{z=1}\mathbf{e} + \frac{d\mathbf{P}(z)}{dz}\big|_{z=1}\mathbf{e} - L^2.
$$

5. Probability that i customers stay in the system and the main server is fault free

$$\alpha_i = \mathbf{p}_i \begin{pmatrix} 1 \\ 0 \end{pmatrix}, i = \overline{0, j_1}, \quad \alpha_i = \mathbf{p}_i \begin{pmatrix} 1 \\ 0 \\ 0 \end{pmatrix}, i = \overline{j_1 + 1, j_2}, \quad \alpha_i = \mathbf{p}_i \begin{pmatrix} 1 \\ 0 \end{pmatrix}, i > j_2.$$

6. Probability that i customers stay in the system, the main server is under repair, and the backup server has not yet connected to the customers service

$$\beta_i = \mathbf{p}_i \begin{pmatrix} 0 \\ 1 \end{pmatrix}, i = \overline{0, j_1}, \quad \beta_i = \mathbf{p}_i \begin{pmatrix} 0 \\ 1 \\ 0 \end{pmatrix}, i = \overline{j_1 + 1, j_2}.$$

7. Probability that i customers stay in the system, the main server is under repair, and the backup server serves a customer

$$\gamma_i = \mathbf{p}_i \begin{pmatrix} 0 \\ 1 \end{pmatrix}, i > j_1.$$

8. Probability that the system is working in mode 0 (the main server is fault free) $P^{(0)} = \sum\limits_{i=0}^{\infty} \alpha_i.$

9. Probability that the system is working in mode 1 (the main server is under repair, and the backup server has not yet connected to the customers service)
$$P^{(1)} = \sum\limits_{i=0}^{j_2} \beta_i.$$

10. Probability that the system is working in mode 2 (the main server is under repair, and the backup server serves a customer) $P^{(2)} = \sum\limits_{i=j_1+1}^{\infty} \gamma_i.$

11. Mean number of switching from the mode 1 to mode 2 (from the mode 2 to mode 1) per unit of time

$$\chi = \beta_{j_2} \lambda = \gamma_{j_1+1} \mu_2.$$

The optimization problem is formulated as the problem of choosing the optimal thresholds j_1, j_2 that minimize the economic criterion of the quality of the system operation of the form

$$E = aL + c_1 P^{(2)} + 2c_2 \chi, \tag{5}$$

where L is a mean number of customers in the system at an arbitrary time; a is a penalty charged per unit of time for each customer that stay in the system; $P^{(2)}$ is the average fraction of the time during which the main server is under repair and the backup server serves customers; c_1 is a unit cost of operation time of the backup server, χ is the mean number of switching from one server to another one per unit of time, c_2 is a cost of one switch.

Such a criterion is an average penalty per unit of time during the steady state operation of the system.

Below we present two examples of numerical optimization. Our aim is to find by direct search the optimal set of values of thresholds j_1, j_2 that provides the minimum value to the cost criterion (5).

In the first example, we consider the following input data: the input rate $\lambda = 25$, the service rates of the main server and the backup server are $\mu_1 = 30$ and $\mu_2 = 20$, respectively, the rate of breakdowns is $h = 0.1$, the repair rate is $\tau = 1$. Under these parameters the load coefficient $\rho = 0.859375$. The cost coefficient are as follows: $a = 1$, $c_1 = 69$, $c_2 = 8.5$.

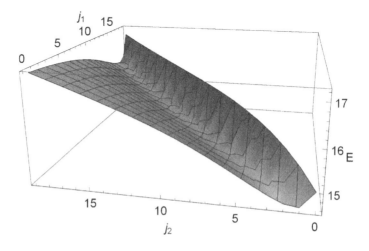

Fig. 1. Criterion E as a function of the thresholds j_1 and j_2

The dependence of the criterion E on the threshold values j_1, j_2 is shown in Fig. 1. The criterion values for different threshold values are also given in Table 1. From the figure and the table it can be seen that the least favorable situation is achieved in two situations: when the thresholds are equal and high and when the difference in the values of thresholds is high. The more favorable values of thresholds, including the optimum values, lie in the region where the difference $0 < j_2 - j_1 \leq 2$. The maximum of the criterion E is reached at the point (0,1). With j_1 and j_2 increasing the value of the criterion also increases. It should be also noted that in the region $j_1, j_2 \subset [0, 19]$ the relative gain from the use of the optimal strategy (0,1) is 20% comparing with the use the most non-optimal strategy (3,19).

In the second example, we consider the following input data: the input rate $\lambda = 7$, the service rates of the main server and the backup server are $\mu_1 = 30$ and $\mu_2 = 15$, respectively, the rate of breakdowns is $h = 0.1$, the repair rate is $\tau = 3$. Under these parameters the load coefficient $\rho = 0.24$. The cost coefficient are as follows: $a = 1$, $c_1 = 3$, $c_2 = 5$. The cost criterion as a function of the

Table 1. Criterion E as a function of the thresholds j_1 and j_2

j_1	j_2	E	j_1	j_2	E	j_1	j_2	E	j_1	j_2	E	j_1	j_2	E
0	0	14,95780	2	5	15,01159	4	14	16,38788	7	13	15,93764	11	13	15,61730
0	1	**14,35471**	2	6	15,17670	j_1	j_2	E	7	14	16,12324	11	14	15,72523
0	2	14,60754	2	7	15,35916	4	15	16,54278	7	15	16,30469	11	15	15,89458
0	3	14,76822	2	8	15,54449	4	16	16,69165	7	16	16,48037	11	16	16,08656
0	4	14,96300	2	9	15,72600	4	17	16,83527	7	17	16,64977	11	17	16,28511
0	5	15,16582	2	10	15,90062	4	18	16,97446	7	18	16,81301	11	18	16,48283
0	6	15,36542	2	11	16,06723	4	19	17,11005	7	19	16,97057	11	19	16,67620
0	7	15,55665	2	12	16,22574	5	5	16,11257	8	8	16,36308	12	12	16,57641
0	8	15,73745	2	13	16,37667	5	6	15,26902	8	9	15,48888	12	13	15,75826
0	9	15,90735	2	14	16,52083	5	7	15,18019	8	10	15,39819	12	14	15,69584
0	10	16,06683	2	15	16,65918	5	8	15,27019	8	11	15,49180	12	15	15,80941
0	11	16,21684	2	16	16,79271	5	9	15,42248	8	12	15,64936	12	16	15,98312
0	12	16,35859	2	17	16,92239	5	10	15,59792	8	13	15,83089	12	17	16,17880
0	13	16,49334	2	18	17,04910	5	11	15,78009	8	14	16,01973	12	18	16,38068
0	14	16,62238	2	19	17,17366	5	12	15,96129	8	15	16,20815	12	19	16,58148
0	15	16,74689	3	3	15,81789	5	13	16,13790	8	16	16,39247	13	13	16,62784
0	16	16,86798	3	4	15,06800	5	14	16,30836	8	17	16,57106	13	14	15,83133
0	17	16,98661	3	5	15,00571	5	15	16,47224	8	18	16,74341	13	15	15,77802
0	18	17,10362	3	6	15,10695	5	16	16,62974	8	19	16,90963	13	16	15,89737
0	19	17,21974	3	7	15,26464	5	17	16,78139	9	9	16,42185	13	17	16,07545
1	1	15,32200	3	8	15,44245	5	18	16,92793	9	10	15,55487	13	18	16,27479
1	2	14,76688	3	9	15,62495	5	19	17,07011	9	11	15,46925	13	19	16,47991
1	3	14,76776	3	10	15,80498	6	6	16,21381	9	12	15,56695	14	14	16,68152
1	4	14,90081	3	11	15,97919	6	7	15,34886	9	13	15,72807	14	15	15,90806
1	5	15,07784	3	12	16,14626	6	8	15,25565	9	14	15,91285	14	16	15,86412
1	6	15,26828	3	13	16,30593	6	9	15,34489	9	15	16,10480	14	17	15,98925
1	7	15,45904	3	14	16,45858	6	10	15,49767	9	16	16,29625	14	18	16,17164
1	8	15,64407	3	15	16,60490	6	11	15,67426	9	17	16,48357	14	19	16,37453
1	9	15,82070	3	16	16,74576	6	12	15,85800	9	18	16,66514	15	15	16,73830
1	10	15,98811	3	17	16,88203	6	13	16,04113	9	19	16,84046	15	16	15,98876
1	11	16,14644	3	18	17,01462	6	14	16,21996	10	10	16,47525	15	17	15,95426
1	12	16,29639	3	19	17,21435	6	15	16,39289	10	11	15,62089	15	18	16,08510
1	13	16,43895	4	4	15,98385	6	16	16,55944	10	12	15,54198	15	19	16,27167
1	14	16,57520	4	5	15,17725	6	17	16,71975	10	13	15,64455	16	16	16,79875
1	15	16,70627	4	6	15,09818	6	18	16,87433	10	14	15,80966	16	17	16,07359
1	16	16,83322	4	7	15,19187	6	19	17,02385	10	15	15,99797	16	18	16,04849
1	17	16,95702	4	8	15,34556	7	7	16,29530	10	16	16,19319	16	19	16,18490
1	18	17,07856	4	9	15,52125	7	8	15,42105	10	17	16,38777	17	17	16,86325
1	19	17,19860	4	10	15,70287	7	9	15,32756	10	18	16,57812	17	18	16,16264
2	2	15,60238	4	11	15,88293	7	10	15,41820	10	19	16,76267	17	19	16,14680
2	3	14,93421	4	12	16,05791	7	11	15,57287	11	11	16,52607	18	18	16,93204
2	4	14,89774	4	13	16,22635	7	12	15,75159	11	12	15,68835	18	19	16,25589
												19	19	17,00522

threshold j_1, j_2 is shown in Fig. 2. In this example in the region $j_1, j_2 \subset [0, 19]$ the minimum value of the cost criterion $E = 5.59596$ is reached at the point (0,2) and the maximum value $E = 6.31951$ is reached at the point (5.16). The relative gain from the use of the optimal strategy (0,2) is 13% comparing with the use the most non-optimal strategy (5,16).

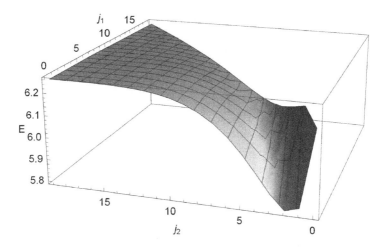

Fig. 2. Criterion E as a function of the thresholds j_1 and j_2

6   Conclusion

In this paper, we consider an unreliable queueing system with high-speed unreliable main server and low speed reliable backup server. The backup server is connected to the service of customers when the main server is not available due to a breakdown. In order to save energy, a hysteresis strategy for connecting a backup server is used. We described the system operation by two-dimensional Markov chain, calculated the steady state distribution and the main performance characteristics of the system. We introduced the cost function and presented the example of numerical optimisation consisting in choosing the threshold values minimizing the cost function. The results can be used for performance evaluations and optimization of real world hybrid communication system consisting the Free Space Optics channel and radio wave channel.

Acknowledgments. This work has been financially supported by the joint grant of Belarusian Republican Foundation for Fundamental Research (No F18R-136) and Russian Foundation for Fundamental Research (No 18-57-00002).

References

1. Arnon, S., Barry, J., Karagiannidis, G., Schober, R., Uysal, M.: Advanced Optical Wireless Communication Systems. Cambridge University Press, Cambridge (2012)
2. Nadeem, F., Kvicera, V., Awan, M.S.: Weather effects on hubrid FSO/RF communication link. IEEE J. Sel. Areas Commun. **27**, 1687–1697 (2009)
3. Wong, D., Abouzeida, A.A.: Throughput and delay analysis for hybrid radio fregnency and free-space-optical (RF/FSO) networks. J. Wirel. Netw. **17**, 877–892 (2011)
4. Nadeem, F., Geiger, B., Leitgeb, E., Awan, M.S., Kandus, G.: Evaluation of switchover algorithms for hybrid FSO-WLAN systems. In: Wireless VITAE, pp. 565–570 (2009)
5. Eslami, A., Vangala, S., Pishro-Nik, H.: Hybrid channel codes for efficient FSO/RF communication systems. IEEE Trans. Commun. **58**, 2926–2938 (2010)
6. Akbulut, A., Gokhan, H., Ari, F.: Design, availability and reliability analysis on an experimental outdoor FSO/RF communication system. In: International Conference ICTON, pp. 403–406 (2005)
7. Letzepis, N., i Fàbregas, A.G., Cowley, W.G.: Outage analysis of the hybrid freespace optical and radio-frequency channel. IEEE J. Sel. Areas Commun. **27**, 1709–1719 (2009)
8. Vishnevsky, V., Kozyrev, D., Semenova, O.V.: Redundant queueing system with unreliable servers. In: Proceedings of the 6th International Congress on Ultra Modern Telecommunications and Control Systems and Workshops (ICUMT), Moscow, pp. 383–386 (2014)
9. Vishnevsky, V.M., Semenova, O.V., Sharov, S.Y.: Modeling and analysis of a hybrid communication channel based on free-space optical and radio-frequency technologies. Autom. Remote Control **72**, 345–352 (2013)
10. Sharov, S.Y., Semenova, O.V.: Simulation model of wireless channel based on FSO and RF technologies. In: Distributed Computer and Communication Networks. Theory and Applications (DCCN 2010), pp. 368–374 (2010)
11. Dudin, A., Klimenok, V., Vishnevsky, V.: Analysis of unreliable single server queueing system with hot back-up server. In: Plakhov, A., Tchemisova, T., Freitas, A. (eds.) EmC-ONS 2014. CCIS, vol. 499, pp. 149–161. Springer, Cham (2015). https://doi.org/10.1007/978-3-319-20352-2_10
12. Klimenok, V., Vishnevsky, V.: Unreliable queueing system with cold redundancy. In: Gaj, P., Kwiecień, A., Stera, P. (eds.) CN 2015. CCIS, vol. 522, pp. 336–346. Springer, Cham (2015). https://doi.org/10.1007/978-3-319-19419-6_32
13. Klimenok, V.: Two-server queueing system with unreliable servers and Markovian arrival process. In: Dudin, A., Nazarov, A., Kirpichnikov, A. (eds.) ITMM 2017. CCIS, vol. 800, pp. 42–55. Springer, Cham (2017). https://doi.org/10.1007/978-3-319-68069-9_4
14. Klimenok, V.I., Dudin, A.N.: Multi-dimensional asymptotically quasi-Toeplitz Markov chains and their application in queueing theory. Queueing Syst. **54**, 245–259 (2006)
15. Dudin, A., Klimenok, V., Lee, M.H.: Recursive formulas for the moments of queue length in the $BMAP/G/1$ queue. IEEE Commun. Lett. **13**, 351–353 (2009)

The Laws of Conservation of Flows in Acyclic Queueing Networks

Gurami Tsitsiashvili[1,2]($\boxtimes$) and Marina Osipova[1,2]

[1] IAM FEB RAS, Vladivostok, Russia
guram@iam.dvo.ru, mao1975@list.ru
[2] Far Eastern Federal University, Vladivostok, Russia

Abstract. In this paper we consider open acyclic queuing network with few input flows, multi-server nodes and service discipline when a presence of customers in a node leads to a work of some its server. Input flows are Poisson and service times in all nodes have exponential distribution. It is proved that in stationary regime input and output flows coincide by their distributions. These results are based on generalization of Burke theorem and on rearrangement of acyclic open network nodes into some classes of nodes J_p, $p = 1, \ldots, s$, so that transition of customer may be only from node of class J_p to node of class J_{p+1}, $p = 1, \ldots, s - 1$.

Keywords: Acyclic queuing network · Output flow · Law of conservation

Introduction

In [1] we consider a single-server queuing system with several independent Poisson input streams and service discipline, in which the device is always busy, if the system has any applications. It is proved that in the stationary regime the flows leaving this system coincide in distribution with the input flows. This statement can be considered as a kind of analogue of the law of conservation of flows. In the present work, the aim is to generalize this conservation law for open acyclic queueing network, the nodes of which are multi-server queuing systems.

This generalization of the law of conservation of flows is proposed to carry out by mathematical induction along the length of the maximum path from the input node of the acyclic graph to the remaining nodes [2]. With this factorization of the nodes of an acyclic graph, the oriented edges connect only the nodes with smaller factors to the nodes with larger factors.

Therefore, the implementation of the proposed research program of the law of conservation of flow distributions consists of the following parts: (1) consideration of a multi-server queuing system with several flows, (2) application of the method of mathematical induction on the above factor.

© Springer Nature Switzerland AG 2019
A. Dudin et al. (Eds.): ITMM 2019, CCIS 1109, pp. 98–108, 2019.
https://doi.org/10.1007/978-3-030-33388-1_9

1   Output Flows in a Multi-server Queuing System

The system with one flow is described by a discrete Markov process characterizing the number of customers at the current time. To describe the Markov process system with multiple flows and even the simplest discipline of service FIFO "first come-first served" is required to characterize the entire queue, i.e, to specify to which flows the customers in queue belong, and to arrange them by the time of receipt.

As a result, the phase space of a discrete Markov process describing such a system consists of vectors whose dimension coincides with the number of customers in the system and therefore is not constant. Each component of this vector is a number of the flow to which the customer in the system belongs. This creates significant analytical and computational difficulties in the study of Markov processes describing systems with multiple flows. Such description in many ways resembles the construction of a piecewise linear Markov process [3,4], which is used not in analytical calculations, but in numerical modelling of queuing systems.

In this paper, we consider a multi-server queuing system with exponentially distributed service times and intervals between the arrivals of customers from several flows. It is proved that stationary output flows coincide in distribution with independent Poisson input flows provided that the servers work if the system has customers. Using the technique of [5], in which a jump of one of the components of the Markov process described above corresponds to the exit from the server to the next customer, we checked the condition of independence of the number of customers of output flows at disjoint time intervals [6]. Closed methods of customers flows analysis are used in [7–9] with different applications to retrial queues [10–14], and systems with feedback [15,16].

However, to determine the intensities of the output flows it is required to use the ergodicity theorem of the analysed Markov process [2]. To establish the ergodicity of the Markov process describing the membership of the customers in the system to different input flows, the stochastic monotonicity conditions [17] and the ergodicity theorem for the regenerating process [18] are used. The theorem of the ergodicity has allowed to establish the equality of the intensities of the input and output flows without calculating marginal distributions.

2   Law of Conservation of Flow Intensities

Let m independent Poisson flows of intensity λ^k, $k = 1, \ldots, m$, enter n – server queuing system $A(n, m)$. Customers of these flows are served on the system servers with an intensity of μ so that if there is at least one customer in the system, the servers work. Let's denote $x_k(t)$ the number of customers k of the input flow that came to the system in the time interval $[0, t]$, $x_k(0) = 0$. Define $y_k(t)$ the number of customers that have been logged out in the time interval $[0, t]$, $y_k(0) = 0$.

Lemma 1. *If the following condition is true*

$$\lambda = \sum_{k=1}^{m} \lambda^k < n\mu, \tag{1}$$

then the convergence in probability is valid

$$\frac{y_k(t)}{t} \to \lambda^k, \ t \to \infty, \ k = 1, \ldots, m. \tag{2}$$

Proof. The number of customers in the k - th input flow that are in the system at the time t satisfies the equality

$$z_k(t) \equiv x_k(t) - y_k(t), \ t \geq 0. \tag{3}$$

It is obvious that the number of customers in the flow $z(t) = \sum_{k=1}^{m} z_k(t)$ and so

$$z_k(t) \leq z(t), \ t \geq 0, \ k = 1, \ldots, m. \tag{4}$$

Since the Markov process $z(t)$ is ergodic, the convergence in probability is fair

$$\frac{z(t)}{t} \to 0, \ t \to \infty. \tag{5}$$

Indeed, fix arbitrary $\varepsilon > 0$ and define z_ε so that

$$\lim_{t \to \infty} P(z(t) \geq z_\varepsilon) < \varepsilon.$$

Then there exists $t_\varepsilon > 0$ so that for any $t \geq t_\varepsilon$

$$\left| \lim_{t \to \infty} P(z(t) \geq z_\varepsilon) - P(z(t) \geq z_\varepsilon) \right| < \varepsilon,$$

and consequently for any $t \geq t_\varepsilon$

$$P(z(t) \geq z_\varepsilon) \leq 2\varepsilon.$$

Using these inequalities we obtain for any $t \geq t_\varepsilon$, $t \geq \dfrac{z_\varepsilon}{2\varepsilon}$

$$2\varepsilon \geq P\left(\frac{z(t)}{t} \geq \frac{z_\varepsilon}{t} \right) \geq P\left(\frac{z(t)}{t} \geq 2\varepsilon \right).$$

So the convergence in probability (5) is proved.

From the relations (4), (5) the convergence in probability follows

$$\frac{z_k(t)}{t} \to 0, \ t \to \infty, \ k = 1, \ldots m. \tag{6}$$

Since the random variable $x_k(t)$ has Poisson distribution with the parameter $\lambda^k t$, it is easy to establish the convergence in probability

$$\frac{x_k(t)}{t} \to \lambda^k, \ t \to \infty. \tag{7}$$

From the formulas (3), (6), (7) we obtain the limit convergence in probability (2). Thus, the intensity of the output flow consisting of customers of the input flow k, coincides with the intensity of λ^k of the input low, $k = 1, \ldots, m$.

3   Construction of Markov Process Describing Queuing System $A(n, m)$

Everywhere further we assume that in the queuing system $A(n, m)$ with m independent input Poisson flows of intensity λ^k, $k = 1, \ldots, m$, the server operates in the presence of at least one customer with the intensity μ.

Let us first describe the operation of the system $A(1, m)$ by Markov process $Z(t)$ with states of the form

$$Z = (z, a_1, \ldots, a_z).$$

Here z is the number of customers in the system, $a_1, \ldots, a_z$ is the sequence of numbers of flows to which the customers in the system belong, $a_k \in \{1, \ldots, m\}$, $k = 1, \ldots, z$.

Customers in the set $a_1, \ldots, a_z$ are ordered by the sequence of their service on the device: the first number on the device is the customer described by the flow number a_1, second number - the customer described by the flow number a_2, etc. After the end of the service the process $Z(t)$ with intensity μ goes from the state $Z = (z, a_1, \ldots, a_z)$ to the state

$$Z' = (z - 1, a_2, \ldots, a_z). \tag{8}$$

When a new customer enters the system from the flow with the number a_{z+1}, the transition from the state Z to the state

$$Z'' = (z + 1, \pi(a_1, \ldots, a_{z+1}))$$

with the intensity $\lambda_{a_{z+1}}$, where π is a permutation of characters $a_1, \ldots, a_{z+1}$ (see Fig. 1). In such a transition, the order of service of the customers in it may change depending on the discipline of service.

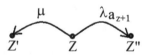

Fig. 1. Transition intensities of the process $Z(t)$.

Thus, for the FIFO service discipline (first come - first served), the arrival of a new customer from the flow with the number a_{z+1} leads to the transition to the state

$$Z'' = (z + 1, a_1, \ldots, a_{z+1}).$$

For the discipline of relative priority between flows the arrival of a new customer from the flow with the number a_{z+1} leads to the transition from the state Z to the state Z'' of the following form:

$$Z'' = (z + 1, a_1, \ldots, a_i, a_{z+1}, a_{i+1}, \ldots, a_z),$$

where $i > 1$ is the maximum of numbers such that $a_i \geq a_{z+1}$. If for all $i : 1 < i \leq z$ the inequality $a_i < a_{z+1}$ is satisfied, then

$$Z'' = (z + 1, a_1, a_{z+1}, a_2, \ldots, a_z). \tag{9}$$

If in the system $A(1, m)$ the absolute priority by flows is set then under the same conditions instead of the ratio (9) the ratio is executed

$$Z'' = (z + 1, a_{z+1}, a_1, a_2, \ldots, a_z).$$

Other service disciplines are possible also and, as a consequence, other ways to set the state of Z'' by rearranging the components of the vector $a_1, \ldots, a_{z+1}$.

Remark 1. In the transition from system $A(1, m)$ to system $A(n, m)$ it is enough to replace the formula (8) with

$$Z' = (z - 1, a_1, \ldots, a_{i-1}, a_{i+1}, \ldots, a_z), \ 1 \leq a_i \leq \min(n, z).$$

This means that one of the customers on its servers leaves the system. The intensity of the transition from the state Z to the state Z' remains the same and is equal to the intensity of service μ.

4   Ergodicity of Markov Process $Z(t)$ and Equality of Distributions of Input and Output Stationary Flows

Since the task of Z', Z'' defines a random process $Z(t)$ as Markov, an important problem is to find out the conditions of ergodicity of this process and consequently that it has a stationary distribution.

Lemma 2. *If the condition is met that there is at least one customer in the system $A(n, m)$ and so one of the servers works, and the inequality (1) is fair, then the Markov process $Z(t)$ is ergodic.*

Proof. The process $z(t)$, $t \geq 0$, describing the total number of customers in the system $A(n, m)$ at the time t, coincides with the process describing the total number of customers in the system $M|M|n|\infty$ with the Poisson input flow with the intensity λ and the exponential distribution of service times with the intensity μ. Therefore, the process $z(t)$ is Markov and ergodic, moreover, the moments $T_1, T_2, \ldots,$ of zeroing the process $z(t)$ are the moments of its regeneration, i.e. the pieces of a random process

$$\{z(t), \ T_n < t \leq T_{n+1}\}, \ n = 1, 2, \ldots,$$

are independent and equally distributed. Then it follows from the condition of this theorem that the average time between adjacent regeneration moments

$$M(T_{n+1} - T_n) < \infty.$$

However, the points $T_1, T_2, \ldots$, are also regeneration points of the process $Z(t)$. Therefore, it follows from the ergodicity theorem for regenerating processes [18] that the Markov process $Z(t)$ is also ergodic and its stationary state is determined.

Theorem 1. *Under the conditions of Lemma 2, the stationary output flows coincide in distribution with the Poisson independent input flows.*

Proof. Since output of a customer of the input flow k from the system corresponds to the jump of the process $Z(t)$ from the state $Z : z > 0$, $a_1 = k$ to the state Z', then the intensity b_k of the stationary output flow consisting of customers of the input flow k, satisfies the equality

$$b_k = \sum_{Z:\, z>0,\, a_i=k,\, 1 \leq i \leq \min(n,z)} \mu P(Z),$$

where $P(Z)$ is the stationary distribution of the random process $Z(t)$. Hence, using a generalization of Burke's theorem [1], we obtain that the stationary output flow in the system $A(n, m)$, consisting of customerss of the input flow k, is Poisson with intensity λ^k. And since every point of the output flow in the system $A(n, m)$ with a probability

$$\frac{\lambda^k}{\sum_{i=1}^m \lambda^i}$$

belongs to the k flow, all output flows are independent [19].

5   Invariance of Stationary Flows in Acyclic Open Queuing Networks

Consider an open queuing network S with a finite number of nodes $1, \ldots, N$ and independent Poisson intensity input flows $\lambda_1^1, \ldots, \lambda_1^m$ (exiting node 1). Motion of customers in the network S is determined by the route matrix

$$\Theta = ||\theta_{i,j}||_{i,j=1}^N$$

consisting of transition probabilities $\theta_{i,j}$ of the customer to node j after completing the service at node i. Moreover, $\theta_{1,i}$ is the probability of the customer of input flow to enter the service node i and $\theta_{i,N}$ is the probability of withdrawal of the customer from the network after service at node i, $1 \leq i < N$. In node k of network S there is n_k of identical servers with exponential distribution of service time of customers having parameter

$$\mu_k, \ 0 < \mu_k \leq \infty, \ k = 2, \ldots, N-1.$$

We compare the transition probability matrix Θ with a directed graph G with a set of nodes $1, \ldots, N$ and a set of V edges $(i, \ j)$, for which the transition

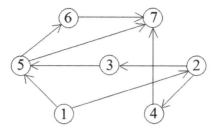

Fig. 2. Acyclic open queuing network with $N = 7$.

probability $\theta_{i,j} > 0$. Suppose that the matrix Θ is arranged in such a way that the directed graph G is acyclic. Let's call a certain open queueing network acyclic (see Fig. 2).

Divide the set of nodes of the graph $1, \ldots, N$ into disjoint subsets

$$J_1, \ldots, J_s, \; J_1 = \{1\}, \; J_s = \{N\}$$

(here J_1 is the node from which the customers come to the network, J_s is the node to which the customers come out of the network) so that any edge of the graph S comes from the node j_p of the class J_p to the node j_q of the class J_q, $p < q$.

Everywhere further we assume that in the graph G for any node $i \in U$ there exists a path from node 1 to node i. For each node i of the graph G we determine the maximum path length l_i from node 1 to node i, $l_1 = 0$. Matrix of the maximums of the paths lengths may be determined by a modification of the Floyd-Warshall [20, P. 1296] algorithm, which is commonly used to compute the matrix of minimal paths lengths between nodes of a graph.

For constructive computation l_i, $i = 2, \ldots, N$, we introduce the following matrix $D^1 = ||d_{i,j}^1||_{i,j=1}^N$:

$$d_{i,j}^1 = \begin{cases} 0, & i = j, \; i = 1, \ldots, N, \\ \infty, & (i, j) \notin V, \\ 1, & (i, j) \in V. \end{cases}$$

Thus for any pair of nodes that are not connected by an edge (by length one), the value $d_{i,j}^1$ is equal to infinity.

Let

$$D^k = ||d_{i,j}^k||_{i,j=1}^N, \; k = 1, \ldots, N,$$

where the value $d_{i,j}^k$ is infinite, if and only if the graph G has no path connecting nodes i, j which in addition to nodes i, j can only pass through nodes $1, \ldots, k$. If such paths exist, then the value of $d_{i,j}^k$ is equal to the maximum length of such paths. It is easy to prove that the following theorem is true.

Theorem 2. *Matrices D^k, $k = 2, \ldots, N$, satisfy recurrent relations*

$$d_{i,j}^k = \begin{cases} \max(d_{i,j}^{k-1}, \; d_{i,k}^{k-1} + d_{k,j}^{k-1}), \text{ if } \max(d_{i,j}^{k-1}, \; d_{i,k}^{k-1} + d_{k,j}^{k-1}) < \infty, \\ \\ \min(d_{i,j}^{k-1}, \; d_{i,k}^{k-1} + d_{k,j}^{k-1}), \text{ otherwise.} \end{cases}$$

Moreover, the matrix $D^N = ||d_{i,j}^N||_{i,j=1}^N$ defines the maximum lengths of paths between the vertices of the graph G, if such paths exist. In the absence of such paths, the corresponding elements of the matrix are infinite.

Thus, the Theorem 2 can be used to calculate the maximum lengths of paths from node 1 to nodes $i = 2, \ldots, N$:

$$l_i = d_{1,i}^N.$$

Let us now proceed to the classification of the nodes of the graph G. Define

$$J_l = \{t : \; d_{1,t} = l\}.$$

Obviously, after this classification, any edge of the graph S goes from the vertex j_p of class J_p to the vertex j_q of class J_q, $p < q$. The selection of node sets $J_1, \ldots, J_s$ is illustrated in Fig. 3.

If $p+1 < q$, then you can enter intermediate dummy nodes in the graph G so that the edges (j_p, j_q) pass through the dummy nodes. Then the flow of output customers from the node j_p to the node j_q can be passed through the fictitious nodes in which the servers in them process the customers instantly (see Fig. 4).

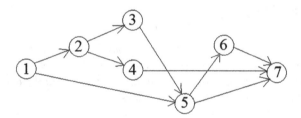

Fig. 3. Transformed acyclic open queuing network.

Denote $\widehat{G}$ a graph (network) G overridden in this way. Put $\widehat{J}_p$, $1 \le p \le s$, sets of nodes of the graph $\widehat{G}$ (see Fig. 5) and transition probabilities between nodes of the graph $\widehat{G}$:

$$\widehat{\theta}_{j_p,j_{p+1}}, \; j_p \in \widehat{J}_p, \; j_{p+1} \in \widehat{J}_{p+1}, \; 1 \le p < s.$$

By induction of p it is possible to calculate the intensity of the input flows to all nodes of the network. Let λ_{j_p} be the total intensity of the input flow to the node $j_p \in \widehat{J}_p$, then

$$\lambda_{j_{p+1}} = \sum_{j_p \in \widehat{J}_p} \lambda_{j_p} \widehat{\theta}_{j_p,j_{p+1}}.$$

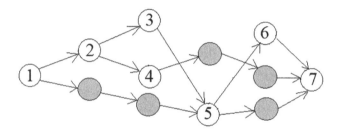

Fig. 4. The transformed acyclic open network with dummy nodes.

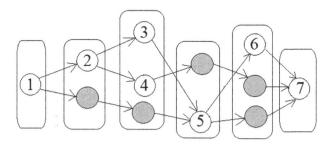

Fig. 5. Flow passing through a transformed acyclic open network with dummy nodes.

Theorem 3. *Let each node of the acyclic open queuing network $\widehat{G}$ meet the condition of ergodicity*

$$\lambda_{j_p} < n_{j_p}\mu_{j_p}$$

and the discipline of servicing customers of flows $1, \ldots, m$ in the network is such that if in any node there is at least one customer then some of servers of this node works. Then the steady-state output flows at individual nodes coincide in distribution with their steady-state input flows.

Proof. The proof is based on the Theorem 1, on the induction of p for nodes from the set $\widehat{J}_p$ (see Fig. 5) and on the following approval. The union of two independent Poisson flows of intensities a, b is a Poisson flow with the intensity $a + b$. If each point of Poisson flow with the intensity $a + b$, regardless of the other events with probability $\dfrac{a}{a + b}$ goes to one flow, and with probability $\dfrac{b}{a + b}$ goes to the other flow, then the resulting flows are independent and Poisson with intensities a, b [19].

6   Conclusion

Considered in Fig. 5 the scheme corresponds to the so-called queuing network of multiphase type [21]. In this network, we can assume that the flow of customers of different types may have different transition probabilities between the nodes of the network.

Assume that the network S has the set of nodes $\{1,\ldots,N\}$ devided into subsets $J_1 = \{1\}$, $J_2,\ldots,J_{s-1},J_s = \{N\}$. Poisson input flow consists of m independent flows $1,\ldots,m$. For them the following transition probabilities are defined: $\theta^k_{j_p,j_{p+1}}$, $j_p \in J_p$, $j_{p+1} \in J_{p+1}$, $k = 1,\ldots,m$. Then total intensity of input flow of the node $j_{p+1} \in J_{p+1}$ is following

$$\lambda_{j_{p+1}} = \sum_{j_p \in J_p} \sum_{k=1}^{m} \lambda^k_{j_p} \theta^k_{j_p,j_{p+1}}$$

and the intensity of the customers from input flow k are

$$\lambda^k_{j_{p+1}} = \sum_{j_p \in J_p} \lambda^k_{j_p} \theta^k_{j_p,j_{p+1}}.$$

Then it is possible to rewrite conditions of ergodicity and laws of flows conservation for all nodes of the network. Also there are other generalizations of queuing network of multiphase type that involve integrating of customers of different flows in the separate units and which divide customers during their passage through the nodes.

This paper is partially supported by Russian Fund for Basic Researches, project 17-07-00177.

References

1. Tsitsiashvili, G.S.: Invariant properties of queuing systems with few flows. DVMJ **2**, 267–270 (2018). (In Russian)
2. Tsitsiashvili, G.: Algorithm of balance equations decomposition and investigation of poisson flows in Jackson networks. In: Dudin, A., Nazarov, A., Kirpichnikov, A. (eds.) ITMM 2017. CCIS, vol. 800, pp. 336–346. Springer, Cham (2017). https://doi.org/10.1007/978-3-319-68069-9_27
3. Gnedenko, B.V., Kovalenko, I.N.: Introduction to Queuing Theory. Nauka, Moscow (1966). (In Russian)
4. Buslenko, N.P.: Modelling of Complex Systems. Nauka, Moscow (1968). (In Russian)
5. Tsitsiashvili, G., Osipova, M.: Modelling of output flows in queuing systems and networks. In: Dudin, A., Nazarov, A., Moiseev, A. (eds.) ITMM/WRQ -2018. CCIS, vol. 912, pp. 106–116. Springer, Cham (2018). https://doi.org/10.1007/978-3-319-97595-5_9
6. Khinchin A.Y. Works on the mathematical queuing theory. Physmatlit, Moscow (1963). (in Russian)
7. Nazarov, A., Dammer, D.: Methods of limiting decomposition and Markovian summation in queueing system with infinite number of servers. In: Dudin, A., Nazarov, A., Moiseev, A. (eds.) ITMM/WRQ -2018. CCIS, vol. 912, pp. 71–82. Springer, Cham (2018). https://doi.org/10.1007/978-3-319-97595-5_6
8. Tananko, I.E., Fokina, N.P.: An analysis method of queueing networks with a degradable structure and non-zero repair times of systems. In: Dudin, A., Nazarov, A., Moiseev, A. (eds.) ITMM/WRQ -2018. CCIS, vol. 912, pp. 184–194. Springer, Cham (2018). https://doi.org/10.1007/978-3-319-97595-5_15

9. Mikheev, P., Pichugina, A., Suschenko, S.: Modeling of a multi-link transport connection by a network of queuing systems. In: Dudin, A., Nazarov, A., Moiseev, A. (eds.) ITMM/WRQ -2018. CCIS, vol. 912, pp. 274–289. Springer, Cham (2018). https://doi.org/10.1007/978-3-319-97595-5_22

10. Nazarov, A., Sztrik, J., Kvach, A.: A survey of recent results in finite-source retrial queues with collisions. In: Dudin, A., Nazarov, A., Moiseev, A. (eds.) ITMM/WRQ -2018. CCIS, vol. 912, pp. 1–15. Springer, Cham (2018). https://doi.org/10.1007/978-3-319-97595-5_1

11. Nazarov, A., Sztrik, J., Kvach, A.: Comparative analysis of methods of residual and elapsed service time in the study of the closed retrial queuing system M/GI/1//N with collision of the customers and unreliable server. In: Dudin, A., Nazarov, A., Kirpichnikov, A. (eds.) ITMM 2017. CCIS, vol. 800, pp. 97–110. Springer, Cham (2017). https://doi.org/10.1007/978-3-319-68069-9_8

12. Lisovskaya, E., Moiseeva, S., Pagano, M.: On the total customers' capacity in multi-server queues. In: Dudin, A., Nazarov, A., Kirpichnikov, A. (eds.) ITMM 2017. CCIS, vol. 800, pp. 56–67. Springer, Cham (2017). https://doi.org/10.1007/978-3-319-68069-9_5

13. Dragieva, V.: System state distribution of a finite-source retrial queue with subscribed customers. In: Dudin, A., Nazarov, A., Moiseev, A. (eds.) ITMM/WRQ -2018. CCIS, vol. 912, pp. 263–273. Springer, Cham (2018). https://doi.org/10.1007/978-3-319-97595-5_21

14. Dudin, A., Deepak, T.G., Joshua, V.C., Krishnamoorthy, A., Vishnevsky, V.: On a $BMAP/G/1$ retrial system with two types of search of customers from the orbit. In: Dudin, A., Nazarov, A., Kirpichnikov, A. (eds.) ITMM 2017. CCIS, vol. 800, pp. 1–12. Springer, Cham (2017). https://doi.org/10.1007/978-3-319-68069-9_1

15. Leskela, L.: Stabilization of an overloaded queueing network using measurement-based admission control. J. Appl. Probab. **43**(1), 231–244 (2006)

16. Shklennik, M., Moiseeva, S., Moiseev, A.: Optimization of two-level discount values using queueing tandem model with feedback. In: Dudin, A., Nazarov, A., Moiseev, A. (eds.) ITMM/WRQ -2018. CCIS, vol. 912, pp. 321–332. Springer, Cham (2018). https://doi.org/10.1007/978-3-319-97595-5_25

17. Shtoyan, D.: Quality Properties and Estimates of Stochastic Models. Mir, Moscow (1979). (In Russian)

18. Klimov, G.P.: Ergodic theorem for regenerative processes. Probab. Theory Its Appl. **21**(2), 402–405 (1976). (in Russian)

19. Tikhonov, V.I., Mironov, A.A.: Markov Processes. Soviet radio, Moscow (1977). (in Russian)

20. Cormen, T.H., Leiserson, C.E., Rivest, R.L.: Introduction to Algorithms, 1st edn. MIT Press and McGraw-Hill (1990). ISBN 0-262-03141-8

21. Tsitsiashvili, G.S., Osipova, M.A., Losev, A.S., Kharchenko, Y.N.: Jackson network as multiphase type network. Int. Math. Forum **12**(7), 303–310 (2017)

Performance Analysis and Optimal Control for Queueing System with a Reserve Unreliable Server Pool

Dmitry Efrosinin[1,2(✉)], Irina Gudkova[1], and Natalia Stepanova[3]

[1] Peoples' Friendship University of Russia (RUDN University),
Miklukho-Maklaya Street 6, 117198 Moscow, Russia
dmitry.efrosinin@jku.at, igudkova@sci.pfu.edu.ru
[2] Johannes Kepler University Linz, Altenbergerstrasse 69, 4040 Linz, Austria
[3] Institute of Control Sciences, RAS,
Profsoyuznaya Street 65, 117997 Moscow, Russia
http://www.rudn.ru, http://www.jku.at, http://www.ipu.ru

Abstract. In this paper a Markovian queueing system supplied with main and reserve unreliable service facility which we refer to as pools is introduced. Usage of the reserve pool is controlled by a hysteretic policy that depends on upper and lower threshold levels of queue length to increase and decrease the total service rate. The system is analysed as a process of type quasi-birth-and-death (QBD), and expressions for the stationary state probabilities are derived. For the cost structure we evaluate the long-run average cost per unit of time and determine the optimal hysteretic policy by implementing genetic algorithm. The sensitivity analysis to study the effect of system parameters and threshold levels on the average cost is provided by a number of numerical examples.

Keywords: Unreliable queueing system · Hysteretic policy · Quasi-birth-and-death process · Long-run average cost · Genetic algorithm

1 Introduction

Hysteretic control policies are used in various types of queueing and reliability models in which the available resources are to be optimally used to minimize the average cost per unit of time. This paper addresses a Markovian queueing system supplied with main and unreliable reserve service facilities, which are hereafter referred to as server pools. The reserve pool is activated in accordance with hysteretic control policy, whose underlying control principle can be easily explained by the following real-world application: Consider a single cell of a cellular (3GPP LTE) network with a Licence Shared Access (LSA) technology

The publication has been prepared with the support of the RUDN University Program 5-100 and funded by RFBR according to the research projects No. 17-07-00142.

(for details see Gudkova et al. [4]), which assumes that the band can be used when the owner does not need it, modelled as a queueing system. The main cellular and LSA bands correspond to the main and the reserve server pools, respectively. If the main pool is completely occupied by users, incoming service requests must be transmitted to the waiting queue. When the number of requests exceeds a certain upper threshold level a reserve LSA pool must be activated. If the number of requests drops below the lower threshold level, the reserve pool must be switched off. Since during service the reserve pool may be required by the band's owner, it is assumed to be unreliable. The central task is to find the optimal value for upper and lower threshold levels to minimize the specified cost function.

Numerous papers on queueing systems with controllable server activation have been published. Among the most popular is a paper by Yadin and Naor [10], who introduced the N-policy for turning the server on when the number of customers in the system reaches a threshold level N and turning it off only when the system becomes idle. This model was generalized by Wang [11] to the case of a unreliable server and random start-up time. Lu and Serfozo [7] presented a control policy for dynamic service rate choice based on hysteresis (retardation) due to the switching costs. Use of such a policy is justified, as it reduces the frequency of transitions between server on and off states, which in turn reduces average cost. Ibe and Keilson [5] studied a multi-server queueing system with hysteretic policy, achieving a stationary state distribution and mean delay of customers in the system in closed form. The problem of finding a balance between high system performance and low power consumption of the reserve servers was addressed by Mitrani [8]. The author has proposed a deterministic fluid approximation of the queueing process to derive some heuristics for the optimal policy. To the best of our knowledge, the problem of optimizing the switching rule for a multi-server queueing system with a multi-server reserve pool subject to failure has not yet been analysed. This work seeks to fill this gap, concentrating mostly on the evaluation of the optimal hysteretic policy. The application of a multi-server queueing system to estimate the impact of the LSA band unreliability to the LSA licensee within the busy period when some interruptions are possible was proposed in [1].

Our performance analysis of the queueing system with a reserve unreliable server pool includes the following contributions. We model the system as controllable queueing system with a hysteretic control policy. It is shown that for a fixed policy the Markov process belongs to a class of quasi-birth-and-death processes with a large number of boundary states according to a two-level threshold policy. We derive expressions for calculation the stationary state distribution and the long-run average cost in explicit form as function of threshold levels. Since a direct optimization of the average cost function over two discrete unknowns by a simple enumerative technique seems to be computationally quite expensive, we have proposed an alternative algorithm based on a genetic optimization methodology.

The rest of the paper is organized as follows: The mathematical model and discussion of the control problem are given in Sect. 2. Section 3 presents the algorithm for calculating the stationary state probabilities and evaluating the average cost function. A sensitivity analysis of the average cost function are presented in Sect. 4.

We use the notations $\mathbf{e}$, $\mathbf{0}$ and I respectively for the column vector of 1's, the raw vector or the matrix of 0's and the identity matrix with appropriate sizes. The notation $1_{\{A\}}$ specifies the indicator function, which takes the value 1 if the event A holds, and 0 otherwise.

2   Mathematical Model

Consider a markovian queueing system with main and reserve server pools of sizes k_1 and k_2, respectively. The total number of identical servers is fixed and denoted by $k = k_1 + k_2$. The requests arrive to the system in accordance to a Poisson process with parameter λ. The service times are exponentially distributed with parameter μ. The main pool is absolutely reliable, whereas the reserve is subject to failures that results in simultaneous failure of all servers in the pool. The service at the reserve pool is assumed to be preemptive. This means that, if there are no waiting customers at the moment of main pool service completion, but some reserve pool servers are busy, those customers must be transmitted to the empty servers of the main pool. In other words, the main pool servers must be kept busy whenever there are more than k_1 customers in the system. This assumption is very important because it renders a separate description of the state of the main server pool in the corresponding Markov chain unnecessary. The lifetime of the reserve pool is exponentially distributed with parameter α. In failure state the repair starts immediately and requires time exponentially distributed with parameter β. The reserve pool can fail even if no customers are present on some of its servers. Requests that cannot be serviced at the moment of arrival form one unbounded queue. The customers whose service was interrupted by failure of the reserve pool when all servers in the main pool were busy go to the head of the queue.

The reserve pool is activated by a two-level threshold policy (hysteretic policy) defined as $f = (q_1, q_2), 0 \le q_2 < q_1 < \infty$. If the reserve pool is switched off and the number of requests exceeds the level $q_1 \ge k_1$, all k_2 of the reserve pool can be treated as activated. After activation, the reserve pool can be found in an operational state with probability p and in a failure state with probability $1 - p$. In the former case all k servers are available for service, whereas in the latter case only k_1 are available. If the reserve servers are in operational or failure mode and the number of requests decreases to level q_2, they are switched off as a group and are no longer observable. When all servers are on, the servers of the reserve pool are indistinguishable from the servers of the main pool and the system behaves like an $M/M/k$ queueing system with unreliable servers. After a failure the system becomes an $M/M/k_1$ queueing system. When the time between deactivation and the next activation of the reserve pool very long, the probability p can be set to be equal to $\beta/(\alpha+\beta)$, that is, to the probability to be at a random epoch in state $s = 1$. The requests in a second pool whose service was interrupted are transferred to the first pool, and occupy the servers if they are available or go to the head of the queue otherwise.

The system state at time t for the fixed hysteresis control policy $f = (q_1, q_2)$ is described by a continuous-time Markov chain

$$\{X(t)\}_{t\geq 0} = \{S(t), N(t)\}_{t\geq 0}$$

with a threshold-dependent state space

$$E = \left\{ x = (s,n) : s = \begin{cases} 0, & 0 \leq n \leq q_1 - 1, \\ 1, 2, & n \geq q_2 + 1 \end{cases} \right\},$$

where $S(t) \in \{0, 1, 2\}$ stands for the states of the reserve pool at time t:

$$S(t) = \begin{cases} 0 & \text{if reserve pool is off;} \\ 1 & \text{if reserve pool is on and operational;} \\ 2 & \text{if reserve pool is on and failed.} \end{cases}$$

$N(t) \in \mathbb{N}_0$ is the number of requests in the system. The transitions of the Markov chain $\{X(t)\}_{t\geq 0}$ are defined by a threshold-dependent infinitesimal matrix $\Lambda^f = [\lambda^f_{xy}]_{x,y\in E}$. The transition diagram for the given Markov process $\{X(t)\}_{t\geq 0}$ is illustrated in Fig. 1, where the service rates are denoted by $\mu_{1,n} = \min\{n, k_1\}\mu$ and $\mu_{2,n} = \min\{n, k\}\mu$.

For the controllable model associated with a Markov process $\{X(t)\}_{t\geq 0}$, we define the following cost structure: c_1 – the holding cost per unit of time for each customer present in the system; c_2 – the pool exploitation cost per unit of time per server; c_3, c_4 – the fixed costs for turning the reserve pool on and off, respectively. The immediate cost $c(x)$ in state $x \in E$ is then of the form

$$c(x) = c_1 n(x) + c_2[k_1 + k_2 1_{\{s(x)\in\{1,2\}\}}] + c_3\lambda 1_{\{x=(0,q_1-1)\}}$$
$$+ c_4[\mu_{2,q_2+1}1_{\{x=(1,q_2+1)\}} + \mu_{1,q_2+1}1_{\{x=(2,q_2+1)\}}],$$

where $s(x)$ and $n(x)$ denote respectively the state of a reserve pool and the number of customers in the system in state $x \in E$. For the given cost structure the long-run average cost per unit of time for the fixed policy f can be written as the gain

$$g^f = \sum_{x\in E} c(x)\pi_x^f = c_1\bar{L}^f + c_2\bar{C}^f + c_3\lambda\pi^f_{(0,q_1-1)} \qquad (1)$$
$$+ c_4\left[\mu_{2,q_2+1}\pi^f_{(1,q_2+1)} + \mu_{1,q_2+1}\pi^f_{(2,q_2+1)}\right],$$

where

$$\bar{L}^f = \sum_{x\in E} n(x)\pi_x^f$$

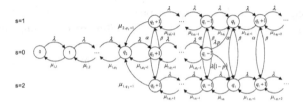

Fig. 1. Transition diagram of the Markov chain

is the mean number of requests in the system and

$$\bar{C}^f = k_1 + k_2 \sum_{\substack{x \in E \\ s(x) \in \{1,2\}}} \pi_x^f = k - k_2 \sum_{\substack{x \in E \\ s(x)=0}} \pi_x^f$$

stands for the mean number of activated servers in the main and reserve pools. These performance characteristics depend on the stationary state distribution, which in turn depends on the threshold policy $f = (q_1, q_2)$ specified above and on the specific choice of k_2. The main objective is to evaluate the optimal hysteretic policy $f^* = (q_1^*, q_2^*)$ and the optimal number of servers k_2^* at the reserve pool such that

$$f^* = \arg \min_{f, k_2} g^f, \tag{2}$$

where $0 \le k_2 \le k-1$, $0 \le q_2 \le q_1 - 1$, $q_1 \ge \max\{q_2 + 1, k_1\}$. The costs c_3 and c_4 have an impact on the average cost, but only if they take large values compared to c_1 and c_2.

3   Stationary State Distribution and Average Cost Function

The stationary state probability vector $\boldsymbol{\pi} = (\pi_x : x \in E)$ for some fixed policy f is a unique solution to $\boldsymbol{\pi} \Lambda^f = \mathbf{0}$ and $\boldsymbol{\pi} \mathbf{e} = 1$. To simplify the notation, we omit the upper index f. By $\boldsymbol{\pi}_n = (\pi_{(1,n)}, \pi_{(2,n)}), n \ge q_2 + 1$ we denote the sub-vectors of stationary probabilities for level n. To describe the transitions within a set of states $\{x = (n, s) : n \ge q_2 + 1, s \in \{1, 2\}\}$, we define the sub-matrices,

$$Q_{0,n} = \lambda I, \ Q_{1,n} = \begin{pmatrix} -(\lambda + \mu_{2,n} + \alpha) & \alpha \\ \beta & -(\lambda + \mu_{1,n} + \beta) \end{pmatrix},$$

$$Q_{2,n+1} = \begin{pmatrix} \mu_{2,n+1} & 0 \\ 0 & \mu_{1,n+1} \end{pmatrix},$$

which specify respectively the transitions from level $n - 1$ to n due to new arrivals, the transitions for remaining to stay at a given level n, and the transitions upon service completion at level $n + 1$. The Markov process $\{X(t)\}_{t \ge 0}$

obviously belongs to the class of quasi-birth-and-death processes (QBD) with a large number of boundary states. If the number of customers in the system exceeds the level $u = \max\{q_1 + 1, k\}$, the corresponding transitions defined by the matrices $\lambda I, Q_{1,u}$ and $Q_{2,u+1} = Q_{2,u}$ form the homogeneous part of the QBD process. It is well known (see, e.g., Neuts [9, Chapter 3]) that the stationary state probability vector $\boldsymbol{\pi}$ exists if and only if

$$\mathbf{p}\lambda I \mathbf{e} < \mathbf{p}Q_{2,u}\mathbf{e},$$

where $\mathbf{p} = (p_1, p_2)$ represents the invariant probability vector of the matrix $Q = \lambda I + Q_{1,u} + Q_{2,u}$. Solving the system $\mathbf{p}Q = \mathbf{0}$ and $\mathbf{p}\mathbf{e} = 1$ easily yields the stability condition of the queueing system under study in the form,

$$\hat{\rho} = \rho \frac{\alpha + \beta}{k_1 \alpha + k\beta} < 1, \quad \text{where } \rho = \frac{\lambda}{\mu}. \tag{3}$$

Theorem 1. *Under the stability condition the stationary state probabilities can be calculated by the following recursive solver:*

$$\pi_{(0,n)} = \frac{\rho^n}{n!}\pi_{(0,0)},\ 0 \le n \le \min\{q_2, k_1\}, \tag{4}$$

$$\pi_{(0,n)} = \left(\frac{\lambda}{k_1\mu}\right)^{n-k_1} \frac{\rho^{k_1}}{k_1!}\pi_{(0,0)},\ k_1 + 1 \le n \le q_2,\ q_2 > k_1, \tag{5}$$

$$\pi_{(0,n)} = \frac{\tau_{q_1-1-n}}{\tau_{q_1-1-q_2}}\pi_{(q_2,0)},\ q_2 + 1 \le n \le q_1 - 1,\ q_2 \le q_1 - 2, \tag{6}$$

$$\boldsymbol{\pi}_n = \sum_{i=q_1}^{u} L_i \prod_{j=0}^{i-(n+1)} M_{i-j-1}\pi_{(0,q_1-1)},\ q_2 + 1 \le n \le q_1 - 1,\ q_2 \le q_1 - 2, \tag{7}$$

$$\boldsymbol{\pi}_n = \sum_{i=n}^{u} L_i \prod_{j=0}^{i-(n+1)} M_{i-j-1}\pi_{(0,q_1-1)},\ q_1 \le n \le u, \tag{8}$$

$$\boldsymbol{\pi}_n = \boldsymbol{\pi}_u R^{n-u},\ n \ge u + 1, \tag{9}$$

where $\tau_0 = 1, \tau_i = 1 + \frac{1}{\rho_{1,q_1-1}}\tau_{i-1}, 1 \le i \le q_1 - q_2 - 1$, and the matrices $M_i, q_2 + 1 \le i \le u - 1$, and $L_i, q_1 \le i \le u$, satisfy the recursive relations,

$$M_{q_2+1} = - Q_{2,q_2+2}Q_{1,q_2+1}^{-1}, \tag{10}$$

$$M_i = - Q_{2,i+1}(Q_{1,i} + \lambda M_{i-1})^{-1},\ q_2 + 2 \le i \le k - 1,\ q_1 \le k - 2,$$

$$L_{q_1} = - \lambda(p, 1 - p)(Q_{1,q_1} + \lambda M_{q_1-1})^{-1},\ q_2 \le q_1 - 2,$$

$$L_{q_1} = - \lambda(p, 1 - p)Q_{1,q_1}^{-1},\ q_2 = q_1 - 1,$$

$$L_i = - \lambda L_{i-1}(Q_{1,i} + \lambda M_{i-1})^{-1},\ q_1 + 1 \le i \le k - 1,\ q_1 \le k - 2,$$

$$L_u = - \lambda L_{u-1}(Q_{1,u} + \lambda M_{u-1} + RQ_{2,u})^{-1}.$$

The probability $\pi_{(0,0)}$ is calculated from the normalizing condition

$$\sum_{n=0}^{q_1-1} \pi_{(0,n)} + \sum_{n=q_2+1}^{u-1} \pi_n \mathbf{e} + \pi_u (I - R)^{-1} \mathbf{e} = 1. \tag{11}$$

The matrix R is the unique non-negative solution with spectral radius less than one of the equation

$$R^2 Q_{2,u} + R Q_{1,u} + \lambda I = \mathbf{0}. \tag{12}$$

Proof. Under the stability condition we obtain the following results by solving the equation $\pi \Lambda^f = \mathbf{0}$ with normalization condition: The stationary state probabilities $\pi_{(0,n)}$ for $0 \le n \le q_2$ can obviously be expressed as functions of $\pi_{(0,0)}$ in the same way as in the $M/M/k_1$ queueing system, giving expressions (4) and (5). For the states $x = (0,n), q_2+1 \le n \le q_1-1$, the system of balance equations can be rewritten in the form

$$\lambda \pi_{(0,q_1-1)} + \mu_{1,q_2+1} \pi_{(0,q_2+1)} = \lambda \pi_{(0,q_2)}, \tag{13}$$
$$\lambda \pi_{(0,q_1-1)} + \mu_{1,q_2+2} \pi_{(0,q_2+2)} = \lambda \pi_{(0,q_2+1)},$$
$$\cdots$$
$$\lambda \pi_{(0,q_1-1)} + \mu_{1,q_1-2} \pi_{(0,q_1-2)} = \lambda \pi_{(0,q_1-3)},$$
$$\lambda \pi_{(0,q_1-1)} + \mu_{1,q_1-1} \pi_{(0,q_1-1)} = \lambda \pi_{(0,q_1-2)}.$$

The last system implies the relations

$$\pi_{(0,n)} = \tau_{q_1-1-n} \pi_{(0,q_1-1)}, \; q_2 \le n \le q_1 - 2.$$

Equation (6) results from $\pi_{(0,q_2)} = \tau_{q_1-1-q_2} \pi_{(0,q_1-1)}$. For the sub-vectors $\pi_n, n \ge q_2 + 1$, we obtain for $q_2 \le q_1 - 2$:

$$\pi_{q_2+1} Q_{1,q_2+1} + \pi_{q_2+2} Q_{2,q_2+2} = \mathbf{0}, \tag{14}$$
$$\lambda \pi_{n-1} + \pi_n Q_{1,n} + \pi_{n+1} Q_{2,n+1} = \mathbf{0}, \; q_2 + 2 \le n \le u, \; n \ne q_1,$$
$$\lambda \pi_{q_1-1} + \pi_{q_1} Q_{1,q_1} + \pi_{q_1+1} Q_{2,q_1+1} + \lambda (p, 1-p) \pi_{(0,q_1-1)},$$
$$\lambda \pi_u R^{n-u-1} + \pi_u R^{n-u} Q_{1,u} + \pi_u R^{n-u+1} Q_{2,u} = \mathbf{0}, \; n \ge u + 1.$$

Routine substitution in system (14) yields

$$\pi_n = \pi_{n+1} M_n, \; q_2 + 1 \le n \le q_1 - 1, \tag{15}$$
$$\pi_n = \pi_{n+1} M_n + \pi_{(0,q_1-1)} L_n, \; q_1 \le n \le u - 1,$$
$$\pi_u = \pi_{(0,q_1-1)} L_u,$$
$$\pi_n = \pi_u R^{n-u}, \; n \ge u + 1.$$

Solving these recursive relations, we obtain (7)–(9). Since all probabilities can be expressed as functions of $\pi_{0,0}$, it is evaluated by means of a normalizing condition (11).

Remark 1. The matrix R can be evaluated in closed form by solving (12) given the relation $RQ_{2,u}\mathbf{e} = \lambda I\mathbf{e}$. The resulting formulas are very complex, but an algorithm of successive substitution can be used instead:

$$R_0 = \mathbf{0}, \; R_{n+1} = -\lambda Q_{1,u}^{-1} - R_n^2 Q_{2,u} Q_{1,u}^{-1}, \; n \geq 0, \tag{16}$$

until $\mathbf{e}'(R_{n+1}^2 Q_{2,u} + R_{n+1}Q_{1,u} + \lambda I)\mathbf{e} \leq \varepsilon$. The sequence $\{R_n\}$ is monotonic and converges to R.

Corollary 1. *The mean performance measures needed for calculating the gain g are given by*

$$\bar{L} = \sum_{n=0}^{q_1-1} n\pi_{(n,0)} + \sum_{n=q_2+1}^{u-1} n\boldsymbol{\pi}_n \mathbf{e} + \pi_{(0,q_1-1)} L_u (uI - (u-1)R)(I - R)^{-2}\mathbf{e},$$

$$\bar{C} = k_1 + k_2 \sum_{n=q_2+1}^{\infty} \boldsymbol{\pi}_n \mathbf{e} = k - k_2 \sum_{n=0}^{q_1-1} \pi_{(n,0)}. \tag{17}$$

According to the average cost function defined in (1), it is a hard task to evaluate analytically the optimal values $\{k_2^*, f^* = (q_1^*, q_2^*)\}$ due to highly non-linear structure and recursive elements in the expression. A direct search method using a simple enumeration of possible values $\{k_2, q_1, q_2\}$ requires enormous CPU time and can therefore not be applied for numerical tests of the sensitivity analysis. This problem can be solved by a dynamic programming approach, although this has some limitations associated with the necessity to truncate the queue length, which in turn can influence the optimal policy and the value of the average cost function especially in heavy-traffic cases, when $\hat{\rho} \approx 1$. Other algorithms can be used to minimize the function g^f, such as the Tabu search algorithm (Glover and Laguna [2]) and particle swarm optimization (see Kennedy and Eberhart [6]). To minimize the function g, we implemented a genetic algorithm (GA, see e.g. Goldberg [3]), which requires reasonable evaluation time and belongs to the class of non-traditional search and optimization techniques.

The main idea of the GA consists of a random realization of the initial population of solutions and subsequent trials to improve the solution through a number of new generations. Each individual solution has an impact on the next generation. A selection of individuals for crossover and mutation is used to create the next population of solutions. The performance of the solution in the optimization problem is evaluated by a so-called fitness function, which in our case coincides with the function g. Below we give a short description of the algorithm implemented for our optimization problem. Recall that the task is to minimize the function

$$g^f := g(k_2, q_1, q_2), \; 0 \leq k_2 \leq k - 1, \; 0 \leq q_2 \leq q_1 - 1, \; q_1 \geq \max\{q_2 + 1, k_1\}.$$

To transform the minimization into a maximization problem, we define a fitness function

$$y(k_2, q_1, q_2) = 200 - g(k_2, q_1, q_2).$$

In the GA, the values of the three unknown variables $x_1 = k_2, x_2 = q_1$ and $x_3 = q_2$ must be represented as a bit string (chromosome) with three segments. The bit length l_j of a segment j for an accuracy up to 10^{-4} is calculated by

$$2^{l_j} < (b_j - a_j) \cdot 10^4 < 2^{l_j} - 1,$$

where $x_j \in [a_j, b_j]$.

Algorithm 1 (Genetic algorithm)
Input: $\lambda, \mu, \alpha, \beta$, total number of servers $k = 20$, population size $n_p = 20$, generation number $n_g = 200$, crossover rate $p_c = 0.25$, mutation rate $p_m = 0.01$, bit length of a variable $l_j = 18, j = 1, 2, 3$. Generate randomly n_p bit strings. Set $n = 0$.
Step 1: Evaluate for the k-th generated bit string the corresponding real value $x_j^{(k)}$,

$$x_j^{(k)} = a_j + N_j^{(k)} \frac{b_j - a_j}{2^{l_j} - 1}, j = 1, 2, 3, \tag{18}$$

where N_j is a decimal number encoded within the k-th bit string.
Step 2: If $n < n_g$, then generate a new population by selecting the best strings as parents according to the roulette wheel, terminate the loop otherwise. This step consists of the following operations:

2a: Evaluate the general compliance population function,

$$F = \sum_{k=1}^{n_p} y(x_1^{(k)}, x_2^{(k)}, x_3^{(k)}) - \min_{j=1,\dots,n_p} y(x_1^{(j)}, x_2^{(j)}, x_3^{(j)}).$$

2b: Evaluate the choice probability p_k of each bit string,

$$p_k = \frac{y(x_1^{(k)}, x_2^{(k)}, x_3^{(k)}) - \min\limits_{j=1,\dots,n_p} y(x_1^{(j)}, x_2^{(j)}, x_3^{(j)})}{F}, k = 1, 2, \dots, n_p.$$

2c: Evaluate the aggregate probability q_k for each bit string,

$$q_k = \sum_{j=1}^{k} p_j, k = 1, 2, \dots, n_p.$$

2d: Generate a random value $r \in [0, 1]$. If $r \leq q_1$, then the first chromosome must be selected; otherwise, select a chromosome $k, 1 \leq k \leq n_p$, such that $q_{k-1} < r \leq q_k$.

Step 3: Select from the population the candidates with probability p_c. Perform a double-point crossover between two parents, exchanging a part of a bit string between the randomly selected crossover points.
Step 4: Perform a mutation of each bit in the population with probability p_m.
Step 5. Evaluate new bit strings as chromosomes of a new population. Set $n = n + 1$, and go to Step 2.

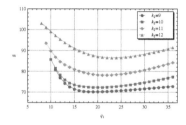

Fig. 2. The function g given optimal q_2^* versus q_1 for various values of k_2

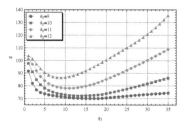

Fig. 3. The function g given optimal q_1^* versus q_2 for various values of k_2

4   Numerical Examples

The genetic algorithm was coded using *Mathematica* 11. The computation time for each optimization run is about 20–40 s, depending on the parameter selection. The numerical examples presented in this section were calculated for a queueing system with a total number of servers $k = 20$. The following system parameter values were chosen:

$$\lambda = 10,\ \mu = 1,\ \alpha = 0.01,\ \beta = 0.1,\ p = 0.9,\ c_1 = 2,\ c_2 = 4,\ c_3 = c_4 = 0. \quad (19)$$

Figure 2 illustrates the sensitivity of the cost function given an optimal value q_2^* for varying q_1 and k_2. The figure shows that the optimal number q_1 of customers needed for activation of the reserve pool increases with increasing k_2. For smaller values of k_2, the function g increases more gently after reaching the optimal value then for larger values k_2. Hence, the overestimation of q_1 has a relatively small impact on the average cost increase, if the number of servers in the reserve pool is considerably smaller than that in the main pool.

Figure 3 plots the cost function given an optimal value q_1^* for varying values of q_2 and k_2. It can be seen from this figure that the optimal number q_2 for deactivating the reserve pool decreases with increasing k_2. In the case of small values k_2, the function g yields a significantly flatter curve after reaching its optimal value.

In Fig. 4, we show the dependence of the function g under the optimal policy $f^* = (q_1^*, q_2^*)$ on the parameter k_2 for various values of λ (Fig. 4a) and μ

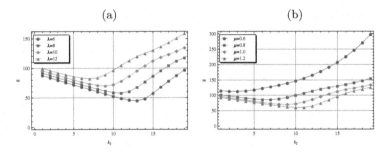

Fig. 4. The function g under optimal policy $f^* = (q_1^*, q_2^*)$ versus k_2 for various values of λ (a) and μ (b)

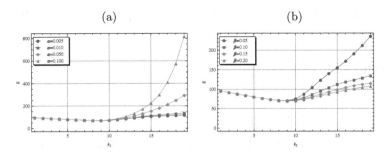

Fig. 5. The function g under optimal policy $f^* = (q_1^*, q_2^*)$ versus k_2 for various values of α (a) and β (b)

(Fig. 4b). Other parameters take the values from (19). We observe that the optimal number of reserve pool servers k_2 increases with decreasing values of λ and with increasing values of μ. Hence, in the case of light traffic (i.e. $\hat{\rho} \ll 1$), the service capacity of the reserve pool, should be higher than in a heavy traffic case ($\hat{\rho} \approx 1$). This can be explained by the choice of pool exploitation cost, which is twice as high as the holding cost. The differences between the curves for average costs and varying arrival and service rates become ,more distinct with increasing k_2, since the increase in the reserve pool's service capacity leads to higher pool exploitation costs.

Figure 5 plots the average cost g under the optimal policy $f^* = (q_1^*, q_2^*)$ for various values α (Fig. 5a) and β (Fig. 5b). We observe that for large α and small β, g rapidly increases with increasing k_2 after reaching the point at $k_2 = 10$. Further, the functions g for different values of α and β coincide when $k \leq 10$. The optimal value k_2 is insensitive to changes of α and β under the optimal policy $f^* = (q_1^*, q_2^*)$. We observe that usage of the reserve pool under the optimal policy outperforms the system in which all $k = k_1 = 20$ reliable servers are permanently in use. Reduction of the pool exploitation cost reduces the cost advantage of the controllable model compared to the non-controllable equivalent model.

5   Conclusions

We have analysed an $M/M/k$ queueing system with k_1 main pool servers and k_2 reserve pool servers, where $k = k_1 + k_2$. The mechanism for activating and consuming reserve service capacity was characterised by hysteretic control policy $f = (q_1, q_2)$ which prescribes the increase/decrease in service rate depending on two queue threshold levels q_1 and q_2. For the given control policy, the corresponding Markov process belongs to the class of quasi-birth-and-death processes. A matrix-analytic approach was used to obtain the stationary state probabilities and to derive an average cost per unit of time as a function of threshold levels and costs. Rather than a direct search method, we employed a genetic optimization algorithm to minimize the average cost function. Numerical analyses showed that the insensitivity of the optimal number k_2 to changes in failure and repair rates under the optimal policy.

References

1. Efrosinin, D., Samouylov, K., Gudkova, I.: Busy period analysis of a queueing system with breakdowns and its application to wireless network under licensed shared access regime. In: Galinina, O., Balandin, S., Koucheryavy, Y. (eds.) NEW2AN/ruSMART -2016. LNCS, vol. 9870, pp. 426–439. Springer, Cham (2016). https://doi.org/10.1007/978-3-319-46301-8_36
2. Glover, F., Laguna, M.: Tabu Search. Kluwer Academic Publishers, Norwell (1997)
3. Goldberg, D.E.: Genetic Algorithm in Search, Optimization and Machine Learning. Addison-Wesley, MA (1989)
4. Gudkova, I., et al.: Service failure and interruption probability analysis for Licensed Shared Access regulatory framework. In: Proceedings of the 7th International Congress on Ultra Modern Telecommunications and Control Systems and Workshops (ICUMT), pp. 123–131 (2015)
5. Ibe, O.C., Keilson, J.: Multi-server threshold queues with hysteresis. Perform. Eval. **21**, 185–213 (1995)
6. Kennedy, J., Eberhart, R.C.: Particle swarm optimization. In: Proceedings International Conference on Neutral Networks, Piscataway, N.J., pp. 1942–1948. IEEE (1995)
7. Lu, F.V., Serfozo, R.F.: $M/M/1$ queueing decision processes with monotone hysteretic optimal policies. Oper. Res. **32**(5), 1116–1132 (1984)
8. Mitrani, I.: Managing performance and power consumption in a server farm. Ann. Oper. Res. **202**, 121–134 (2013)
9. Neuts, M.F.: Matrix-Geometric Solutions in Stochastic Models. The Johns Hopkins University Press, Baltimore (1981)
10. Yadin, M., Naor, P.: Queueing systems with a removable service station. J. Oper. Res. Soc. **14**, 393–405 (1963)
11. Wang, K.-H.: Optimal control of a removable and non-reliable server in an $M/M/1$ queueing system with exponential startup time. Math. Methods Oper. Res. **58**, 29–39 (2003)

Study of a Service Process by a Loop Algorithm by Means of a Stopped Random Walk

Andrei V. Zorine[(✉)] [iD]

Lobachevsky State University of Nizhni Novgorod,
23, Gagarina Prospekt, 603950 Nizhni Novgorod, Russian Federation
andrei.zorine@itmm.unn.ru

Abstract. A queueing system with two Poisson input flows, infinite capacity queues, and a single server is studied. If the first queue is empty at the service time termination of the second queue, the server makes the loop by prolongation of the service time for the second queue by the same amount until the first queue is entered by a customer. Otherwise, a cyclic switching is used. We employ the fact that during certain intervals the second queue is described by a random walk stopped at a random time. In result the queueing system is modeled by a multidimensional discrete Markov chain, the server state and queues' lengths its elements. A necessary condition for the stationary probability distribution existence is found.

Keywords: Conflicting queueing system · Loop control algorithm · Stopped random walk · Stationary probability distribution · Necessary condition

1 Introduction

In queueing systems with several conflicting input flows different control algorithms are used. Some control algorithms allow for a relatively easy study of conditions for the stationary probability distribution existence of the modelling stochastic process, e.g. cyclic algorithms. On the contrary, control algorithms relying on system state dependent switching of the server can lead to analytically intractable models [1–4]. Stability analysis of queueing systems like these is complicated by varying server regimes.

Consider a queueing system with two input flows controlled by an algorithm with a loop (a d-limited polling system [5–7] with two stations and state-dependent switching as will be described below). Assume that the server spends a fixed (nonrandom) amount of time at each queue in turn. If the first queue is empty when the server finishes its service time at the second queue, the server prolongates the service time for the second queue by the same amount of time (this prolongation can be drawn as a loop on a server state-switching graph).

© Springer Nature Switzerland AG 2019
A. Dudin et al. (Eds.): ITMM 2019, CCIS 1109, pp. 121–135, 2019.
https://doi.org/10.1007/978-3-030-33388-1_11

Prolongations take place until a customer arrives to the first queue. After service of the first queue no prolongation is possible. In other words, the first queue has a higher priority.

A control algorithm of this kind can be used for traffic control at intersections governed by traffic-light signals as well as for automated microchip production machines. Depending on the inputs intensities, the typical behaviour of such a system is more like that of a purely cyclic system (in a heavy load case), or an M/G/1/∞ batch system (the case of light first flow), or a mixture of the two. In turn, the mixture coefficients depend on all system parameters in quite a complicated way.

A natural mathematical model for the queuing system is an embedded denumerable Markov chain describing the server state and queue lengths. The focus of the present paper is on the stability conditions, i.e. conditions under which a stationary probability distribution exists for a stochastic process describing the queueing system. Stability conditions for polling systems with state-dependent routing were obtained using different methods [8–11]. They don't seem to be directly applicable to the present queuing system. In the present paper we consider another embedded process, such that the server state changes at every observation instant. This can be done at cost of skipping a random number of steps in the original Markov chain. In Sect. 2 we define the model precisely and find transition probabilities for the embedded chain by solving a purely abstract problem about a stopped random walk. Then in Sect. 3 we explore a representation of a multi-queue system governed by the loop algorithm as a cyclic algorithm with random durations of server states and apply the method from [3].

2   The Problem Statement and the Stochastic Model

Consider a queueing system with two conflicting Poisson inputs Π_1, Π_2. The intensity of Π_j is λ_j, $j = 1, 2$. Customers from Π_j join a queue O_j of unlimited capacity. A server has two states, $\Gamma^{(1)}$ and $\Gamma^{(2)}$. Only customers from O_j get serviced in the state $\Gamma^{(j)}$. The server spends a constant time T_j in the state $\Gamma^{(j)}$. When this time elapses, the server instantly switches to the state $\Gamma^{(2)}$ if $j = 1$, but if $j = 2$ then $\Gamma^{(1)}$ becomes the new server state only when the queue O_1 is non-empty, otherwise a new time slot in the state $\Gamma^{(2)}$ takes place. The server loops in $\Gamma^{(2)}$ until new arrivals from Π_1. Epochs of these T_1- and T_2-time endings will be called the control epochs, and denoted τ_i, $i = 0, 1, \ldots$. To define the service process we use the notion of a saturation flow [12]. Instead of specifying probability distribution functions for service time of a single customer in every queue we set the large number of customers who can be served during the server sojourn in its state. The reader may assume that the actual service times can have different distributions and even be dependent as long as the total amounts of serviced customers per a server visit. In the state $\Gamma^{(j)}$ the saturation flow Π_j^{sat} holds $\ell_j > 0$ customers during the time T_j, and the other saturation flow Π_r^{sat}, $r \neq j$, holds no customers.

An example of a real-life queueing situation satisfying the above assumptions is an intersection with state-dependent traffic light switching. If yellow light signals (when cars may pass) can be adjoined to green light signals, and if a lower-priority direction can be let through only when there are no vehicles in a perpendicular high-priority direction, then our assumptions are fulfilled.

We will consider a stochastic sequence

$$\{(\varGamma_i, \kappa_{1,i}, \kappa_{2,i}); i = 0, 1, \ldots\} \tag{1}$$

on a probability space $(\Omega, \mathfrak{F}, \mathbf{P})$, where $\varGamma_0 \in \{\varGamma^{(1)}, \varGamma^{(2)}\}$ is the initial server state at time $\tau_0 = 0$, $\varGamma_i \in \{\varGamma^{(1)}, \varGamma^{(2)}\}$ is the server state during the time slot $(\tau_{i-1}, \tau_i]$, $\kappa_{j,i}$ is the number in O_j at time τ_i, $i = 1, 2, \ldots$, dependence on the elementary outcome ω implied but omitted in notation, as usual. We have functional relations

$$\varGamma_{i+1} = \varGamma^{(1)}, \quad \kappa_{1,i+1} = \max\{0, \kappa_{1,i} + \eta_{1,i} - \ell_1\}, \quad \kappa_{2,i+1} = \kappa_{2,i} + \eta_{2,i},$$

on the set $\{\omega \colon \varGamma_i = \varGamma^{(2)}, \kappa_{1,i} > 0\}$, $\eta_{j,i}$ possessing a Poisson probability distribution with parameter $\lambda_j T_1$, $j = 1, 2$, and we have

$$\kappa_{1,i+1} = \kappa_{1,i} + \eta_{1,i}, \quad \kappa_{2,i+1} = \max\{0, \kappa_{2,i} + \eta_{2,i} - \ell_2\}, \quad \varGamma_{i+1} = \varGamma^{(2)},$$

on the set $\{\omega \colon \varGamma_i = \varGamma^{(1)}\} \cup \{\omega \colon \varGamma_i = \varGamma^{(2)}, \kappa_{1,i} = 0\}$, $\eta_{j,i}$ having a Poisson probability distribution with parameter $\lambda_j T_2$, $j = 1, 2$. In effect, sequence (1) is a homogeneous irreducible Markov chain.

Let $S' = \{\varGamma^{(1)}, \varGamma^{(2)}\} \times \{0, 1, \ldots\} \times \{0, 1, \ldots\}$ be the state space of the process (1), put $S_0 = \{(\varGamma^{(2)}, 0, x_2); x_2 = 0, 1, \ldots\}$. Introduce stopping moments

$$\theta_0 = 0, \quad \theta_{i+1} = \min\{k \colon k > \theta_i, (\varGamma_k, \kappa_{1,k}, \kappa_{2,k}) \notin S_0\}.$$

Set $\hat{\varGamma}_i = \varGamma_{\theta_i}$, $\hat{\kappa}_{j,i} = \kappa_{j,\theta_i}$, $i = 0, 1, \ldots$. The new sequence

$$\{(\hat{\varGamma}_i, \hat{\kappa}_{1,i}, \hat{\kappa}_{2,i}); i = 0, 1, \ldots\} \tag{2}$$

is another Markov chain. To find its transition probabilities, let us note that one has $\theta_{i+1} = \theta_i + 1$ on a set

$$\{\omega \colon \hat{\varGamma}_i = \varGamma^{(2)}, \hat{\kappa}_{1,i} > 0\} \cup \{\omega \colon \hat{\varGamma}_i = \varGamma^{(1)}, \hat{\kappa}_{1,i} > 0\},$$

while on a set

$$\{\omega \colon \hat{\varGamma}_i = \varGamma^{(1)}, \hat{\kappa}_{1,i} = 0\}$$

the number of prolongations is $\nu_{i+1} = \theta_{i+1} - \theta_i$, geometrical random variable taking on value $k = 1, 2, \ldots$ with probability $(1 - p)p^{k-1}$, $p = e^{-\lambda_1 T_2}$. Also on the set $\{\omega \colon \hat{\varGamma}_i = \varGamma^{(1)}, \hat{\kappa}_{1,i} = 0, \hat{\kappa}_{2,i} = x_2\}$ quantities κ_{2,θ_i}, κ_{2,θ_i+1}, $\ldots$, $\kappa_{2,\theta_{i+1}}$ behave as the number in an $M/G/1/\infty$ queue with batch service and initial queue length $\hat{\kappa}_{2,i}$, and $\hat{\kappa}_{2,i+1}$ is the number at the stopping time ν_{i+1}. To employ this observation, let us introduce auxiliary i.i.d. Poisson variables η_i', $i = 1, 2, \ldots$ with parameter $\lambda_2 T_2$, and variables

$$\kappa_0' = b, \quad \kappa_{i+1}' = \max\{0, \kappa_i' + \eta_i' - \ell_2\}.$$

Further, let us introduce a family of probability generating functions ($|z| \leqslant 1$, $k = 0, 1, \ldots$):

$$\Phi_k(z, b) = \mathbf{E}(z^{\kappa_k'}), \quad \Phi(p, z; b) = \sum_{k=0}^{\infty} p^k \Phi_k(z; b), \quad q_j(z; t) = \exp\{\lambda_j t(z-1)\}.$$

Then, in law,

$$\mathbf{E}(z_1^{\hat{\kappa}_{1,i+1}} z_2^{\hat{\kappa}_{2,i+1}} \mid \{\omega \colon \hat{\Gamma}_i = \Gamma^{(1)}, \hat{\kappa}_{1,i} = 0, \hat{\kappa}_{2,i} = x_2\})$$

$$= \sum_{k=1}^{\infty} (e^{-\lambda_1 T_2})^{k-1} \left(\sum_{b=1}^{\infty} z_1^b \frac{(\lambda_1 T_2)^b)}{b!} e^{-\lambda_1 T_2} \right) \Phi_k(z_2; x_2)$$

$$= (e^{\lambda_1 T_2 z_1} - 1)(\Phi(p, z_2; x_2) - z_2^{x_2}). \quad (3)$$

We claim the following:

Lemma 1. *Let $\beta_j = \beta_j(p)$, $j = 1, 2, \ldots, \ell_2$, be the (distinct) zeroes of an equation*

$$z^{\ell_2} - pq_2(z; T_2) = 0$$

lying inside a unit disk $|z| < 1$. Then

$$\Phi(p, z; b) = \frac{(z - \beta_1) \times \ldots \times (z - \beta_{\ell_2})}{z^{\ell_2} - pq_2(z; T_2)} \left(\frac{1}{(1 - \beta_1) \times \ldots \times (1 - \beta_{\ell_2})} \right.$$

$$\left. + \sum_{j=1}^{\ell_2} \frac{(\beta_j)^{\ell_2 - 1}}{\prod_{s \neq j} (\beta_j - \beta_s)} \left(\frac{z^{b+1} - (\beta_j)^{b+1}}{z - \beta_j} - \frac{1 - (\beta_j)^{b+1}}{1 - \beta_j} \right) \right). \quad (4)$$

Proof. Conditioning on κ_i' gives

$$\mathbf{E}(z^{\kappa_{k+1}'}) = z^{-\ell_2} q_2(z; T_2) \mathbf{E}(z^{\kappa_k'})$$

$$+ \sum_{x=0}^{\ell_2 - 1} \mathbf{P}(\{\omega \colon \kappa_k' = x\}) \sum_{w=0}^{\ell_2 - 1 - x} \varphi_2(w; T_2)(1 - z^{x+w-\ell_2}). \quad (5)$$

Recalling the definition of $\Phi_{k+1}(z; b)$ and $\Phi_k(z; b)$, by multiplying (5) by p^{k+1} and then summing up for $k = 0, 1, \ldots$ one gets

$$\sum_{k=0}^{\infty} p^{k+1} \Phi_{k+1}(z; b) = z^{-\ell_2} q_2(z; T_2) p \sum_{k=0}^{\infty} p^k \Phi_k(z; b)$$

$$+ p \sum_{x=0}^{\ell_2 - 1} \left(\sum_{k=0}^{\infty} p^k \mathbf{P}(\{\omega \colon \kappa_k' = x\}) \right) \sum_{w=0}^{\ell_2 - x} \varphi_2(w; T_2)(1 - z^{x+w-\ell_2}).$$

Since $\mathbf{E}(z^{\kappa_0'}) = z^b$, one finally has

$$(z^{\ell_2} - pq_2(z; T_2))\Phi(p, z; b) = z^{b+\ell_2} + p \sum_{x=0}^{\ell_2 - 1} \hat{\Phi}(p; b, x) \sum_{w=0}^{\ell_2 - x} \varphi_2(w; T_2)(z^{\ell_2} - z^{x+w}).$$

By virtue of Rouché's theorem, an equation $z^{\ell_2} - pq_2(z; T_2) = 0$ has exactly ℓ_2 zeros $\beta_j = \beta_j(p)$, $j = 1, 2, \ldots, \ell_2$ in the open unit disk $|z| < 1$. Positions of the zeros don't depend on the initial value $\kappa'_0 = b$. The numbers $\beta_1, \beta_2, \ldots, \beta_{\ell_2}$ are all distinct. Indeed, β_j is the unique zero of

$$z = (\zeta_{\ell_2})^j p^{1/\ell_2} e^{\lambda_2 T_2(z-1)/\ell_2},$$

lying in the unit circle, where ζ_{ℓ_2} is the ℓ_2th primitive root of unity. Let us consider a factorization

$$(z^{\ell_2} - pq_2(z; T_2)) = (z - \beta_1)(z - \beta_2) \times \ldots \times (z - \beta_{\ell_2})$$
$$\times \frac{z^{\ell_2} - pq_2(z; T_2)}{(z - \beta_1)(z - \beta_2) \times \ldots \times (z - \beta_{\ell_2})} = u_e(z; p)u_i(z; p).$$

The function

$$u_i(z; p) = \frac{z^{\ell_2} - pq_2(z; T_2)}{(z - \beta_1)(z - \beta_2) \times \ldots \times (z - \beta_{\ell_2})}$$

is analytic in the open unit disk $|z| < 1$, continuous up to the boundary, and non-vanishing. So,

$$\Phi(p, z; b) = \frac{1}{u_i(z; p)} \cdot \frac{z^{b+\ell_2} + p \sum_{x=0}^{\ell_2-1} \hat{\Phi}(p; b, x) \sum_{w=0}^{\ell_2-1-x} \varphi_2(w; T_2)(z^{\ell_2} - z^{x+w})}{u_e(z; p)}.$$

In the numerator we have a polynomial of degree $\ell_2 + b$ in the variable z, at the same time the numbers $\beta_1, \beta_2, \ldots, \beta_{\ell_2}$ should be among its zeros. In other words, one has

$$z^{b+\ell_2} + p \sum_{x=0}^{\ell_2-1} \hat{\Phi}(p; b, x) \sum_{w=0}^{\ell_2-1-x} \varphi_2(w; T_2)(z^{\ell_2} - z^{x+w})$$
$$= (z^b + A_1(p)z^{b-1} + \ldots + A_{b-1}(p)z + A_b(p) + \hat{A}(p))$$
$$\times (z - \beta_1)(z - \beta_2) \times \ldots \times (z - \beta_{\ell_2}) \quad (6)$$

with some yet undefined coefficients $A_1(p), \ldots, A_b(p)$. are independent of the unknown $\hat{\Psi}_x(p)$, $x = 0, 1, \ldots, \ell_2 - 1$. Interestingly, they are exactly the coefficients obtained after division of a monomial z^{ℓ_2+b} by $(z - \beta_1)(z - \beta_2) \times \ldots \times (z - \beta_{\ell_2})$. In particular, they are the symmetric polynomials

$$A_1(p) = \beta_1 + \beta_2 + \ldots + \beta_{\ell_2},$$
$$A_2(p) = \beta_1^2 + \beta_2^2 + \ldots + \beta_{\ell_2}^2 + \beta_1\beta_2 + \beta_1\beta_2 + \ldots + \beta_{\ell_2-1}\beta_{\ell_2},$$
$$\ldots,$$
$$A_j(p) = \sum_{\substack{k_1+\ldots+k_{\ell_2}=j \\ k_1 \geqslant 0, \ldots, k_{\ell_2} \geqslant 0}} (\beta_1)^{k_1} \times \ldots \times (\beta_{\ell_2})^{k_{\ell_2}}, \qquad j = 1, 2, \ldots, b.$$

Considering the partial fractions decomposition of a generating function $\sum_{j=1}^{\infty} t^j A_j(p)$ we quickly prove a concise (fixed-length) formula

$$A_j(p) = \frac{(\beta_1)^{j+\ell_2-1}}{(\beta_1 - \beta_2) \times \ldots \times (\beta_1 - \beta_{\ell_2})}$$

$$+ \frac{(\beta_2)^{j+\ell_2-1}}{(\beta_2 - \beta_1) \times \ldots \times (\beta_2 - \beta_{\ell_2})} + \frac{(\beta_{\ell_2})^{j+\ell_2-1}}{(\beta_{\ell_2} - \beta_1) \times \ldots \times (\beta_{\ell_2-1} - \beta_{\ell_2})}.$$

In particular, for $j = 0$ we have $A_0(p) = 1$. Now let us find the unknown $\hat{A}(p)$. We put $z = 1$ into both sides of equality (6). We have an equation

$$1 = (1 - \beta_1) \times \ldots \times (1 - \beta_{\ell_2})(1 + A_1(p) + \ldots + A_b(p) + \hat{A}(p)).$$

Hence,

$$\hat{A}(p) = (1 - \beta_1)^{-1} \times \ldots \times (1 - \beta_{\ell_2})^{-1} - 1 - A_1(p) - \ldots - A_b(p).$$

Then,

$$A_0(p)(z^b - 1) + A_1(p)(z^{b-1} - 1) + \ldots + A_{b-1}(p)(z - 1)$$

$$+ \big((1 - \beta_1) \times \ldots \times (1 - \beta_{\ell_2})\big)^{-1} = \frac{1}{(1 - \beta_1) \times \ldots \times (1 - \beta_{\ell_2})}$$

$$+ \sum_{j=1}^{\ell_2} \frac{(\beta_j)^{\ell_2-1}}{\prod_{s \neq j}(\beta_j - \beta_s)} \left(\frac{z^{b+1} - (\beta_j)^{b+1}}{z - \beta_j} - \frac{1 - (\beta_j)^{b+1}}{1 - \beta_j} \right).$$

This proves the lemma.   □

Lemma 1 is an extension of known facts firstly because as a rule only the case $b = 0$ is studied in the majority of researches. Secondly, they were more interested in the limit of $(1 - p)\Phi(p, z; b)$ as $p \to 1$ as it gives the stationary probability distribution for the random walk.

Let $Q_i(r; x_1, x_2) = \mathbf{P}(\{\omega : \hat{\Gamma}_i = \Gamma^{(r)}, \hat{\kappa}_{1,i} = x_1, \hat{\kappa}_{2,i} = x_2\})$,

$$\Psi_i(z_1, z_2; r) = \sum_{x_1=1}^{\infty} \sum_{x_2=0}^{\infty} z_1^{x_1} z_2^{x_2} Q_i(r; x_1, x_2), \quad r = 1, 2.$$

Lemma 2. *One has* $(p = e^{-\lambda_1 T_2})$

$$\Psi_{i+1}(z_1, z_2; 1) + \sum_{x_2=0}^{\infty} Q_{i+1}(1; 0, x_2) z_2^{x_2}$$

$$= q_2(z_2; T_1) \sum_{x_1=1}^{\ell_1-1} \sum_{x_2=0}^{\infty} Q_i(2; x_1, x_2) z_2^{x_2} \sum_{b=0}^{\ell_1-1-x_1} \varphi_1(b; T_1)(1 - z_1^{x_1+b-\ell_1})$$

$$+ z_1^{-\ell_1} q_1(z_1; T_1) q_2(z_2; T_1) \Psi_i(z_1, z_2; 2), \quad (7)$$

$$\Psi_{i+1}(z_1, z_2; 2) = q_1(z_1; T_2)z_2^{-\ell_2}q_2(z_2; T_2)\Psi_i(z_1, z_2; 1)$$

$$+ \sum_{x_1=1}^{\infty}\sum_{x_2=0}^{\ell_2-1} Q_i(1; x_1, x_2)z_1^{x_1}q_1(z_1; T_2)\sum_{b=0}^{\ell_2-1-x_2}\varphi_2(b; T_2)(1 - z_2^{x_2+b-\ell_2})$$

$$+ (p^{-1}q_1(z_1; T_2) - 1)\sum_{x_2=0}^{\infty} Q_i(1; 0, x_2)z_2^{x_2}\left(z_2^{-x_2}\Phi(p, z_2; x_2) - 1\right). \tag{8}$$

Proof. Since

$$\Psi_{i+1}(z_1, z_2; 1) + \sum_{x_2=0}^{\infty} Q_{i+1}(1; 0, x_2)z_2^{x_2} = \mathbf{E}\big[z_1^{\hat{\kappa}_{1,i+1}}z_2^{\hat{\kappa}_{2,i+1}}I(\{\omega\colon \hat{\Gamma}_{i+1} = \Gamma^{(1)}\})\big],$$

by conditioning on $\hat{\Gamma}_i$, $\hat{\kappa}_{1,i}$, and $\hat{\kappa}_{2,i}$, one gets:

$$\Psi_{i+1}(z_1, z_2; 1) + \sum_{x_2=0}^{\infty} Q_{i+1}(1; 0, x_2)z_2^{x_2} = \sum_{x_1=1}^{\infty}\sum_{x_2=0}^{\infty} Q_i(2; x_1, x_2)$$

$$\times \mathbf{E}(z_1^{\max\{0, x_1+\eta_{1,\theta_i}-\ell_1\}}z_2^{x_2+\eta_{2,\theta_i}} \mid \{\omega\colon \hat{\Gamma}_i = \Gamma^{(2)}, \hat{\kappa}_{1,i} = x_1, \hat{\kappa}_{2,i} = x_2\})$$

$$= \sum_{x_1=1}^{\infty}\sum_{x_2=0}^{\infty} Q_i(2; x_1, x_2)z_2^{x_2}q_2(z_2; T_1)$$

$$\times \bigg(\mathbf{E}(z_1^{x_1+\eta_{1,\theta_i}-\ell_1} \mid \{\omega\colon \hat{\Gamma}_i = \Gamma^{(2)}, \hat{\kappa}_{1,i} = x_1, \hat{\kappa}_{2,i} = x_2\})$$

$$+ \mathbf{E}\big(z_1^{\max\{0, x_1+\eta_{1,\theta_i}-\ell_1\}} - z_1^{x_1+\eta_{1,\theta_i}-\ell_1}\big|\{\omega\colon \hat{\Gamma}_i = \Gamma^{(2)}, \hat{\kappa}_{1,i} = x_1, \hat{\kappa}_{2,i} = x_2\})\bigg),$$

which leads to (7).

Now let us prove equality (8). We have $\nu_{i+1} = 1$ on the set $\{\omega\colon \hat{\Gamma}_i = \Gamma^{(1)}, \hat{\kappa}_{1,i} = x_1, \hat{\kappa}_{2,i} = x_2\}$, $x_1 > 0$. Thus

$$\Psi_{i+1}(z_1, z_2; 2) = \mathbf{E}\big[z_1^{\hat{\kappa}_{1,i+1}}z_2^{\hat{\kappa}_{2,i+1}}I(\{\omega\colon \hat{\Gamma}_{i+1} = \Gamma^{(2)}\})\big]$$

$$= \bigg(\sum_{x_1=1}^{\infty}\sum_{x_2=0}^{\infty} Q_i(1; x_1, x_2)$$

$$\times \mathbf{E}(z_1^{x_1+\eta_{1,\theta_i}}z_2^{\max\{0,x_2+\eta_{2,\theta_i}-\ell_2\}} \mid \{\omega\colon \hat{\Gamma}_i = \Gamma^{(1)}, \hat{\kappa}_{1,i} = x_1, \hat{\kappa}_{2,i} = x_2\})\bigg)$$

$$+ \sum_{x_2=0}^{\infty} Q_i(1; 0, x_2)\mathbf{E}(z_1^{\hat{\kappa}_{1,i+1}}z_2^{\hat{\kappa}_{2,i+1}} \mid \{\omega\colon \Gamma_i = \Gamma^{(2)}, \kappa_{1,i} = 0, \kappa_{2,i} = x_2\}).$$

The sum in parentheses is

$$\sum_{x_1=1}^{\infty}\sum_{x_2=0}^{\infty} Q_i(1,x_1,x_2)z_1^{x_1}q_1(z_1;T_2)$$

$$\times\left(\mathbf{E}(z_2^{x_2+\eta_{2,\theta_i}-\ell_2}\mid\{\hat{\Gamma}_i=\Gamma^{(1)},\hat{\kappa}_{1,i}=x_1,\hat{\kappa}_{2,i}=x_2\})\right.$$

$$\left.+\mathbf{E}(z_2^{\max\{0,x_2+\eta_{2,\theta_i}-\ell_2\}}-z_2^{x_2+\eta_{2,\theta_i}-\ell_2}\mid\{\omega\colon\hat{\Gamma}_i=\Gamma^{(1)},\hat{\kappa}_{1,i}=x_1,\hat{\kappa}_{2,i}=x_2\})\right)$$

$$=q_1(z_1;T_2)z_2^{-\ell_2}q_2(z_2;T_2)\Psi_i(z_1,z_2;1)$$

$$+\sum_{x_1=1}^{\infty}\sum_{x_2=0}^{\ell_2-1} Q_i(1,x_1,x_2)z_1^{x_1}q_1(z_1;T_2)\sum_{b=0}^{\ell_2-1-x_2}\varphi_2(b;T_2)(1-z_2^{x_2+b-\ell_2}).$$

The second sum is transformed using formulae (3), (4) to prove (8).   □

3 Some Stationarity Conditions

Under the assumptions above, a sequence

$$\{(\hat{\Gamma}_i,\hat{\kappa}_{1,i}); i=0,1,\dots\} \tag{9}$$

is also an irreducible periodic Markov chain. Let us set,

$$Q_{1,i}(r,x)=\mathbf{P}(\{\omega\colon\hat{\Gamma}_i=\Gamma^{(r)},\hat{\kappa}_{1,i}=x\}),$$

$$\Psi_{1,i}(z;1)=\sum_{x=0}^{\infty}Q_{1,i}(r,x)z^x,\qquad \Psi_{1,i}(z;2)=\sum_{x=1}^{\infty}Q_{1,i}(2,x)z^x,\quad |z|\leqslant 1.$$

The notations for a stationary probability distribution resemble this one but omit i and $(i+1)$ in the subscripts.

Theorem 1. *For the existence of a stationary probability distribution of the Markov chain* (9) *it is necessary and sufficient that* $\lambda_1(T_1+T_2)-\ell_1<0$.

Proof. Substitution $z_1=z$, $z_2=1$ into recurrence relations (7), (8) produces

$$\Psi_{1,i+1}(z;1)=z^{-\ell_1}q_1(z;T_1)\Psi_{1,i}(z;2)$$

$$+\sum_{x=1}^{\ell_1-1}Q_{1,i}(2,x)\sum_{b=0}^{\ell_1-1-x}\frac{(\lambda_1T_1)^b}{b!}e^{-\lambda_1T_1}(1-z^{x+b-\ell_1}), \tag{10}$$

$$\Psi_{1,i+1}(z;2)=q_1(z;T_2)\Psi_{1,i}(z;1)+Q_{1,i}(1,0)e^{-\lambda_1T_2}\frac{q_1(z;T_2)-1}{1-e^{-\lambda_1T_2}}. \tag{11}$$

If a stationary probability distribution exists, the corresponding probability generating functions must satisfy equations

$$\Psi_1(z;1)=z^{-\ell_1}q_1(z;T_1)\Psi_1(z;2)$$

$$+\sum_{x=1}^{\ell_1-1}Q_1(2,x)\sum_{b=0}^{\ell_1-1-x}\frac{(\lambda_1T_1)^b}{b!}e^{-\lambda_1T_1}(1-z^{x+b-\ell_1}), \tag{12}$$

$$\Psi_1(z;2)=q_1(z;T_2)\Psi_1(z;1)+Q_1(1,0)e^{-\lambda_1T_2}\frac{q_1(z;T_2)-1}{1-e^{-\lambda_1T_2}}. \tag{13}$$

It's trivial to see that $\Psi_1(1;1) = \Psi_1(1;2) = 1/2$. Substituting Taylor expansions in the left neighborhood of $z = 1$

$$z^{-\ell_1} q_1(z; T_1) = 1 + (\lambda_1 T_1 - \ell_1)(z - 1) + o(z - 1),$$
$$q_1(z; T_2) = 1 + \lambda_1 T_2(z - 1) + o(z - 1),$$
$$1 - z^{x+b-\ell_1} = (\ell_1 - x - b)(z - 1) + o(z - 1)$$

into (12), (13), summing up the two equations, combining similar terms, and sending $z \to 1$ from the left we get in effect

$$\sum_{x=1}^{\ell_1-1} Q_1(2,x) \sum_{b=0}^{\ell_1-1-x} \frac{(\lambda_1 T_1)^b}{b!} e^{-\lambda_1 T_1} (\ell_1 - x - b)$$
$$+ Q_1(1,0) e^{-\lambda_1 T_2} \frac{\lambda_1 T_2}{1 - e^{-\lambda_1 T_2}} = \frac{\ell_1 - \lambda_1(T_1 + T_2)}{2}. \quad (14)$$

Now, assumption $\ell_1 \leqslant \lambda_1(T_1 + T_2)$ leads to an impossible conclusion that all the stationary probabilities $Q_1(1,0)$, $Q_1(2,b)$, $b = 2, 3, \ldots, \ell_1 - 1$ are non-positive.

To prove that the inequality $\lambda_1(T_1 + T_2) - \ell_1 < 0$ is sufficient for the existence of the stationary probability distribution, assume the converse, that given this inequality no stationary probabilities exist. Then it's easy to prove that a sequence of mathematical expectations $\{\mathbf{E}\hat{\kappa}_{1,i}; i = 0, 1, \ldots\}$ grows to infinity independently of the initial probability distribution. But we plan to demonstrate that these mathematical expectations are bounded uniformly in $i = 0, 1, \ldots$. Indeed, recurrence relations (10), (11) extend the probability generating functions in analytic way into a disk $\{z: |z| < 1 + \varepsilon\}$ for some $\varepsilon > 0$. We can select a real z_1, $1 < z_1 < 1 + \varepsilon$ and define a real-valued sequence

$$\Psi_0^+ = \Psi_{1,0}(z_1; 1), \qquad \Psi_1^+ = \Psi_{1,1}(z_1; 1),$$
$$\Psi_{i+1}^+ = z_1^{-\ell_1} q_1(z_1; T_1) q_1(z; T_2) \Psi_i^+ + z_1^{-\ell_1} q_1(z_1; T_1) e^{-\lambda_1 T_2} \frac{q_1(z_1; T_2) - 1}{1 - e^{-\lambda_1 T_2}}$$
$$+ \sum_{x=1}^{\ell_1-1} \sum_{b=0}^{\ell_1-1-x} \frac{(\lambda_1 T_1)^b}{b!} e^{-\lambda_1 T_1} (1 - z_1^{x+b-\ell_1}), \quad i = 0, 1, \ldots$$

Since the derivative

$$\frac{d}{dz} \left(z_1^{-\ell_1} q_1(z_1; T_1) q_1(z; T_2) \right) \Big|_{z=1} = \lambda_1(T_1 + T_2) - \ell_1$$

is negative by assumption, the sequence $\{\Psi_i^+; i = 0, 1, \ldots\}$ converges, hence it's bounded. At the same time the sequence dominates $\{\Psi_{1,i}(z_1; 1); i = 0, 1, \ldots\}$, i.e. $\Psi_{1,i}(z_1; 1) \leqslant \Psi_i^+$ for all $i = 0, 1, \ldots$. It follows then that both the sequence $\{\Psi_{1,i}(z_1; 1); i = 0, 1, \ldots\}$ and the sequence $\{\Psi_{1,i}(z_1; 2); i = 0, 1, \ldots\}$ are bounded, and the above-mentioned mathematical expectations get bounded due to the Chauchy's integral formula. The contradiction proves that a stationary probability distribution exists.

Theorem 2. *Let $\alpha_1 = 1$, α_2, ..., α_{l_1} be the (distinct) zeroes of the equation*

$$z^{\ell_1} - q_1(z; T_1 + T_2) = 0$$

lying in the unit disk $|z| \leqslant 1$. Then the stationary probabilities $Q_1(1,0)$, $Q_1(2,b)$, $b = 2, 3, \ldots, \ell_1 - 1$ are the solution to the linear algebraic system

$$\sum_{x=1}^{\ell_1-1} Q_1(2,x) \sum_{b=0}^{\ell_1-1-x} \frac{(\lambda_1 T_1)^b}{b!} e^{-\lambda_1 T_1}(\alpha_w^{\ell_1} - \alpha_w^{x+b})$$

$$+ Q_1(1,0)e^{-\lambda_1 T_2} \cdot \frac{e^{\lambda_1(T_1+T_2)(\alpha_w-1)} - e^{\lambda_1 T_1(\alpha_w-1)}}{1 - e^{-\lambda_1 T_2}} = 0, \quad w = 2,3,\ldots,\ell_1,(15)$$

$$\sum_{x=1}^{\ell_1-1} Q_1(2,x) \sum_{b=0}^{\ell_1-1-x} \frac{(\lambda_1 T_1)^b}{b!} e^{-\lambda_1 T_1}(\ell_1 - x - b)$$

$$+ Q_1(1,0)e^{-\lambda_1 T_2} \frac{\lambda_1 T_2}{1 - e^{-\lambda_1 T_2}} = \frac{\ell_1 - \lambda_1(T_1 + T_2)}{2}. \qquad (16)$$

Proof. Stationary Eqs. (12), (13) in matrix form can be written as

$$\begin{pmatrix} z^{\ell_1} & -q_1(z; T_1) \\ -q_1(z; T_2) & 1 \end{pmatrix} \cdot \begin{pmatrix} \Psi_1(z; 1) \\ \Psi_1(z; 2) \end{pmatrix}$$

$$= \begin{pmatrix} \displaystyle\sum_{x=1}^{\ell_1-1} Q_1(2,x) \sum_{b=0}^{\ell_1-1-x} \frac{(\lambda_1 T_1)^b}{b!} e^{-\lambda_1 T_1}(z^{\ell_1} - z^{x+b}) \\ Q_1(1,0)e^{-\lambda_1 T_2} \dfrac{q_1(z; T_2) - 1}{1 - e^{-\lambda_1 T_2}} \end{pmatrix}.$$

Let's multiply the above equation from left by the conjugate matrix

$$\begin{pmatrix} 1 & q_1(z; T_1) \\ q_1(z; T_2) & z^{\ell_1} \end{pmatrix}.$$

We get (back in scalar form)

$$\left(z^{\ell_1} - q_1(z; T_1 + T_2)\right)\Psi_1(z; 1) = Q_1(1,0)e^{-\lambda_1 T_2}q_1(z; T_1)\frac{q_1(z; T_2) - 1}{1 - e^{-\lambda_1 T_2}}$$

$$+ \sum_{x=1}^{\ell_1-1} Q_1(2,x) \sum_{b=0}^{\ell_1-1-x} \frac{(\lambda_1 T_1)^b}{b!} e^{-\lambda_1 T_1}(z^{\ell_1} - z^{x+b})$$

and

$$\left(z^{\ell_1} - q_1(z; T_1 + T_2)\right)\Psi_1(z; 2) = Q_1(1,0)e^{-\lambda_1 T_2}z^{\ell_1}\frac{q_1(z; T_2) - 1}{1 - e^{-\lambda_1 T_2}}$$

$$+ q_1(z; T_2) \sum_{x=1}^{\ell_1-1} Q_1(2,x) \sum_{b=0}^{\ell_1-1-x} \frac{(\lambda_1 T_1)^b}{b!} e^{-\lambda_1 T_1}(z^{\ell_1} - z^{x+b}).$$

By virtue of Rouché's theorem, an equation $z^{\ell_1} - q_1(z; T_1 + T_2) = 0$ has exactly ℓ_1 zero in a disk $|z| \leqslant 1$. Indeed, one has

$$|q_1(z; T_1 + T_2)| = \left|\exp\{\lambda_1(T_1 + T_2)(z - 1)\}\right| \leqslant 1 = |z^{\ell_1}| \qquad \text{for } |z| = 1.$$

We shall call them $z_1 = 1$, z_2, ..., z_{ℓ_1}. They should also zero out the right-hand sides of the two equations above. So, we have $(\ell_1 - 1)$ equations like (15). Finally, Eq. (16) is the same as (14). □

Equations (15), (16) are linear in the unknown stationary probabilities, but involve quantities (the zeros α_2, ..., α_{ℓ_1}) which can't be written in a simple form in terms of the system's parameters. But they can be determined numerically to any desired accuracy. Then the stationary probability generating functions $\Psi_1(z; 1)$ and $\Psi_1(z; 2)$ will have only numerical parameters. It's worth noting that the stability condition for the first queue coincides with the stability condition for a purely cyclic service algorithm.

Theorem 3. *Let $\ell_2 = 1$ or 2. For the existence of the stationary probability distribution of the Markov chain (2) it is necessary that*

$$\frac{1}{2}(\lambda_2(T_1 + T_2) - \ell_2) + \frac{p}{1-p}(\lambda_2 T_2 - \ell_1)Q_1(1, 0) < 0. \tag{17}$$

Proof. Let's assume that the stationary probability distribution for the Markov chain (2) exists. Selecting it in place of the initial probability distribution we guarantee that the probability generating functions are independent of i. Substitution $z_1 = 1$ into recurrence relations of Lemma 2 yields

$$\Psi(1, z_2; 1) + \sum_{x_2=0}^{\infty} Q(1; 0, x_2)z_2^{x_2} = q_2(z_2; T_1)\Psi(1, z_2; 2), \tag{18}$$

$$\Psi(1, z_2; 2) = z_2^{-\ell_2} q_2(z_2; T_2)\Psi(1, z_2; 1)$$
$$+ \sum_{x_1=1}^{\infty}\sum_{x_2=0}^{\ell_2-1} Q(1; x_1, x_2) \sum_{b=0}^{\ell_2-1-x_2} \varphi_2(b; T_2)(1 - z_2^{x_2+b-\ell_2})$$
$$+ (p^{-1} - 1)\sum_{x_2=0}^{\infty} Q(1; 0, x_2)z_2^{x_2}\left(z_2^{-x_2}\Phi(p, z_2; x_2) - 1\right). \tag{19}$$

To proceed we need to calculate the derivative of $\Phi(p, z_2; x_2)$ w.r.t z_2 at $z_2 = 1$. From Equality (5) we get, after some trivial transformations:

$$\frac{d\Phi}{dz_2}(p, 1; x_2) = \frac{1}{1-p}\left(\frac{1}{1-\beta_1} + \ldots + \frac{1}{1-\beta_{\ell_2}}\right) - \frac{\ell_2 - p\lambda_2 T_2}{(1-p)^2}$$
$$+ \frac{1}{1-p}\left(x_2\sum_{j=1}^{\ell_2}(\beta_j)^{\ell_2-1}\prod_{s\neq j}\frac{1-\beta_s}{\beta_j - \beta_s} - \sum_{j=1}^{\ell_2}\frac{(\beta_j)^{\ell_2} - (\beta_j)^{\ell_2+x_2}}{1-\beta_j}\prod_{s\neq j}\frac{1-\beta_s}{\beta_j - \beta_s}\right).$$

Lagrange's interpolation formula justifies identities

$$\sum_{j=1}^{\ell_2}(\beta_j)^{\ell_2-1}\prod_{s\neq j}\frac{1-\beta_s}{\beta_j-\beta_s}=1$$

and

$$\frac{1}{1-\beta_1}+\ldots+\frac{1}{1-\beta_{\ell_2}}-\sum_{j=1}^{\ell_2}\frac{(\beta_j)^{\ell_2}}{1-\beta_j}\prod_{s\neq j}\frac{1-\beta_s}{\beta_j-\beta_s}=\ell_2.$$

Then the derivative equals

$$\frac{1}{1-p}\Big(\frac{1}{1-\beta_1}+\ldots+\frac{1}{1-\beta_{\ell_2}}\Big)-\frac{\ell_2-p\lambda_2 T_2}{(1-p)^2}$$

$$+\frac{1}{1-p}\Big(x_2+\ell_2-\frac{1}{1-\beta_1}-\ldots-\frac{1}{1-\beta_{\ell_2}}+\sum_{j=1}^{\ell_2}\frac{(\beta_j)^{\ell_2+x_2}}{1-\beta_j}\prod_{s\neq j}\frac{1-\beta_s}{\beta_j-\beta_s}\Big)$$

$$=-\frac{\ell_2-p\lambda_2 T_2}{(1-p)^2}+\frac{1}{1-p}\Big(x_2+\ell_2+\sum_{j=1}^{\ell_2}\frac{(\beta_j)^{\ell_2+x_2}}{1-\beta_j}\prod_{s\neq j}\frac{1-\beta_s}{\beta_j-\beta_s}\Big).$$

With the last formula the following expansions hold:

$$q_2(z_2;T_1)=1+\lambda_2 T_1(z_2-1)+o(z_2-1),$$
$$z_2^{-\ell_2}q_2(z_2;T_2)=1+(\lambda_2 T_2-\ell_2)(z_2-1)+o(z_2-1),$$
$$(p^{-1}-1)(z_2^{-x_2}\Phi(p,z_2;x_2)-1)=(p^{-1}-1)\big[(1-p)^{-1}-1\big]$$

$$+(z_2-1)(p^{-1}-1)\Big(-x_2(1-p)^{-1}+\frac{d\Phi}{dz_2}(p,1;x_2)\Big)+o(z_2-1)$$

$$=1+(z_2-1)\Big(\frac{\lambda_2 T_2-\ell_2}{(1-p)}+\frac{1}{p}\sum_{j=1}^{\ell_2}\frac{(\beta_j)^{\ell_2+x_2}}{1-\beta_j}\prod_{s\neq j}\frac{1-\beta_s}{\beta_j-\beta_s}\Big)+o(z_2-1),$$

Let us add (18), (19) and use the above expansions:

$$\Psi(1,z_2;1)+\sum_{x_2=0}^{\infty}Q(1;0,x_2)z_2^{x_2}+\Psi(1,z_2;2)=(1+\lambda_2 T_1(z_2-1)+o(z_2-1))$$

$$\times\Psi(1,z_2;2)+(1+(\lambda_2 T_2-\ell_2)(z_2-1)+o(z_2-1))\Psi(1,z_2;1)$$

$$+\sum_{x_1=1}^{\infty}\sum_{x_2=0}^{\ell_2-1}Q(1;x_1,x_2)\sum_{b=0}^{\ell_2-1-x_2}\varphi_2(b;T_2)((\ell_2-x_2-b)(z_2-1)+o(z_2-1)$$

$$+\sum_{x_2=0}^{\infty}Q(1;0,x_2)z_2^{x_2}\Big(1+(z_2-1)\Big(\frac{\lambda_2 T_2-\ell_2}{(1-p)}+\frac{1}{p}\sum_{j=1}^{\ell_2}\frac{(\beta_j)^{\ell_2+x_2}}{1-\beta_j}\prod_{s\neq j}\frac{1-\beta_s}{\beta_j-\beta_s}\Big)$$

$$+o(z_2-1)\Big).$$

Then after combining similar terms, dividing by $(z_2 - 1)$, and sending $z_2 \to 1$ $(z_2 < 1)$ we obtain

$$
0 = \lambda_2 T_1 \Psi(1,1;2) + (\lambda_2 T_2 - \ell_2)\Psi(1,1;1)
$$
$$
+ \sum_{x_1=1}^{\infty} \sum_{x_2=0}^{\ell_2-1} Q(1;x_1,x_2) \sum_{b=0}^{\ell_2-1-x_2} \varphi_2(b;T_2)(\ell_2 - x_2 - b)
$$
$$
+ \sum_{x_2=0}^{\infty} Q(1;0,x_2)\left(\frac{\lambda_2 T_2 - \ell_2}{(1-p)} + \frac{1}{p}\sum_{j=1}^{\ell_2} \frac{(\beta_j)^{\ell_2+x_2}}{1-\beta_j} \prod_{s\neq j} \frac{1-\beta_s}{\beta_j - \beta_s}\right). \tag{20}
$$

On the other hand, from formula (18) we get whence

$$
\Psi(1,1;1) + \sum_{x_2=0}^{\infty} Q(1;0,x_2) = \Psi(1,1;2) = \frac{1}{2}.
$$

If we put these values into (20) we get

$$
0 = \frac{1}{2}(\lambda_2(T_1 + T_2) - \ell_2) + \sum_{x_1=1}^{\infty} \sum_{x_2=0}^{\ell_2-1} Q(1;x_1,x_2) \sum_{b=0}^{\ell_2-1-x_2} \varphi_2(b;T_2)(\ell_2 - x_2 - b)
$$
$$
+ \sum_{x_2=0}^{\infty} Q(1;0,x_2)\left(\frac{\lambda_2 T_2 - \ell_2}{(1-p)} + \frac{1}{p}\sum_{j=1}^{\ell_2} \frac{(\beta_j)^{\ell_2+x_2}}{1-\beta_j} \prod_{s\neq j} \frac{1-\beta_s}{\beta_j - \beta_s} - (\lambda_2 T_2 - \ell_2)\right).
$$

In result,

$$
0 = \frac{1}{2}(\lambda_2(T_1 + T_2) - \ell_2) + \frac{p}{1-p}(\lambda_2 T_2 - \ell_2)Q_1(1,0)
$$
$$
+ \sum_{x_1=1}^{\infty} \sum_{x_2=0}^{\ell_2-1} Q(1;x_1,x_2) \sum_{b=0}^{\ell_2-1-x_2} \varphi_2(b;T_2)(\ell_2 - x_2 - b)
$$
$$
+ \sum_{x_2=0}^{\infty} Q(1;0,x_2)\frac{1}{p}\sum_{j=1}^{\ell_2} \frac{(\beta_j)^{\ell_2+x_2}}{1-\beta_j} \prod_{s\neq j} \frac{1-\beta_s}{\beta_j - \beta_s}. \tag{21}
$$

It follows from the definition of $\beta_1, \beta_2, \ldots, \beta_{\ell_2}$ that two at most are real, while the others make pairs of complex conjugates. Then the value

$$
\sum_{j=1}^{\ell_2} \frac{(\beta_j)^{\ell_2+x_2}}{1-\beta_j} \prod_{s\neq j} \frac{1-\beta_s}{\beta_j - \beta_s} \tag{22}
$$

coincides with its complex conjugate, so it's real. Up to this time we have a proof it's positive when $\ell_2 = 1$ or 2. If $\ell_2 = 1$, the unique root β_1 is positive and less than unity, thus $\beta_1^{1+x_2}(1-\beta_1)^{-1} > 0$.

Let's assume now that $\ell_2 = 2$. There are two real solutions, $0 < -\beta_2 < \beta_1 < 1$. The sum of interest equals

$$\frac{\beta_1^{2+x_2}}{1-\beta_1} \cdot \frac{1-\beta_2}{\beta_1-\beta_2} + \frac{\beta_2^{2+x_2}}{1-\beta_2} \cdot \frac{1-\beta_1}{\beta_2-\beta_1}$$

$$= \frac{1-\beta_2}{(1-\beta_1)(\beta_1-\beta_2)} \left(\beta_1^{2+x_2} - \beta_2^{2+x_2} \frac{(1-\beta_1)^2}{(1-\beta_2)^2} \right) > 0.$$

An assumption that Inequality (17) is false leads to a nonsense conclusion that all stationary probabilities in (21) are null or negative. Hence, Inequality (17) is necessary for the existence of a stationary probability distribution of the Markov chain (2). □

Theorems 1–3 characterize the stability regime of the queueing system with loop algorithm in a form suitable for numerical verification. We have to remark that the stationary probability $Q_1(1,0)$ occurring in the necessary stationarity condition for the Markov chain (2) is also a stationary probability of the state $(\Gamma^{(1)}, 0)$ of the Markov chain (9). Its presence in (17) comes from the fact that the first queue, O_1, can be stable on its own, independently of the second queue, and then it 'modulates' the second input flow, Π_2, by controlling the service tact durations (i.e., whether it loops or not). Figure 1 depicts the set of pairs (λ_1, λ_2) for which the condition in Theorem 2 is fulfilled.

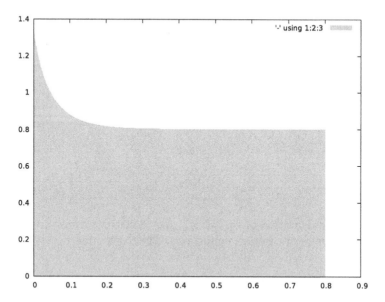

Fig. 1. A stationary probability distribution existence domain for $T_1 = 10$, $T_2 = 15$, $\ell_1 = \ell_2 = 20$ (shaded)

Obviously, the assumption that $\ell_2 = 1$ or 2 plays role only in the last step of the proof. Actually, we computationally verified positivity of the quantities (22) even for larger integer values of the parameter ℓ_2. Also, in our Monte-Carlo simulations we saw that the inequalities from Theorems 1 and 3 provided stability to the both queues.

4 Conclusion

The queueing system under study gives an example where the stationarity conditions have an essentially non-linear form in terms of input parameters such as input intensities. Still, the method for analysis of two conflicting Poisson queues governed by a cyclic algorithm with prolongations which is demonstrated in this paper allows to obtain such conditions. A possible issue when generalizing the method is the necessity to obtain and analyze explicit formulas like (4) and (22).

References

1. Litvak, N.V., Fedotkin, M.A.: A probabilistic model for the adaptive control of conflict flows. Autom. Remote Control 5(61), 777–784 (2000)
2. Proydakova, E.V.: Necessary conditions for the existence of a stationary distribution of output flows in a system with priority direction. Vestnik NNSU 1, 167–172 (2007). (in Russian)
3. Zorine, A.V.: On ergodicity conditions in a polling model with Markov modulated input and state-dependent routing. Queueing Syst. 2(76), 223–241 (2014)
4. Rachinskaya, M., Fedotkin, M.: Stationarity conditions for the control systems that provide service to the conflicting batch poisson flows. In: Rykov, V.V., Singpurwalla, N.D., Zubkov, A.M. (eds.) ACMPT 2017. LNCS, vol. 10684, pp. 43–53. Springer, Cham (2017). https://doi.org/10.1007/978-3-319-71504-9_5
5. Tagaki, H.: Analysis of Polling Systems. MIT Press, Cambridge (1986)
6. Borst, S.C.: Polling Systems, Volume 115 of CWI tract. Centrum voor Wiskunde en Informatica, Amsterdam (1996)
7. Vishnevskii, V.M., Semenova, O.V.: Mathematical methods to study the polling systems. Autom. Remote Control 2(67), 173–220 (2006)
8. Fayolle, G., Lasgouttes, J.: A state-dependent polling model with Markovian routing. INRIA-Report, 2279 (1994)
9. Schassberger, R.: Stability of polling networks with state-dependent server routing. Probab. Eng. Inf. Sci. 9, 539–550 (1995)
10. Foss, S., Last, G.: Stability of polling systems with state-dependent routing and with exhaustive service policies. Ann. Appl. Probab. 1(6), 116–137 (1996)
11. Foss, S., Last, G.: On the stability of greedy polling systems with general service policies. Probab. Eng. Inf. Sci. 1(12), 49–68 (1998)
12. Fedotkin, M.A.: Optimal control for conflict flows and marked point processes with selected discrete component. I. Liet. mat. rinkinys. 4(28), 783–794 (1988). (in Russian)

Resource Queueing System $MMPP^{(2,\nu)}|GI_2|\infty$ with Parallel Service of Multiple Paired Customers

Tatiana Bushkova[1], Ekaterina Pavlova[1], Svetlana Rozhkova[2], Svetlana Moiseeva[1($\boxtimes$)], and Michele Pagano[3]

[1] Tomsk State University, Lenina Avenue, 36, 634050 Tomsk, Russia
{bushkova70,pavlovakatya_2010,smoiseeva}@mail.ru
[2] Tomsk Politechnic University, Lenina Avenue, 30, 634050 Tomsk, Russia
rozhkova@tpu.ru
[3] University of Pisa, Pisa, Italy
m.pagano@iet.unipi.it

Abstract. The article proposes the method for investigating the heterogeneous queuing system of $MMPP^{(2,\nu)}|GI_2|\infty$ type with resource splitting and parallel service. Each customer is characterized by a random total capacity which is independent of the service time. Based on the asymptotic analysis, it is possible to deduce the expressions for characteristic function of the process of the total amount of resource in two-service unit system. The mathematical models of this type could be of great interest in terms of application in telecommunication, for example, for modeling wireless network, enhancing the existing and designing new ones.

Keywords: Markov modulated Poisson process · Infinite server · Paired requests

1 Introduction

The development of 4G mobile broadband technology has contributed to a rapid growth in wireless network traffic. It has affected the Quality of Service (QoS) and resulted in the network capacity crunch.

Today, multiplexing is regarded as a major solution to increase spatial reuse and, consequently, to meet ever-growing traffic demand in 5th Generation (5G) wireless systems [1].

At the same time, resource sharing in wireless network depends on the adequate management of information flows. Multiplexing of two or more heterogeneous networks, for example, combination of traditional LTE operating in licensed spectrum and machine-to-machine traffic operating in new unlicensed spectrum of narrow band internet of things (NB-IoT), could increase reliability and intelligence of decision-taking systems. The analysis of sources, recommendations, and standards of such international organizations as 3GPP, IEEE, ETSI

© Springer Nature Switzerland AG 2019
A. Dudin et al. (Eds.): ITMM 2019, CCIS 1109, pp. 136–149, 2019.
https://doi.org/10.1007/978-3-030-33388-1_12

has revealed that there is a need for end-to-end models which would adequately describe the peculiarities of access control methods and resource redundancy in 5G networks. To enhance the efficiency of redundancy mechanism, it is required to develop new models and methods to analyze the efficiency of heterogeneous broadband wireless networks.

The issues of sharing telecommunication system resources were for the first time addressed by Kleinrock in the work on shared-memory packet switches [2]. The methods of research were further developed by scientific schools of Basharin and Zhozhikashvili. The peculiarity of the models was a multiplicativity of the stationary distribution. Then, such models were investigated with regard to the satellite communication systems. At the same time, some scholars achieved the results in the study of exponential homogenous and heterogeneous queuing networks, with their models sharing the same peculiarity. The key findings on mathematical models for sharing channel resources were obtained by Kelly, and later Ross [3–6]. It is possible to assume that their works triggered the research in resource sharing for multiservice networks of the next 3G and 4G generations. Among the studies on network resource sharing in case of two types of traffic, the works by Bousset and Beylot are worth noting [7].

In the second half of the 20th century, Romm and Skitovich for the first time proposed the generalization of the Erlang problem in terms of the resource system models. According to this generalization, each customer entering the queue possesses a certain information property termed as capacity. Denial of service occurs when capacity of the arriving customer exceeds the difference between the queuing network capacity and the number of customers being served at the moment of a new customer arrival.

Tikhonenko [8,9] together with the colleagues significantly contributed to the development of the research methods for resource queuing network. He considered the queuing systems of random capacity as a class of systems characterized by a certain capacity and service time which is dependent or independent of the number of customers. Samuilov uses the queuing networks with limited resources as models of wireless communication networks of the next generation [10,11]. Almost all research works mentioned above investigate the models which are referred to so-called Markovian models (input flow is the simplest and service time follows the exponential distribution). At the same time, as described in [12], in order to model modern flows of information transmission, it is required to apply non-Poisson models: MAP-flow (Markovian Arrival Process) or MMPP-flow (Markov Modulated Poisson Process). To investigate the models in terms of the mentioned flow types, simulation modeling, matrix and numerical methods are used. The analytical results have been obtained only for peculiar cases. The present article investigates the characteristics of queuing systems with random customer capacities in the wireless network with splitting requests [13–15]. Unlike the well-known models, the considered ones would allow engineers to estimate the required volume of redundant resources for traffic operating on the internet of things and develop the strategy to allocate resources in case of competing traffic.

2   Statement of the Problem

2.1   Mathematical Model

Consider the queueing system with two service units of two different types contains of unlimited number of servers (Fig. 1). Each customer carries a random capacity of some resource. Customers arrive in the system according to Markov Modulated Poisson Process (MMPP), which is governer by Markov chain $k(t) = 1, ..., K$, which determined by the matrix of infinitesimal characteristics $\mathbf{Q}$ and matrix of conditional intensities $\mathbf{\Lambda} = diag[\lambda_1, ..., \lambda_K]$. At the time of occurrence of the event in the arriving flow each customer splits, the customer entering the queue splits, so that there are two customers in the system. Each customer goes to a free server in the first and second service units, where the service is performed during a random time with distribution function $B_i(x), i = 1, 2$ corresponding to the type of unit.

Let each customer requires some random capacity $\nu_i > 0, i = 1, 2$ with distribution function $G_i(y), i = 1, 2$.

The total capacity of each customer on each service unit in the system at time t is described as $\{\nu_1(t), \nu_2(t)\}$. The problem of exploring of two-dimensional stochastic process $\{V_1(t), V_2(t)\}$ is set, where

$$V_j = \sum_{i=1}^{\infty} \nu_i^{(j)}, j = 1, 2.$$

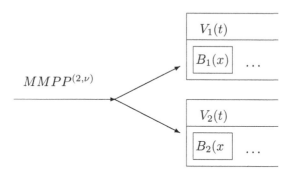

Fig. 1. Heterogeneous queue $MMPP^{(2,v)}|G_{(2,v)}|\infty$ with random customers capacities

This process is not Markovian, therefore, we use the dynamic screening method to investigate it (Fig. 2).

We assume the system is empty at the moment t_0, and let us fix some arbitrary moment T in the future. $S_i(t)$ denotes the probability that a customer arriving at time t will be served in the system within the moment T on i-type of service unit, $S_i(t) = 1 - B_i(T - t), i = 1, 2$, for $t_0 \leq t \leq T$. The total capacity of arrivals screened before the moment t is described by $w_1(t), w_2(t)$. As shown in

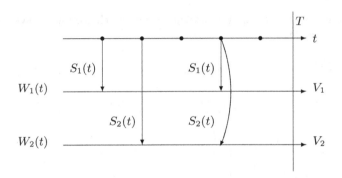

Fig. 2. Screening of customers arrivals

[7], the probability distribution of the capacity of the customer in the system at the moment T equals to the probability distribution of the capacity of screened arrivals on the axis:

$$P\{V_1(T) < w_1, V_2(T) < w_2\} = P\{w_1(T) < w_1, w_2(T) < w_2\}.$$

2.2 Kolmogorov Differential Equations

Let us consider the process $\{k(t), w_1(t), w_2(t)\}$, where $k(t)$ is state of Markov chain and describe it by

$$P(k, w_1, w_2, t) = P\{k(t) = k, w_1(t) < w_1, w_2(t) < w_2\}.$$

According to the formula of the total probability we deduce the equation

$$
\begin{aligned}
P(k, w_1, w_2, t + \Delta t) &= (1 - \lambda_k \Delta t)(1 - q_{\nu k} \Delta t) P(k, w_1, w_2, t) \\
&+ \lambda_k \Delta t \Big[S_1(t)(1 - S_2(t)) \int_0^{w_1} P(k, w_1 - y_1, w_2, t) dG_1(y_1) \\
&+ (1 - S_1(t)) S_2(t) \int_0^{w_2} P(k, w_1, w_2 - y_2, t) dG_2(y_2) \\
&+ S_1(t) S_2(t) \int_0^{w_1} \int_0^{w_2} P(k, w_1 - y_1, w_2 - y_2, t) dG_1(y_1) dG_2(y_2) \\
&+ (1 - S_1(t))(1 - S_2(t)) P(k, w_1, w_2, t) \Big] + \sum_{\nu \neq k} q_{\nu k} P(\nu, w_1, w_2, t) + o(\Delta t)
\end{aligned}
$$

Regarding two-dimensional Markovian process, the following system of Kolmogorov differential equations is written:

$$\frac{\partial P(k, w_1, w_2, t)}{\partial t} = -\lambda_k(S_1(t) + S_2(t) - S_1(t)S_2(t))P(k, w_1, w_2, t))$$

$$+\lambda_k\left[S_1(t)(1 - S_2(t))\int_0^{w_1} P(k, w_1 - y_1, w_2, t)dG_1(y_1)\right.$$

$$+S_2(t)(1 - S_1(t))\int_0^{w_2} P(k, w_1, w_2 - y_2, t)dG_2(y_2) \qquad (1)$$

$$+S_1(t)S_2(t)\int_0^{w_1}\int_0^{w_2} P(k, w_1 - y_1, w_2 - y_2, t)dG_1(y_1)dG_2(y_2)\right]$$

$$+\sum_v q_{vk}P(k, v_1, v_2, t)$$

with initial condition

$$P(k, w_1, w_2, t_0) = \begin{cases} 1, w_1 = w_2 = 0, \\ 0, \text{ otherwise.} \end{cases}$$

We introduce the characteristic function

$$h(k, u_1, u_2, t) = \int_0^\infty e^{ju_1w_1} \int_0^\infty e^{ju_2w_2} P(k, dw_1, dw_2, t),$$

where $j = \sqrt{-1}$ is the imaginary unit. Then we can rewrite system (1)

$$\frac{\partial h(k, u_1, u_2, t)}{\partial t} = \lambda_k h(k, u_1, u_2, t)[S_1(t)(G_1^*(u_1) - 1) + S_2(t)(G_2^*(u_2) - 1)$$

$$+S_1(t)S_2(t)(G_1^*(u_1) - 1)(G_2^*(u_2) - 1)] + \sum_v q_{vk}h(\nu, u_1, u_2, t). \qquad (2)$$

Let us write vector-matrix equation

$$\frac{\partial \mathbf{h}(u_1, u_2, t)}{\partial t} = \mathbf{h}(u_1, u_2, t)[\mathbf{\Lambda}(S_1(t)(G_1^*(u_1) - 1) + S_2(t)(G_2^*(u_2) - 1)$$

$$+ S_1(t)S_2(t)(G_1^*(u_1) - 1)(G_2^*(u_2) - 1)) + \mathbf{Q}]. \qquad (3)$$

with initial condition

$$\mathbf{h}(u_1, u_2, t_0) = \mathbf{r}, \qquad (4)$$

where

$$\mathbf{h}(u_1, u_2, t) = \{h(1, u_1, u_2, t), h(2, u_1, u_2, t), \ldots, h(K, u_1, u_2, t)\}$$

and $\mathbf{r} = [r(1), r(2), ..., r(K)]$ stands for the stationary distribution of the underlying Markov chain. Vector $\mathbf{r}$ satisfies the following linear system

$$\begin{cases} \mathbf{r}\mathbf{Q} = 0, \\ \mathbf{r}\mathbf{e} = 1. \end{cases} \qquad (5)$$

3   Asymptotic Analysis Method

In general the exact solution of Eqs. (3), (4) is not available, but it may be found under asymptotic conditions. In this work we consider the system under the condition of parallel infinitely growing service time.

3.1   First-Order Asymptotic Analysis

Let us denote the mean service time in each unit as

$$b_i = \int_0^\infty (1 - B_i(x)dx, i = 1, 2,$$

then the asymptotic condition is $b_i \to \infty$.

We formulate and prove the following statement.

Theorem 1. *The first-order asymptotic characteristic function of the probability distribution of the process $w_1(t), w_2(t)$ has the form*

$$\mathbf{h}^{(1)}(u_1, u_2, t) = \mathbf{r} \exp\left\{ j\kappa_1 \sum_{i=1}^{2} u_i a_i \int_{t_0}^{t} S(\tau)d\tau \right\},$$

where $\kappa_1 = \mathbf{r}\Lambda\mathbf{e}$ and $a_i = \int_0^\infty y_i dG_i(y_i)$, $i = 1, 2$ is the mean customer capacity.

Proof. By performing the substitutions

$$b_1 = \frac{1}{\varepsilon}, b_2 = \frac{1}{q\varepsilon}, t\varepsilon = \tau, t_0\varepsilon = \tau_0,$$
$$S_1(t) = \bar{S}_1(\tau), S_2(t) = \bar{S}_2(\tau), u_1 = \varepsilon x_1, u_2 = \varepsilon x_2, \qquad (6)$$
$$\mathbf{h}(u_1, u_2, t) = F_1(x_1, x_2, \tau, \varepsilon)$$

in Eqs. (3), (4), we obtain

$$\varepsilon \frac{\partial F_1(x_1, x_2, \tau, \varepsilon)}{\partial \tau} = F_1(x_1, x_2, \tau, \varepsilon)\Big[\Lambda \bar{S}_1(\tau)\Big(\int_0^\infty e^{j\varepsilon x_1 y_1} dG_1 - 1 \Big)$$
$$+ \bar{S}_2(\tau)\Big(\int_0^\infty e^{j\varepsilon x_2 y_2} dG_2 - 1 \Big)$$
$$+ \bar{S}_1(\tau)\bar{S}_2(\tau)\Big(1 - \int_0^\infty e^{j\varepsilon x_1 y_1} dG_1 - \int_0^\infty e^{j\varepsilon x_2 y_2} dG_1 \qquad (7)$$
$$+ \int_0^\infty e^{j\varepsilon x_1 y_1} dG_1 \int_0^\infty e^{j\varepsilon x_2 y_2} dG_2 \Big) + \mathbf{Q} \Big]$$

with initial condition

$$F_1(x_1, x_2, \tau_0, \varepsilon) = \mathbf{r} \qquad (8)$$

We find solution of problems (7), (8) $F_1(x_1, x_2, \tau) = \lim_{\varepsilon \to 0} F_1(x_1, x_2, \tau, \varepsilon)$ in two steps.

Step 1. We consider Eqs. (7), (8) in limit $\varepsilon \to 0$, and taking into account decomposition $e^{j\varepsilon x_i y_i} = 1 + j\varepsilon x_i y_i + o(\varepsilon^2)$, we obtain

$$F_1(x_1, x_2, \tau)\mathbf{Q} = 0 \tag{9}$$

Taking into account (5), we can conclude that $F_1(x_1, x_2, \tau)$ can be expressed as

$$F_1(x_1, x_2, \tau) = \mathbf{r}\Phi(x_1, x_2, \tau), \tag{10}$$

where $\Phi(x_1, x_2, \tau)$ is some scalar function which satisfies the condition

$$\Phi(x_1, x_2, \tau_0) = 1. \tag{11}$$

Step 2. Let us substitute (10) into (7) and multiply by vector $\mathbf{e}$, taking into account system (5) and $\kappa_1 = \mathbf{r}\Lambda\mathbf{e}$

$$\varepsilon \frac{\partial \Phi(x_1, x_2, \tau)}{\partial \tau} = \kappa_1 \Phi(x_1, x_2, \tau)[\bar{S}_1(\tau)j\varepsilon x_1 a_1 \\ + \bar{S}_2(\tau)j\varepsilon x_2 a_2 + \bar{S}_1(\tau)\bar{S}_2(\tau)j^2\varepsilon^2 x_1 x_2 a_1 a_2] + o(\varepsilon^2). \tag{12}$$

Then let us divide the result by ε and perform the asymptotic transition $\varepsilon \to 0$. We obtain the following differential equation for $\Phi(x_1, x_2, \tau)$

$$\frac{\partial \Phi(x_1, x_2, \tau)}{\partial \tau} = \kappa_1 j \Phi(x_1, x_2, \tau)[\bar{S}_1(\tau)x_1 a_1 + \bar{S}_2(\tau)x_2 a_2].$$

The solution of problems (11), (12)

$$\Phi(x_1, x_2, \tau) = \exp\left\{\kappa_1 j \sum_{i=1}^{2} x_i a_i \int_{-\infty}^{\tau} \bar{S}_i(\xi)d\xi\right\}.$$

Substituting this expression into (10), we obtain

$$F_1(x_1, x_2, \tau) = \mathbf{r}\exp\left\{\kappa_1 j \sum_{i=1}^{2} x_i a_i \int_{-\infty}^{\tau} \bar{S}_i(\xi)d\xi\right\}.$$

Therefore, we can write

$$\mathbf{h}(u_1, u_2, \tau) = F_1(x_1, x_2, \tau, \varepsilon) \approx F_1(x_1, x_2, \tau) = \mathbf{r}\exp\left\{\kappa_1 j \sum_{i=1}^{2} x_i a_i \int_{-\infty}^{\tau} \bar{S}_i(\xi)d\xi\right\}$$

$$= \mathbf{r}\exp\left\{\kappa_1 j \sum_{i=1}^{2} u_i a_i \int_{t_0}^{t} \bar{S}_i(\xi)d\xi\right\}$$

Thus, the proof is complete.

We set $t = T$ $t_0 = -\infty$, then we can write for two-dimensional process $\{V_1, V_2\}$

$$\mathbf{h}(u_1, u_2) = M\left\{e^{j(u_1 w_1 + u_2 w_2)}\right\} = \mathbf{h}(u_1, u_2, T)\mathbf{e} = \exp\left\{\kappa_1 j \sum_{i=1}^{2} u_i a_i b_i\right\},$$

where

$$b_i = \int_{0}^{\infty} (1 - B_i(x))dx, i = 1, 2.$$

3.2   Second-Order Asymptotic Analysis

The main result is the following theorem.

Theorem 2. *The second-order asymptotic characteristic function of the probability distribution of the process $w_1(t), w_2(t)$ has the form*

$$
\begin{aligned}
\mathbf{h}^{(2)}(u_1, u_2, t) = \mathbf{r} \exp \Big\{ & j\kappa_1 \sum_{i=1}^{2} u_i a_i \int_{t_0}^{t} S_i(\tau) d\tau \\
& + \frac{(ju_1)^2}{2} \Big[\kappa_1 a_1^{(2)} \int_{t_0}^{t} S_1(\tau) d\tau + 2\kappa_2 a_1^{(2)} \int_{t_0}^{t} S_2^2(\tau) d\tau \Big] \\
& + \frac{(ju_1)^2}{2} \Big[\kappa_1 a_2 \int_{t_0}^{t} S_1(\tau) d\tau + 2\kappa_2 a_2^{(2)} \int_{t_0}^{t} S_2^2(\tau) d\tau \Big] \\
& + ju_1 ju_2 \Big[\kappa_1 a_1 a_2 \int_{t_0}^{t} S_1(\tau) S_2(\tau) d\tau + 2\kappa_2 \int_{t_0}^{t} S_1(\tau) S_2(\tau) d\tau \Big] \Big\},
\end{aligned}
$$

where $\kappa_2 = \mathbf{f}_2 \Lambda \mathbf{e}$, $a_i^{(2)} = \int_0^\infty y_i{}^2 dG_i(y_i), i = 1, 2$ and the row vector $\mathbf{f}_2$ satisfies the linear matrix system

$$
\begin{cases} \mathbf{f}_2 \mathbf{Q} = \mathbf{r}(\kappa_1 \mathbf{I} - \Lambda) \\ \mathbf{f}_2 \mathbf{e} = 0 \end{cases} \tag{13}
$$

Proof. Let $\mathbf{h}_2(x_1, x_2, t)$ be a vector function which satisfies the equation

$$
\frac{\partial \mathbf{h}_2(x_1, x_2, t)}{\partial t} = \mathbf{h}_2(x_1, x_2, t) \exp \Big\{ j\kappa_1 \sum_{i=1}^{2} u_i a_i \int_{t_0}^{t} S(\tau) d\tau \Big\}. \tag{14}
$$

Substituting (14) into (3) and (4), we obtain

$$
\begin{aligned}
\frac{\partial \mathbf{h}_2(x_1, x_2, t)}{\partial t} = \mathbf{h}_2(x_1, x_2, t) \Big[& \Lambda(S_1(t)(G_1^*(u_1) - 1) + S_2(t)(G_2^*(u_2) - 1) \\
& + S_1(t)S_2(t)(G_1^*(u_1) - 1)(G_2^*(u_2) - 1)) \\
& + \mathbf{Q} - j\kappa_1(u_1 a_1 S_1(t) + u_2 a_2 S_2(t)) \Big]
\end{aligned} \tag{15}
$$

with initial condition

$$
\mathbf{h}_2(u_1, u_2, t_0) = \mathbf{r}, \tag{16}
$$

where $\mathbf{I}$ is identity matrix.

Let us make the following substitutions

$$
\begin{aligned}
b_1 &= \frac{1}{\varepsilon^2}, b_2 = \frac{1}{q\varepsilon^2}, t\varepsilon^2 = \tau, t_0 \varepsilon^2 = \tau_0, S_1(t) = \bar{S}_1(\tau), \\
& S_2(t) = \bar{S}_2(\tau), u_1 = \varepsilon x_1, u_2 = \varepsilon x_2, \\
& \mathbf{h}_2(u_1, u_2, t) = F_2(x_1, x_2, \tau, \varepsilon).
\end{aligned} \tag{17}
$$

We rewrite problems $(15), (16)$ considering (17)

$$\varepsilon^2 \frac{\partial F_2(x_1, x_2, \tau, \varepsilon)}{\partial \tau} = F_2(x_1, x_2, \tau, \varepsilon) \Big[\mathbf{\Lambda} \bar{S}_1(\tau) \Big(\int_0^\infty e^{j \varepsilon x_1 y_1} dG_1 - 1 \Big)$$

$$+ \bar{S}_2(\tau) \Big(\int_0^\infty e^{j \varepsilon x_2 y_2} dG_2 - 1 \Big)$$

$$+ \bar{S}_1(\tau) \bar{S}_2(\tau) \Big(1 - \int_0^\infty e^{j \varepsilon x_1 y_1} dG_1 - \int_0^\infty e^{j \varepsilon x_2 y_2} dG_1 \qquad (18)$$

$$+ \int_0^\infty e^{j \varepsilon x_1 y_1} dG_1 \int_0^\infty e^{j \varepsilon x_2 y_2} dG_2 \Big) + \mathbf{Q}$$

$$- j \varepsilon \kappa_1 (x_1 a_1 \bar{S}_1(\tau) + x_2 a_2 \bar{S}_2(\tau)) \mathbf{I} \Big]$$

with initial condition

$$F_2(x_1, x_2, \tau_0, \varepsilon) = \mathbf{r}. \qquad (19)$$

We find the asymptotic solution of problems $(18), (19)$ in three steps.

Step 1. Under limit condition $\varepsilon \to 0$, we obtain system

$$\begin{cases} F_2(x_1, x_2, \tau) \mathbf{Q} = 0 \\ F_2(x_1, x_2, \tau_0) = \mathbf{r}, \end{cases}$$

and taking into account (5), we can write

$$F_2(x_1, x_2, \tau) = \mathbf{r} \Phi_2(x_1, x_2, \tau), \qquad (20)$$

where $\Phi_2(x_1, x_2, \tau)$ is some scalar function which satisfies condition

$$\Phi_2(x_1, x_2, \tau_0) = 1. \qquad (21)$$

Step 2. Using (20), the function $F_2(x_1, x_2, \tau)$ can be represented in the expansion form

$$F_2(x_1, x_2, \tau, \varepsilon) = \Phi_2(x_1, x_2, \tau, \varepsilon)[\mathbf{r} + j\varepsilon(x_1 a_1 \bar{S}_1(\tau) + x_2 a_2 \bar{S}_2(\tau)) \mathbf{f}_2] + O(\varepsilon^2), \quad (22)$$

where $\mathbf{f}_2$ is a row vector that satisfies the condition $\mathbf{f}_2 \mathbf{e} = const$. Let us use (22) and taking into account decomposition $e^{j \varepsilon x_i y_i} = 1 + j \varepsilon x_i y_i + O(\varepsilon^2)$ in (18), we obtain the following equation for the vector function $\mathbf{f}_2$

$$\mathbf{f}_2 \mathbf{Q} + \mathbf{r}(\mathbf{\Lambda} - \kappa_1 \mathbf{I}) = 0.$$

Step 3. We multiply (18) by vector $\mathbf{e}$ and use (22) taking into account

$$e^{j \varepsilon x_i y_i} = 1 + j \varepsilon x_i y_i + \frac{(j \varepsilon x_i y_i)^2}{2} + O(\varepsilon^3)$$

and denoting by $\kappa_2 = \mathbf{f}_2 \mathbf{\Lambda} \mathbf{e}$ we obtain the following equation for $\Phi_2(x_1, x_2, \tau)$

$$\frac{\partial \Phi_2(x_1, x_2, \tau)}{\partial \tau} = \Phi_2(x_1, x_2, \tau) \Big[\frac{(j x_1)^2}{2} (\kappa_1 a_1 \bar{S}_1(\tau) + 2 \kappa_2 a_1^{(2)} \bar{S}_1^2(\tau))$$

$$+ \frac{(j x_2)^2}{2} (\kappa_1 a_2 \bar{S}_2(\tau) + 2 \kappa_2 a_2^{(2)} \bar{S}_2^2(\tau)) \qquad (23)$$

$$+ j x_1 j x_2 (\kappa_1 a_1 a_2 \bar{S}_1(\tau) \bar{S}_2(\tau) + 2 \kappa_2 \bar{S}_1(\tau) \bar{S}_2(\tau)) \Big].$$

The solution of this equation with initial condition (22) is as follows

$$\Phi_2(x_1, x_2, \tau) = \exp\left\{\frac{(jx_1)^2}{2}\left(\kappa_1 a_1 \int_0^\tau \bar{S}_1(\xi)d\xi + 2\kappa_2 {a_1}^{(2)}\int_0^\tau \bar{S}_1^2(\xi)d\xi\right)\right.$$

$$+\frac{(jx_2)^2}{2}\left(\kappa_1 a_2 \int_0^\tau \bar{S}_2(\xi)d\xi + 2\kappa_2 {a_2}^{(2)}\int_0^\tau \bar{S}_2^2(\xi)d\xi\right) \qquad (24)$$

$$\left.+jx_1 jx_2\left(\kappa_1 a_1 a_2 \int_0^\tau \bar{S}_1(\xi)\bar{S}_2(\xi)d\xi + 2\kappa_2 \int_0^\tau \bar{S}_1(\xi)\bar{S}_2(\xi)d\xi\right)\right\}.$$

Substituting this expression in Formula (20) and performing the substitutions which are inverse to (14) and (17), we obtain

$$\mathbf{h}^{(2)}(u_1, u_2, t) = \mathbf{r}\exp\left\{j\kappa_1 \sum_{i=1}^2 u_i a_i \int_{t_0}^t S_i(\tau)d\tau\right.$$

$$+\frac{(ju_1)^2}{2}\left[\kappa_1 {a_1}^{(2)}\int_{t_0}^t S_1(\tau)d\tau + 2\kappa_2 {a_1}^{(2)}\int_{t_0}^t S_2^2(\tau)d\tau\right]$$

$$+\frac{(ju_2)^2}{2}\left[\kappa_1 a_2 \int_{t_0}^t S_1(\tau)d\tau + 2\kappa_2 {a_2}^{(2)}\int_{t_0}^t S_2^2(\tau)d\tau\right]$$

$$\left.+ju_1 ju_2\left[\kappa_1 a_1 a_2 \int_{t_0}^t S_1(\tau)S_2(\tau)d\tau + 2\kappa_2 \int_{t_0}^t S_1(\tau)S_2(\tau)d\tau\right]\right\},$$

The proof is complete.

Corollary. We assume $t = T$ and $t_0 \to -\infty$ and consider that

$$P\{V_1(T) < w_1, V_2(T) < w_2\} = P\{w_1(T) < w_1, w_2(T) < w_2\}.$$

We obtain the steady-state characteristic function of the process under study $\{V_1(t), V_2(t)\}$

$$h^{(2)}(u_1, u_2) = \exp\left\{j\kappa_1(u_1 a_1 b_1 + u_2 a_2 b_2)\right.$$

$$+\frac{(ju_1)^2}{2}(\kappa_1 a_1 b_1 + 2\kappa_2 {a_1}^{(2)}\beta_2)$$

$$\qquad\qquad (25)$$

$$+\frac{(ju_2)^2}{2}(\kappa_1 a_2 b_2 + 2\kappa_2 {a_2}^{(2)}\beta_2)$$

$$\left.+ju_1 ju_2(\kappa_1 a_1 a_2 \beta_{12} + 2\kappa_2 \beta_{12})\right\},$$

which we call the second-order asymptotic characteristic function, where

$$\beta_i = \int_{-\infty}^0 S_i^2(x)dx = \int_{-\infty}^0 (1 - B_i(-x))^2 dx = \int_0^\infty (1 - B_i(x))^2 dx, i = 1, 2,$$

$$\beta_{12} = \int_{-\infty}^0 S_1(x)S_2(x)dx = \int_{-\infty}^0 (1 - B_1(-x))(1 - B_2(-x))dx$$

$$= \int_0^\infty (1 - B_1(x))(1 - B_2(x))dx.$$

From the form of the characteristic function (25), it is clear that the probability distribution of the two-dimensional process $\{V_1(t), V_2(t)\}$ is asymptotically Gaussian with vector of means

$$\mathbf{a} = \begin{bmatrix} \kappa_1 a_1 b_1 & \kappa_1 a_2 b_2 \end{bmatrix}$$

and covariance matrix

$$\mathbf{K} = \begin{bmatrix} \kappa_1 a_1 b_1 + 2\kappa_2 a_1{}^{(2)}\beta_2 & \kappa_1 a_1 a_2 \beta_{12} + 2\kappa_2 \beta_{12} \\ \kappa_1 a_1 a_2 \beta_{12} + 2\kappa_2 \beta_{12} & \kappa_1 a_2 b_2 + 2\kappa_2 a_2{}^{(2)}\beta_2 \end{bmatrix}$$

that is variance of total customer capacity on each service unit has form

$$\sigma_i = \kappa_1 a_i b_i + 2\kappa_2 a_i{}^{(2)}\beta_2, i = 1, 2.$$

Then using (25) and set $u_1 = u_2 = u$ we can obtain the characteristic function for total customer capacity in the system, on both service units

$$h(u) = M\{e^{V_1 + V_2}\} = exp\Big\{ j\kappa_1 u(a_1 b_1 + a_2 b_2) + \frac{(ju)^2}{2}(\kappa_1 u(a_1 b_1 + a_2 b_2) \\ + 2\kappa_2 \beta_2 (a_1{}^{(2)} + a_2{}^{(2)}) + 2\beta_{12}(\kappa_1 a_1 a_2 + 2\kappa_2)) \Big\}.$$

4   Numerical Results

Result (25) is obtained under the asymptotic condition $b_i \to \infty$. Therefore, the result may be used just as an approximation. We make series of simulation experiments and compare asymptotic distributions with empiric ones.

Let us set the matrix of infinitesimal characteristics and matrix of conditional intensities

$$\mathbf{Q} = \begin{bmatrix} -0.8 & 0.4 & 0.4 \\ 0.3 & -0.6 & 0.3 \\ 0.4 & 0.4 & -0.8 \end{bmatrix}, \quad \mathbf{\Lambda} = \begin{bmatrix} 0.5 & 0 & 0 \\ 0 & 1 & 0 \\ 0 & 0 & 1.5 \end{bmatrix}.$$

Then intensity of incoming flow $\kappa_1 = \mathbf{r\Lambda e} = 1$. Let us also assume that customer capacities have uniform distribution in $[0, 1]$ and service time has gamma distribution with shape and inverse scale parameters $\alpha_1 = 0.5$ and $\beta_1 = \alpha_1/N$ on the first service unit, $\alpha_2 = 1.5$ and $\beta_2 = \alpha_2/N$ on the second service unit. There are results of a comparison of the probability distribution of the total customer capacities in different service units obtained using simulation and asymptotic analysis, and the approximation accuracy is clearly demonstrated on Figs. 3, 4.

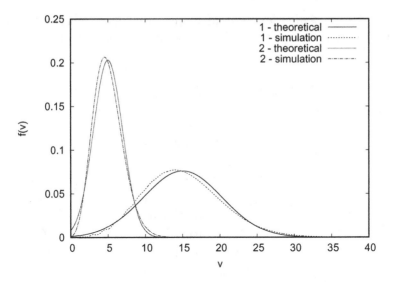

Fig. 3. Distributions of the total resource ($N = 10$) for the first (1) and the second (2) service units

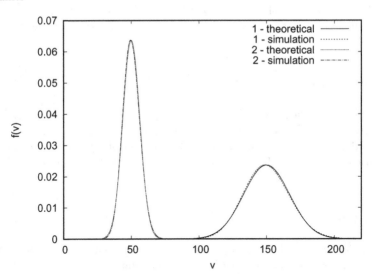

Fig. 4. Distributions of the total resource ($N = 100$) for the first (1) and the second (2) service units

Let us denote

$$\Delta_i = \sup_x |Fas_i(x) - Fim_i(x)|, i = 1, 2,$$

$$\Delta_{12} = \sup_{x,y} |Fas(x, y) - Fim(x, y)|,$$

Where $Fim(x), Fim(x, y)$ is the cumulative distribution function built on the basis of simulation results and $Fas(x), Fas(x, y)$ is the Gaussian approximation based on (25).

Table 1. Distance Kolmogorov

N	Δ_{12}	Δ_1	Δ_2
1	0.373	0.373	0.369
3	0.118	0.111	0.117
5	0.066	0.061	0.063
7	0.043	0.040	0.042
10	0.029	0.027	0.029
20	0.018	0.017	0.018
50	0.011	0.011	0.011
100	0.008	0.007	0.008

As we can see from the (Table 1), the use of a Gaussian approximation is justified when the average service time is 10 times greater than the intensity of the incoming flow.

5   Conclusion

In this paper we considered a $MMPP^{(2,\nu)}/GI_2/\infty$ queueing system with Markov Modulated Poisson Process input, each customer occupying a random resource amount independent of its service time. We have obtained expressions for the asymptotic characteristic function of the process of the total volume of the occupied resource in two-service unit system.

References

1. Moltchanov, D., et al.: Improving session continuity with bandwidth reservation in mmwave communications. IEEE Wirel. Commun. **8**(1), 105–108 (2019)
2. Kleinrock, L.: Resource allocation in computer systems and computer communication networks. In: IFIP Cong. Proc., pp. 11–18, North-Holland (1974)
3. Kelly, F.P.: Reversibility and Stochastic Networks, p. 630. Wiley, New York (1979)
4. Ross, K.W.: Multiservice Loss Models for Broadband Telecommunication Networks, p. 343. Springer, Berlin (1995). https://doi.org/10.1007/978-1-4471-2126-8
5. Ross, K.W., Liu, Y., Wu, D.: Modeling and analysis of multi-channel P2P live video systems. Proc. IEEE/ACM Trans. Netw. **18**(4), 1063–6692 (2010)
6. Boussetta, K., Belyot, A.-L.: Multirate resource sharing for unicast and multicast connections. In: Tsang, D.H.K., Kühn, P.J. (eds.) Broadband Communications. ITIFIP, vol. 30, pp. 561–570. Springer, Boston, MA (2000). https://doi.org/10.1007/978-0-387-35579-5_47

7. Begishev, V., Samuylov, A., Moltchanov, D., Samouylov, K.: Modeling the process of dynamic resource sharing between LTE and NB-IoT services. In: Vishnevskiy, V.M., Samouylov, K.E., Kozyrev, D.V. (eds.) DCCN 2017. CCIS, vol. 700, pp. 1–12. Springer, Cham (2017). https://doi.org/10.1007/978-3-319-66836-9_1

8. Tikhonenko, O., Kempa, W.M.: The generalization of AQM algorithms for queueing systems with bounded capacity. In: Wyrzykowski, R., Dongarra, J., Karczewski, K., Waśniewski, J. (eds.) PPAM 2011. LNCS, vol. 7204, pp. 242–251. Springer, Heidelberg (2012). https://doi.org/10.1007/978-3-642-31500-8_25

9. Tikhonenko, O., Kempa, W.M.: Queue-size distribution in M/G/1-type system with bounded capacity and packet dropping. In: Dudin, A., Klimenok, V., Tsarenkov, G., Dudin, S. (eds.) BWWQT 2013. CCIS, vol. 356, pp. 177–186. Springer, Heidelberg (2013). https://doi.org/10.1007/978-3-642-35980-4_20

10. Naumov, V.A., Samuilov, K.E., Samuilov, A.K.: On the total amount of resources occupied by serviced customers. Autom. Remote Control **77**(8), 1419–1427 (2016)

11. Naumov, V.A., Samuilov, K.E.: Analysis of networks of the resource queueing systems. Autom. Remote Control **79**(5), 822–829 (2018)

12. Paxson, V., Floyd, S.: Wide area traffic: the failure of Poisson modeling. IEEE/ACM Trans. Netw. **3**(3), 226–244 (1995)

13. Moiseev, A., Moiseeva, S., Lisovskaya, E.: Infinite-server queueing tandem with MMPP arrivals and random capacity of customers. In: Proceedings of 31st European Conference on Modelling and Simulation Proceedings, pp. 673–679 (2017)

14. Naumov, V., Samouylov, K., Samouylov, A.: Total amount of resources occupied by serviced customers. Autom. Remote Control **77**(8), 1419–1427 (2016)

15. Lisovskaya, E., Moiseeva, S., Pagano, M., Potatueva, V.: Study of the MMPP/GI/∞ queueing system with random customers' capacities. Inf. Appl. **11**(4), 111–119 (2017)

Modelling of Virtual Radio Resources Slicing in 5G Networks

Kirill Ageev[1], Armen Garibyan[1], Anastasia Golskaya[1],
Yuliya Gaidamaka[1,2], Eduard Sopin[1,2(✉)], Konstantin Samouylov[1,2],
and Luis M. Correia[3]

[1] Applied Informatics and Probability Theory Department, Peoples'
Friendship University of Russia, RUDN University, Miklukho-Maklaya Str. 6,
Moscow 117198, Russian Federation
`{ageev-ka,garibyan-a,golskaya-a,gaidamaka-yv,`
`sopin-es,samuylov-ke}@rudn.ru`
[2] Institute of Informatics Problems, FRC CSC RAS, Vavilova street 44-2,
Moscow 119333, Russian Federation
[3] INESC-ID, Instituto Superior Técnico (IST), University of Lisbon,
Rua Alves Redol 9, 1000-029 Lisbon, Portugal
`luis.m.correia@tecnico.ulisboa.pt`

Abstract. Virtual radio resource management system (VRRM) provides optimal sharing of virtualized resources of an infrastructure provider (InP) between several virtual network operators (VNO). One of the main objectives of the VRRM is to optimize the usage of radio access network (RAN) by dynamic sharing between slices in a fair manner according to their contracted Service Level Agreement (SLA). The paper presents the architecture of the VRRM simulation tool in terms of queuing systems. Besides, using the developed simulator, we analyze a practical scenario with 3 VNOs and different types of SLAs and investigate performance metrics under variation of traffic load and SLAs.

Keywords: Virtual RAN · Resource slicing · SLA · Event-based simulation

1 Introduction

Network slicing is a key capability of modern network systems that allows a single network to simultaneously support a wide range of application scenarios (e.g., automotive, utilities, smart cities, high-tech manufacturing) and business models that impose a wide variety of requirements on network functions and expected performance. This allows operators to create and manage multiple dedicated logical networks with

The publication has been prepared with the support of the "RUDN University Program 5-100" (mathematical model development) and funded by RFBR according to the research projects No. 18-00-01555(18-00-01685) and 19-07-00933 (numerical analysis). This work has been developed within the framework of the COST Action CA15104, Inclusive Radio Communication Networks for 5G and beyond (IRACON).

A. Dudin et al. (Eds.): ITMM 2019, CCIS 1109, pp. 150–161, 2019.
https://doi.org/10.1007/978-3-030-33388-1_13

specific functionality running on top of the overall infrastructure. Each of these logical networks is called a network slice and can be adapted to provide specific system behavior to best support specific service/application domains [1–3].

In 5G Network slicing can be used following cases:

- Advanced mobile broadband (eMBB): this usage scenario applies to high data rates, high user density, high user mobility, highly variable data rates, deployment and coverage. Enhanced mobile broadband will be complemented by new applications and requirements in addition to existing mobile broadband applications to increase productivity and provide increasingly seamless user access;
- Machine type mass communication (mMTC): this usage scenario is characterized by a very large number of connected devices, typically transmitting a relatively low amount of non-latency-sensitive data. Devices should be inexpensive and have a very long battery life;
- Ultra-reliable and low-latency communication (URLLC): this usage scenario has strict requirements for features such as bandwidth, latency, and reliability. Some examples include wireless control of industrial production or manufacturing processes, remote medical surgery, distribution automation in an intelligent network, transportation safety, etc.

Network slicing enables an 5G network operator to provide customized networks by slicing a network into multiple virtual, and end-to-end networks, referred to as network slices. Each network slice can be defined according to different requirements on functionality, performance and specific users.

Network slicing allows the 5G network operator to provide dedicated logical networks (i.e., network slices) with customer specific functionalities. A network slice, spanning all the network segments including radio access network, transport network and core network, can be dedicated to specific types of service [4]. When a UE is only associated to a single dedicated network slice, the 5G network can identify the association of the UE with the network slice based on user subscription, context, service provider's policy, etc. Otherwise, if a UE accesses multiple network slice instances simultaneously, it is recommended that the UE provide information to the network to assist the network slice selection process.

For each network slice, dedicated resources (e.g., virtualized network functions, network bandwidth, QoS) are allocated and an error or fault that occurs in one slice does not cause any effect in other slices [5].

In the paper, we develop mathematical model in terms of queuing system for a standalone network slice. Since the analysis of the whole system with multiple slices governed by the Virtual Radio Resource Management system (VRRM) is too complex, we also develop a simulation model for evaluation of VRRM performance metrics.

The rest of the paper is organized as follows. Section 2 describes network slicing model including service model, while Sect. 3 is devoted to the mathematical model of resource allocation and queuing in a standalone VNO. Section 4 provides description of the simulation tool's architecture. The case study is presented in Sect. 5, while Sect. 6 concludes the paper.

2   Virtual Resources Slicing Model

2.1   RRM Model

One of the main concepts of network slicing is virtualization of physical resources from different Radio Access Technologies (RATs) and managing available virtual resources. Radio resources of each base station are controlled managed by Local RRMs (LRRMs) and the Common RRM system (CRRM) manages all radio resources from different base stations of different RATs. Finally, the VRRM system provides virtualization of available physical resources and divides them into slices by means of bitrate instead of Radio Resource Units (RRUs).

VNOs demand for Capacity-as-a-Service (CaaS) from VRRM. VRRM is in charge of managing the total available capacity provided by CRRM, through aggregating all the RRUs from different RATs, and sharing the capacity by configuring isolated slices associated with different services of VNOs. By providing isolation and abstraction of elements, VRRM allows each VNO to deploy its own protocol stack on the same set of RRUs per RAT (e.g. resource blocks in LTE, codes in UMTS, timeslots in GSM and carriers in Wi-Fi), thus promoting the concept of multi-lease in a heterogeneous virtualized environment with existing multiple access methods. A simplified diagram of the network functionality is presented in Fig. 1 for a better understanding of the mechanisms of interaction among the involved parties.

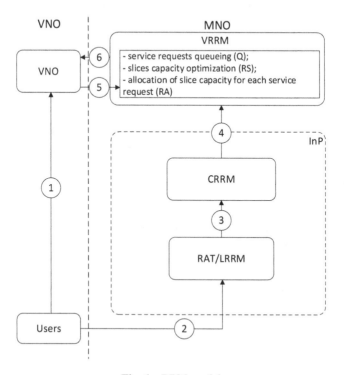

Fig. 1. RRM model

Here (1) is the session request arriving to VNO, (2) is the channel condition information sent to base station of RAT, (3) is the RRU status information gathered by CRRM from all available LRRM. CRRM is informing VRRM of total available resource capacity (4), while VNO informs VRRM of service request parameters (5). Finally, VRRM defines the slice capacity for the VNO (6). It is worth noting that the numbers in Fig. 1 do not necessarily represent the operational order in sequence. The physical infrastructure and network elements managed by the InP are displayed in a dotted box [6, 7].

VRRM, VNOs, and physical resource define the current network structure. Mobile Network Operator locates on the top of this block. MNO is the owner of the resource. A network virtualization tool is offered in this system, which consists of a VRM module and several virtual operators (VNO), which allows the owner of cellular equipment to sell its resources to several operators. At the same time, virtual operators get a cheaper opportunity to enter the market without buying equipment, and at they will be logically separated from each other within the same network, which will allow them to provide their subscribers with any telecommunication services.

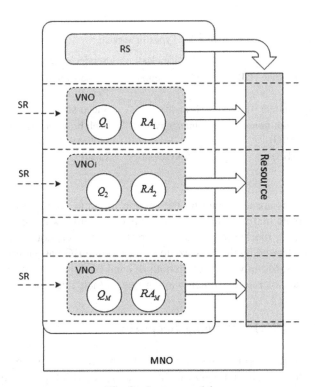

Fig. 2. System model

Figure 2 shows the system model. The request to start a user session (we will refer further as service requests, SR) arrive to VNO and provide information on service parameters. We assume here that each mobile device can handle no more than one session at a time. According to the service parameters and SLA types, each VNO allocates available slice capacity to service requests (Resource Allocation, RA) and places some of them to the queue (Q) if there are not enough resources for all requests. The Resource Sharing (RS) block gathers information on the load at VNOs and determines slice capacities accordingly.

2.2   Service Model

The set of services provided in wireless networks by virtual operators may be different. However, for the correct operation of the network and VRRM, a set of parameters is defined that must be set for each service before it is put into operation: class of service, service priority, violation priority, maximum rate, minimum rate, maximum waiting time for the start of service [8]. Table 1 summarizes the SLA types and their requirements.

Table 1. Types of service.

SLA type	Service class	Service example
Guaranteed bitrate (GB)	Conversational	Voice over IP (VoIP); Real-time games
Guaranteed bitrate (GB)	Streaming	Video streaming
Best effort with minimum guaranteed (BG)	Interactive	Web browsing, file transfer protocol (FTP)
Best effort (BE)	Background	Email

GB type sessions are described by the session duration, minimum and maximum required bitrate. Note that streaming GB sessions can tolerate waiting in a queue with the help of buffering mechanisms, but conversational GB cannot. BG and BE type sessions are characterized by the data size to be sent, while BG has additional minimum bitrate requirements. Moreover, although both BG and BE type sessions can be queued, BG sessions do not tolerate too much waiting due to their interactive nature.

3   Standalone VNO Queueing Scenario

In this section, we develop a mathematical model in terms of queueing systems with impatient customers and retrial group for the analysis of BG type sessions. BG type corresponds to 3rd class of services, interactive, that include file sharing, web browsing and social networking. We consider the scenario, in which the user can use one of the three above services. The system contains R resources, the queue size and the retrial group size are denoted by N_q and N_{rg}. In other words the network has a total capacity

$R_{[Mbps]}$, a queue is a buffer in which session requests are accumulated in case the whole network capacity is occupied. The time that a service request spends in the queue corresponds to the waiting for the start of the session. The maximum number of simultaneously served sessions N depends on the total network capacity and the minimum assignable data rate b_{BG}^{min} to each session, all other sessions are waiting in the queue.

Table 2. Notations of the mathematical model.

R	Volume of resources of a VNO
N_q	Queue size
N_{rg}	Size of the retrial group
θ_S	Parameter of the file size distribution
μ	The average service rate
β	The intensity of customers leaving the queue to the retrial group
α	The intensity of customers moved to the queue from the retrial group
λ	Arrival intensity of customers
$p(n,m)$	The stationary probability that n sessions are served or waiting in the queue and m sessions are in the retrial group
$\sigma_{GB}(t)$	Volume of resources allocated to a GB service request at time t
$\sigma_{BG}(t)$	Volume of resources allocated to a BG service request at time t
$\sigma_{BE}(t)$	Volume of resources allocated to a BE service request at time t
$N_{GB}(t)$	Number of service requests of type GB at time t
$N_{BG}(t)$	Number of service requests of type BG at time t
$N_{BE}(t)$	Number of service requests of type BE at time t
$N(t)$	Total number of all service requests at time t
b_{GB}^{max}	Maximum volume of resources required to a GB type service request
b_{GB}^{min}	Minimum volume of resources required to a GB type service request
b_{BG}^{min}	Minimum volume of resources required to a BG type service request
$R_{FR^*}(t)$	Volume of resources that may be unused if all GB and BG service requests are allocated minimum and there are no BE service requests

Each session implies transferring of a file of random volume. The file size is assumed to be exponentially distributed with the parameter θ_S. Note that the service rate is given by $\mu = R\theta_S$. Since users are impatient, they can leave the queue after an exponentially distributed time with parameter β and reenter the queue after an exponentially distributed time with parameter α. The transition of the user from the queue into the retrial group describes the process when due to the limited waiting time, attempt to start a session is aborted. Thus, the user comes into the retrial group and after a random amount of time returns to the queue, which describes the process of attempting to start a session. All notations are summarized in Table 2.

Let $p(n,m)$ be the stationary probability that n sessions are served or waiting in the queue and m sessions are in the retrial group. The following formulas define set of equilibrium equations:

$$\lambda p(0,0) = \mu p(1,0); \tag{1}$$

$$(\lambda + \mu)p(n,0) = \mu p(n+1,0) + \lambda p(n-1,0) + \beta p(n-1,1), \quad 0 < n \leq N; \tag{2}$$

$$(\lambda + \mu + \alpha(n-N))p(n,0) = \mu p(n+1,0) + \lambda p(n-1,0) + \beta p(n-1,1), \\ N < n < N + N_q; \tag{3}$$

$$(\mu + \alpha N_q)p(N+N_q,m) = \lambda p(N+N_q-1,m) + \beta(m+1)p(N+N_q-1,m+1), \, m < N_{rg}, \tag{4}$$

$$\mu p(N+N_q, N_{rg}) = \lambda p(N+N_q-1, N_{rg}), \tag{5}$$

$$(\lambda + \beta m)p(0,m) = \mu p(1,m), 1 \leq m \leq N_{rg}, \tag{6}$$

$$(\lambda + \mu + m\beta)p(n,m) = \lambda p(n-1,m) + \mu p(n+1,m) + (m+1)\beta p(n-1,m+1), \\ 1 \leq n < N, 1 \leq m < N_{rg}, \tag{7}$$

$$(\lambda + \mu + m\beta)p(N,m) = \lambda p(N-1,m) + + \mu p(N+1,m) \\ + \alpha p(N+1,m-1) + (m+1)\beta p(N-1,m+1), 1 \leq m < N_{rg}, \tag{8}$$

$$(\lambda + \mu + m\beta + (n-N)\alpha)p(n,m) = \lambda p(n-1,m) + \mu p(n+1,m) \\ + (n+1-N)\alpha p(n+1,m-1) + (m+1)\beta p(n-1,m+1), \\ N < n < N + N_q, 1 \leq m < N_{rg}, \tag{9}$$

$$(\lambda + \mu + N_{rg}\beta)p(n,N_{rg}) = \lambda p(n-1,N_{rg}) + \mu p(n+1,N_{rg}), 0 < n < N, \tag{10}$$

$$(\lambda + \mu + N_{rg}\beta)p(n,N_{rg}) = \lambda p(n-1,N_{rg}) + \mu p(n+1,N_{rg}) \\ + (n+1-N)\alpha p(n+1,N_{rg}-1), N \leq n < N + N_b. \tag{11}$$

Solving the system of Eqs. (1–11) we can find blocking probabilities according to the following formulas:

$$B_1 = \sum_{m=0}^{N_{rg}} p(N+N_q,m), \tag{12}$$

$$B_2 = \sum_{n=0}^{N+N_q} p(n,N_{rg}), \tag{13}$$

$$B_1 = \sum_{m=1}^{N_{rg}} p(N+N_q,m), \tag{14}$$

$$B = B_1 + B_2 - p(N+N_q,N_{rg}). \tag{15}$$

Note that in real networks there are no blocking of BG type sessions. Therefore, the blocking probability B corresponds to the probability of SLA violation in the network.

Similar scenario can be developed for each standalone VNO with any SLA type. However, analytical modelling of a VNO with several SLA types is too complex. For that reason, we develop the simulation tool in the next section.

4 Simulation Tool's Architecture

4.1 Event-Based Simulation

The simulation tool models the VRRM system according to the event-based approach. There are three types of events. They are the arrival of a service request, the departure of a service request and the slice reconfiguration event. On the arrival and the departure of a service request, RA function reallocates available capacity to the new number of sessions. At slice reconfiguration, the RS block gathers information on the network load from all slices and assigns new resource volumes for them.

VNO consists of three queues to handle each type of traffic (GB, BG, BE). The length of each queue is set before the simulation begins. The resource is the total bitrate that the VNO can use. VNO is allocated a slice from shared virtual resource. The capacity of slices is optimized and reconfigured every Δ ms by RS block. RA function allocates slice capacity to service requests according to resource allocation rules depending on the type of traffic. Some of the service requests are waiting in block Q. Thus, the input data in the VNO is the flow of service requests with specified service type and parameters. Then by the rules (16)–(19) service requests are allocated some resources.

4.2 TG Architecture

The traffic generator is initiated with a set of services, requests for which it will generate traffic, together with the service parameters, the arrival intensities and distribution of interarrival times for each type of service. The traffic generator runs in several threads according to the number of virtual operators. When the input data is received, the core of the traffic generator determines the next interval of a new service request arrival for each flow and their parameters (file size, service duration, minimum/maximum bitrate, etc.). The generated sequence of arrivals and service parameters is then sent to the event manager block of the simulation tool. The architecture is scalable and allows use in conjunction with various software.

4.3 Resource Allocation Rules

Here we summarize the resource allocation rules for different SLA types. Table 2 explains the notation used to formalize the allocation rules.

The main idea of resource allocation rule is as follows. Firstly, we allocate the highest priority GB and BG service requests their minimum requirements. The rest of resources is denoted $R_{FR^*}(t)$. Then these free resources are distributed equally among all service requests, i.e. all requests receive additional $\frac{R_{FR^*}(t)}{N(t)}$ resources. Finally, if the volume of resources allocated to BG service requests is greater than their maximum requirement $b_{GB}^{\max}$, then their share of resources is decreased to $b_{GB}^{\max}$ and the remaining resources are distributed equally between BG and BE service requests.

Let

$$N(t) = N_{GB}(t) + N_{BG}(t) + N_{BE}(t),$$
$$R_{FR^*}(t) = R - N_{GB}(t) * b_{GB}^{\min} - N_{BG}(t) * b_{BG}^{\min}. \tag{16}$$

Then the formal expressions for the allocation rules are given as follows.

$$\sigma_{GB}(t) = \begin{cases} b_{GB}^{\max}, & if \ \frac{R_{FR^*}(t)}{N(t)} > b_{GB}^{\max} - b_{GB}^{\min}, \\ b_{GB}^{\min} + \frac{R_{FR^*}(t)}{N(t)}, & if \ \frac{R_{FR^*}(t)}{N(t)} < b_{GB}^{\max} - b_{GB}^{\min}, \end{cases} \tag{17}$$

$$\sigma_{BG}(t) = \begin{cases} b_{BG}^{\min} + \frac{R_{FR^*}(t)}{N(t)}, & if \ \frac{R_{FR^*}(t)}{N(t)} < b_{GB}^{\max} - b_{GB}^{\min}, \\ b_{BG}^{\min} + \frac{R_{FR^*}(t) - \left(b_{GB}^{\max} - b_{GB}^{\min}\right) * N_{GB}(t)}{N_{BG}(t) + N_{BE}(t)}, & if \ \frac{R_{FR^*}(t)}{N(t)} > b_{GB}^{\max} - b_{GB}^{\min}, \end{cases} \tag{18}$$

$$\sigma_{BE}(t) = \begin{cases} \frac{R_{FR^*}(t)}{N(t)}, & \frac{R_{FR^*}(t)}{N(t)} < b_{GB}^{\max} - b_{GB}^{\min}, \\ \frac{R_{FR^*}(t) - \left(b_{GB}^{\max} - b_{GB}^{\min}\right) * N_{GB}(t)}{N_{BG}(t) + N_{BE}(t)}, & \frac{R_{FR^*}(t)}{N(t)} > b_{GB}^{\max} - b_{GB}^{\min}. \end{cases} \tag{19}$$

5   Case Study

In this section, we present the results of resource slicing case study. Assume that the system capacity is 200 Mbps (resource volume) and three VNOs. The arrival rates of session requests are given below.

$$\lambda_{GB}^1 = 1.8; \quad \lambda_{GB}^2 = 2.4; \quad \lambda_{GB}^3 = 3.0;$$

$$\lambda_{BG}^1 = 1.2; \quad \lambda_{BG}^2 = 1.6; \quad \lambda_{BG}^3 = 2.0.$$

Service rates of GB type sessions in each VNO are assumed to be equal to each other:

$$\mu_{GB}^1 = \mu_{GB}^2 = \mu_{GB}^3 = \frac{1}{60} \ [sessions/s]$$

Thus, the average service time of GB type sessions is 60 s. File sizes of BG type sessions θ_{BG}^1, θ_{BG}^2, θ_{BG}^3 are assumed to have uniform distribution on the interval [4 MB; 20 MB].

Minimum bitrate requirements $R_{GB}^{i,\min}$ for GB type sessions are identical for each of the VNOs and equal to 0.3 Mbps, while maximum $R_{GB}^{i,\max}$ is equal to 5 Mbps. For BG type sessions, the minimum bitrate $R_{BG}^{i,\min}$ is equal to 1 Mbps.

The maximum queue length for each type of traffic in each VNO is 50.

At the beginning, the resource is divided equally between all operators, but during simulations, the resource distribution between slices become more stable and oscillates about the values that are proportional to the arrival rates to each VNO (see in Figs. 3 and 4).

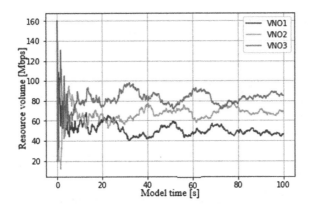

Fig. 3. Dynamics of the resource volume allocated to virtual operators at model times

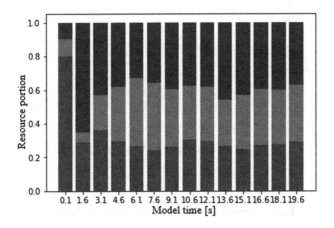

Fig. 4. Dynamics of the resource volume shares allocated to virtual operators

In the case of a fixed resource volume for a separate VNO, with given arrival rates and resource requirements, the first VNO will have an excess of the resource, while the third VNO will suffer from the shortage of resources. Resource reconfiguration allows to solve this problem and reduces the number of blocked requests of the third VNO due to a slight increase in the probability of blocking on the first. Note that resource reconfiguration of slices affects mainly the highest priority GB type sessions (see Fig. 5).

Also, with certain delta, similar improvements can be seen in the average queue length, the average service duration of BG type requests, as well as the average waiting time (Fig. 6).

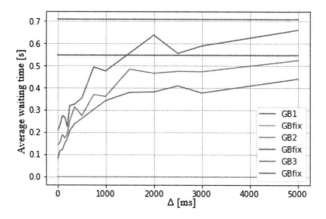

Fig. 5. Average waiting time in each GB applications

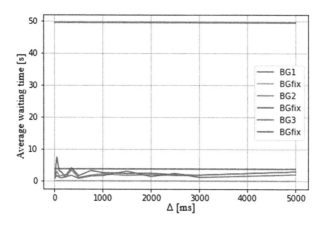

Fig. 6. Average waiting time in each BG applications

6   Conclusion

This paper proposes a general software tool architecture for developing and analyzing mechanisms of SLA-based network slicing procedures for virtual RANs. The architecture shown in Fig. 3 reflects the two main components of the software tool - a module for sharing and optimizing resources between tenants (VNOs) in accordance with the SLA and a module for managing tenant resources in accordance with the algorithms for servicing user requests for the service provisioning. A distinctive feature of the tool architecture is its scalability in terms of the main parameters of the network slicing policies as well as the approach to the development of a tenant resource management module, based on the application of queues with random requirements for system resources.

To achieve these goals, the proposed architecture software is based on the VRRM model, developed on the basis of the hierarchical architecture that was presented in a previous paper [6], and formulated as a convex optimization problem to cope with the concept of proportional fairness of slicing resources for tenants. Recall that the model is based on three types of SLA contracts (GB, BG and BE), and it is for these three types of contracts that models of queuing systems with random requirements for network resources are developed and incorporated into the software tool architecture.

We have already conducted preliminary studies and built models of queuing systems for all types of SLA contracts, and also developed a preliminary version of the tenant resource allocation management module. Some results of numerical are shown in Sects. 4 and 5 of current paper.

In further studies, we are to analyze the options for formalizing the optimization problem in the resource sharing module, taking into account the results of such works as [9, 10], as well as the work of [6], where some improvements and clarifications will be proposed. Particular attention will be paid to the development of the traffic generator module, which should receive the parameters of user requests for services on the front-end, and offer random flows customers to the queuing systems, which are elements of the tenant resource allocation management module, at the back-end.

The goal of all the work is to develop methods and algorithms for the network slicing in the target architectures proposed in the 3GPP standards.

References

1. Pérez-Romero, O., Sallent, R., Ferrús, R., Agustí, R.: On the configuration of radio resource management in a sliced RAN. In: Proceedings of NOMS 2018 IEEE/IFIP Network Operations and Management Symposium, pp. 1–6, Taipei (2018)
2. 3GPP TR 28.801: Study on management and orchestration of network slicing for next generation network, December 2017
3. 3GPP TS 23.501 V15.4.0 – system architecture for the 5G system
4. ITU-T Rec. Y.3101: Requirements of the IMT-2020 network, January 2018
5. ITU-T Rec. Y.3112 - framework for the support of network slicing in the IMT-2020 network, December 2018
6. Rouzbehani, B., Correia, L.M., Caeiro, L.: An SLA-based method for radio resource slicing and allocation in virtual RANs. In: EURO-COST, University of Lisbon (2018)
7. Rouzbehani, B., Correia, L.M., Caeiro, L.: A real-time computational resource management in C-RAN. In: EURO-COST, University of Lisbon (2018)
8. Khatibi, S.: Radio resource management strategies in virtual networks. Ph.D., p. 196, University of Lisbon (2016)
9. Sciancalepore, V., Zanzi, L., Costa-Perez, X., Capone, A.: ONETS: online network slice broker from theory to practice. arXiv:1801.03484 (2018)
10. Wang, G., Feng, G., Quek, T.Q.S., Qin, S., Wen, R., Tan, W.: Reconfiguration in network slicing: optimizing profit and performance. IEEE Trans. Netw. Serv. Manag. **16**(2), 591–605 (2019)

A Multiphase Queueing Model for Performance Analysis of a Multi-hop IEEE 802.11 Wireless Network with DCF Channel Access

Andrey Larionov[1]([⊠])(iD), Vladimir Vishnevsky[1], Olga Semenova[1], and Alexander Dudin[2](iD)

[1] V.A. Trapeznikov Institute of Control Sciences of RAS, Profsoyuznaya 65, Moscow, Russia
larioandr@gmail.com, vishn@inbox.ru, olgasmnv@gmail.com
[2] Belarusian State University, Nezavisimosti Avenue 4, Minsk, Belarus

Abstract. Methods of queueing theory are often used for telecommunication systems performance evaluation. According to this approach, all MAC and PHY layer protocols details are modelled with a given service time distribution, and the precision of the estimated properties depends on this distribution function selection. While for Ethernet and wireless relay networks the service time is roughly equal to the sum of packet and headers sizes divided by the channel bitrate, it is not so easy to estimate the service time for wireless networks channels based on CSMA/CA including IEEE 802.11 (WiFi) and IEEE 802.15.4 (ZigBee) due to random exponential backoff and collisions. In this paper we use a simple model of the saturated channel described as a semi-Markov random absorbing process, which is based on Bianchi model [2]. We use this process to find a service time phase-type (PH) distribution matching the first two moments of the semi-Markov process. We also use channel simulation model to sample transmission delays in unsaturated mode and find another PH-distribution using G-FIT [12] algorithm. Then we use these PH-distributions to estimate end-to-end delays, queue sizes and nodes utilization in a multi-hop wireless network with linear topology using $M/PH/1/N \to \bullet/PH/1/N \bullet \cdots \bullet/PH/1/N$ queueing network. All results are compared to wireless network simulation model, and demonstrate the cases in which the queueing network tends to provide results close to real network performance.

Keywords: CSMA/CA · DCF · Semi-Markov random absorbing process · PH-distribution · Queueing networks · Multi-hop wireless networks

---

This work was partly financially supported by the Russian Foundation for Basic Research, grant No. 18-57-00002.

A. Dudin et al. (Eds.): ITMM 2019, CCIS 1109, pp. 162–176, 2019.
https://doi.org/10.1007/978-3-030-33388-1_14

1   Introduction

When analyzing the performance characteristics of various telecommunication networks, the queuing theory methods are often used. The packet transmission time through the communication channels is modelled by a random service process, the input data flows are modelled by random arrival flows, and the memory of routers in which the packets are buffered while waiting for transmission is modelled with finite or infinite queues. In the simplest case, both the service time and the interarrival intervals are modelled by the exponential distribution, and the queues are assumed to have unlimited capacity. The queueing network model in this case has nodes of the $M/M/1$-type, and the performance characteristics of such network can be easily calculated. However, the estimates obtained may have low accuracy.

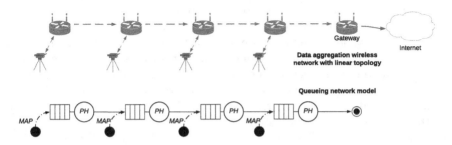

Fig. 1. An example of modelling a wireless network with linear topology using a tandem queuing network with the arrival MAP flows.

For greater certainty, let us focus on the consideration of a particular case of telecommunication networks, namely, the linear topology networks. Traffic is assumed to be transmitted through the network mainly in one direction. An illustration of such a network is shown in Fig. 1. Examples of such networks include sensor networks along pipelines or communication networks for connecting cameras along the highways. Moreover, many results obtained for networks with a linear topology are easily scaled to the case of networks with an arbitrary acyclic topology; therefore, the class of networks under consideration turns out to be quite large.

Insufficient accuracy of the model $M/M/1 \to \bullet/M/1 \to \cdots \bullet /M/1$ for both wired and wireless networks follows from three reasons. First, packets on a network hardly ever have an exponential size distribution. Moreover, the various channel and network layer protocols have their own limitations on the maximum packet size, and the minimum size is due to the presence of the headers and a preamble. Because of this, even in the case of a wired network, the usage of exponential service time can introduce significant error. Secondly, the intervals between packets also do not always have an exponential distribution. For example, if data from the sensors or cameras is transmitted within a fixed interval, then the intervals distribution will be closer to a constant. Thirdly, when transmitting packets over the network, their size does not change which means that their service time should not change either.

The first and second problems can be solved using more appropriate service time distributions and the input flow models. For example, a good generalization is the queuing networks with the $MAP/PH/1/N$-type nodes where the traffic is modelled by the Markovian arrival processes (MAPs), and the service time has the phase-type distribution (PH). Queuing networks of this type are well studied, see for instance [15]. Solving the third problem leads to systems with correlated services which are beyond the scope of this paper.

For effective modelling of the real telecommunication networks using the queuing networks $MAP/PH/1/N \rightarrow \bullet/PH/1/N \rightarrow \cdots \bullet /PH/1/N$, it is necessary to have the service time distribution functions that model a distribution of packets transmission time through a communication channel with sufficient accuracy. In the case of a wireless network using CSMA/CA as an access scheme, the transmission duration is rather complicated. In particular, in addition to the size of the transmitted data and headers, it is necessary to take into account the random waiting time for channel listening (backoff), fixed inter-frame intervals, waiting time, receipt of acknowledgement and re-transmissions due to collisions.

For a mathematical description of the packet transmission process in a wireless network channel, we will use a semi-Markov process with absorbing state, in which the probabilities of state transitions are determined by a Markov chain, and the waiting times in each state are independent (can have arbitrary distributions, not necessarily exponential). Also in such processes one or several absorbing states are selected, having reached any of which the process stops. The time spent in the process from the initial state to hitting any of the absorbing states is a random variable that is proposed to be used as a model of packet transmission time.

The paper is organized as follows. In Sect. 2 we will briefly observe papers related to transmission time modelling. Then in Sect. 3 we will describe the channel we model and provide numerical parameters used in experiments. In Sect. 4 service time modelling with a semi-Markov absorbing process and then PH distributions fitting will be discussed, and in Sect. 5 the results of a numerical experiment of a multi-hop wireless network performance evaluation using the queueing model with PH distributions found earlier will be presented. Finally, Sect. 6 discusses the results and concludes the paper.

2   Previous Work

Markov and semi-Markov processes for performance analysis of wireless networks with DCF channels are studied in a variety of papers. One of the key papers by Bianchi [2] proposes a Markov chain to analyse the operation of an IEEE 802.11 network under the DCF channel access. The states of the chain correspond to backoff slots and transmission attempts. The key assumptions of the model are the saturated network conditions and the fact that the conditional probability of a collision during any transmission is constant. On the base of the proposed

Markov chain, the network throughput is estimated. The semi-Markov processes are also often used to describe the operation of wireless networks. For example, in [10] a semi-Markov process was proposed to analyse the operation of the data link layer of the IEEE 802.15.4-2006 network; authors estimate the network bandwidth, distribution of intervals between successful transmissions and the distribution of the waiting time in backoff.

A large number of papers, e.g. [1, 3–9, 11, 13, 14], are devoted to the analysis of the packet transmission time (access delay) in IEEE 802.11 networks with DCF channel access. Chatzimisios et.al. in [3] describe a simple model for calculating the average packet transmission time based on the Bianchi model suggesting a limited number of transmission attempts. [1] presents an analytical model for calculating the packet transmission delays in saturated network. In [11] authors consider the DCF ad-hoc network in a saturated mode, derive the analytical relations for the expectation and variance of the packet service time taking into account the limited number of transmission attempts. The authors also give the expression for the generation function of the packet transmission time. The paper [14] considers unsaturated network with queues. In [5], Dong et.al. study the ad-hoc network in an unsaturated mode assuming an exponential distribution of time intervals between packet arrivals and, apart from analyzing the service time on individual nodes, consider the delay in transmission over multi-step routes taking into account the hidden stations. The paper [8] is also devoted to the problem of estimating the transmission time in the presence of hidden stations. [13] presents not only the detailed analysis of the packet transmission time in an ad-hoc network in the unsaturated conditions but also considers the case of transmitting a group of packets using the IEEE 802.11e EDCA mechanism. Felemban and Ekici in [6] analyze the transmission time of packets in one-step ad-hoc networks refining the Bianchi model by introducing additional Markov chains for backoff states, they also analyze unsaturated network performance. Another detailed analysis of the performance of ad-hoc networks including packet transmission times is presented in [4]. We should also mention the paper [9] where the packet transmission time is described as the PH distribution.

Note that the primary goal of our work is not in modelling the transmission time of packets on individual channels, but in using the models of transmission time for the tandem queuing networks such as $MAP/PH/1/N \rightarrow \cdots \rightarrow \bullet/PH/1/N$ for evaluating the characteristics of wireless multi-hop networks with linear topology. Although we further propose to use a simple semi-Markov model for analyzing the packet transmission time in a network operating in saturated mode, to clarify the results one can use the transmission time estimates from the works listed above especially concerning the network in unsaturated mode.

3   Wireless DCF Channel

As a channel access method, we consider the simplest implementation of the IEEE 802.11 DCF. This access method is one of the earliest and is a basis for a significant number of wireless LANs and sensor networks. On the one hand, this mechanism is simple to be analyzed and well studied but, on the other hand, it allows taking into account the key features of wireless networks that affect data transfer time without the need to simulate complex protocols and methods.

The DCF mechanism is described in detail in literature (see, e.g., [2]), and the most detailed description can be found in the IEEE 802.11 standard. In this paper we simplify the mechanism by ignoring RTS/CTS transmissions, assuming infinite transmission attempts and not taking into account post-backoff which leads to correlated service times and the need to use MSP (Markov Service Process) instead of PH distributions.

In the following, we use the DCF parameters values given in Table 1. These parameters coincide with [2] which simplifies the model validation. However, they can be easily modified to analyze the higher-speed, modern versions of the IEEE 802.11 channels. In addition, the table shows the traffic parameters that we will use when modeling a network in unsaturated mode.

Table 1. The DCF mechanism parameters and traffic transfer modes used in the simulation of a wireless network.

Parameter	Value	Unit
PHY header	128	bit
MAC DATA header	272	bit
MAC ACK frame	112 (+ PHY header)	bit
Slot duration (σ)	50	μs
DIFS	128	μs
SIFS	28	μs
CWmin	16	
CWmax	1024	
Bitrate	1000	kbps
IP header	160	bit
Payload size	10688	bit
Low traffic rate	1	kbps
Medium traffic rate	20	kbps
High traffic rate	180	kbps

4   Simulation of the Packet Transmission Time over the Wireless Channel

To represent the packet transmission time in the DCF channel as a PH distribution, we consider the classical IEEE 802.11 network in which all stations transmit in a common area (also called collision domain) where all stations "hear" each other. Thus, the simultaneous transmissions of any two stations interfere with each other and lead to collisions. An example of a network is shown in Fig. 2.

Consider two modes of network operation: saturated mode when each station always has a packet to transmit and unsaturated mode in which the packets arrive at the station in accordance with some random distribution and there are intervals when the station doesn't transmit anything. We will select the parameters of this distribution so that on average each station needs to transmit user data (payload) with a given bitrate (low, medium or high, see Table 1).

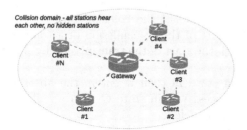

Fig. 2. The structure of the wireless network which is the model for service time estimation: N clients within the same collision domain transmit data to the gateway.

The time of a packet transmission (service) is referred to as the time elapsed from the arrival of the packet to the transmitter until the successful transmission is confirmed. For the analysis of packet transmission time, we use both the simulation and the analytical models: for a saturated mode, the service time is obtained as the distribution of time to reach the absorbing state in a semi-Markovian random process based on the Bianchi [2] model, and for an unsaturated mode, the service time sets are obtained using the network simulation. Then using this data, we apply the methods of constructing PH-distributions by moments or samples set.

4.1   The Structure of the Semi-Markov Process for the Saturated Mode

A large number of models based on [2] were proposed to analyze various performance parameters of IEEE 802.11 networks. The Bianchi model [2] itself is primarily intended for analyzing the network bandwidth in a saturated mode. For the purposes of this paper, we construct a semi-Markov absorption process that simulates the service time of a packet which is convenient because, on the one hand, it is easy to calculate the expectation and variance of the packet service time, and on the other hand, it is convenient to generate a samples set for subsequent use of the EM procedure [12].

The structure of the semi-Markov process is shown in Fig. 3. Here $CW = $ CWmin, $m = log_2(\text{CWmax/CWmin})$ is the number of times the size of the content window may increase. The structure of states and transition probabilities almost completely coincides with the chain described in [2]. However, here we divide the zero slot in which the transmission is carried out into two different states D and C: if the process is in D then no collisions occurred and the packet described by a random variable $\tau_D = (\xi + \text{HDR})/B + \text{SIFS} + \text{ACK}/B + 2\delta$ is successfully transmitted, where ξ is a random variable describing the packet size, HDR is the fixed headers sizes, B is a channel bitrate, ACK is a bit size of the frame Ack along with the headers, and δ is the propagation delay. If the process is in state C then a collision occurred during the transfer and the duration of our stay in this state is modeled using the random variable $\tau_C = (\xi + \text{HDR})/B + T_W$, where T_W is a constant equal to the maximum acknowledgment waiting time (for a more accurate estimate of the collision duration, we should also take into account that the start of the transmission of a competing frame does not necessarily coincide with the start of the transmission by the station in question, see, for example, [2]).

If the process is in state D then it enters the absorbing state next with probability 1. If the process was in state C then a collision occurred and the next transfer attempt is modeled.

The time spent in state E depends on whether there were transmission attempts and whether there was a collision among the rest of stations. This time is described using a random variable τ_E:

$$\mathbb{P}\{\tau_E = \sigma\} = 1 - P_{\text{tr}}(n-1)$$
$$\mathbb{P}\{\tau_E = \tau_D\} = P_{\text{tr}}(n-1)P_s(n-1)$$
$$\mathbb{P}\{\tau_E = \tau_C\} = P_{\text{tr}}(n-1)(1 - P_s(n-1))$$

Here the expressions for P_{tr}, P_s are calculated similarly as formulas (10) and (11) in [2] by using $n-1$ instead of n. The expressions like $\tau_E = \tau_D$ mean that with the given probability τ_E has the same distribution as τ_D. By calculating the moment generating function one can easily obtain the values of the expectation and variance for the semi-Markov random process we have constructed.

Note that to calculate the expectation and variance of packet transmission time, you can use the more accurate formulas given in paper [11].

4.2   Transmission Time Analysis

To validate the semi-Markov process described above, transmission time mean and standard deviation values were compared with the results obtained from a simulation model of a wireless network operating in saturated mode. We considered two different payload size distributions: $\xi \equiv \text{const} \equiv P$ and $\xi \sim Exp(1/P)$, where P – mean value of payload bit size. Figure 4 shows the simulation and analytic results for a network with up to five clients. We also considered unsaturated mode with three types of traffic (see Table 1).

As Fig. 4 shows, semi-Markov process provides very precise results on saturated networks channels with up to four stations. If the network operates in unsaturated mode, then under low traffic both mean and deviation values are much smaller then in saturated case. However, it is obvious that under heavy traffic the conditions are more close to saturated mode. This effect is due to a smaller number of collisions in unsaturated network under low traffic load.

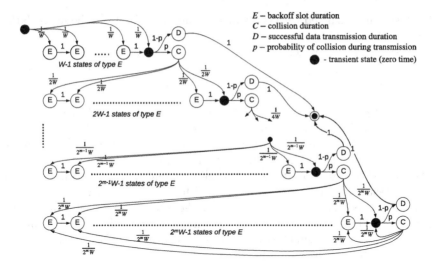

Fig. 3. The structure of the semi-Markov process with absorption built on the basis of the Bianchi model which model the transmission time in the collision domain in the saturated mode

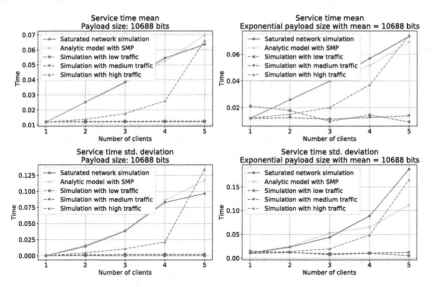

Fig. 4. Mean and standard deviation values for a channel in saturated and unsaturated network, estimated using network simulation modelling and semi-Markov analytic model.

4.3 PH-Fitting of Packet Transmission Time

Based on the results above, we will fit PH-distributions into three sets:

1. PH-distributions matching two moments of semi-Markov process ($\mathcal{S}_1$);

2. PH-distributions fitted with transmission time sampled from the semi-Markov process ($\mathcal{S}_2$);
3. PH-distributions for unsaturated network channel, fitted with samples obtained from a network simulation model when transmitting low intensity traffic ($\mathcal{S}_3$)

To fit PH distribution by the first two moments, one need to solve a non-linear optimization problem of loss function minimization with constraints on moments values, like it is described in [16]. When fitting PH-distribution with a samples set, we use G-FIT algorithm [12] based on EM-procedure.

We use PH distributions to model wireless channels in a network with linear topology. Assuming that stations are placed far enough from each other, collisions arise between neighbour stations only. We also assume ideal radio channel with BER equal zero, neglecting errors in frames transmission due to multipath propagation, signal attenuation and low signal-to-noise (SNR) ratio. In these assumptions, three cases may take place:

1. there are no collisions in a single-hop network, so it is equal to a collision domain network with a single client;
2. each boundary station in a multi-hop network collide with only one neighbour, which is equal to a collision domain network with two clients;
3. each middle station in a multi-hop network collide with two neighbours, which is equal to a collision domain network with three clients.

It should be noticed, that if the stations are placed closer to each other, collisions with two-hop neighbours may arise. However, we are not going to consider this scenario.

Thus, we need to fit the PH distributions for networks that have from one to three stations. Each distributions set $\mathcal{S}_i$ contains six PH distributions: $\mathcal{S}_i = \{\beta_{i,j}^{(const)}, \beta_{i,j}^{(exp)} : j = 1, 2, 3\}$, matching collision domain networks with one, two or three stations, and either constant or exponentially distributed payload size.

5   Numerical Estimations for a Multi-hop Wireless Network with Linear Topology

To study the performance of a multi-hop wireless network with linear topology we make use of a tandem open queueing network $MAP/PH/1/N \rightarrow \bullet PH/1/N \rightarrow \cdots \rightarrow \bullet/PH/1/N$. We use an iterative procedure defined in [15] to compute mean values of response times, end-to-end delays, queue sizes and utilization. The key problem in applying this iterative procedure is the exponential growth of the state space. To get approximate results, one can use departure MAP flows approximations with MAP flows of smaller order [16], or use Monte-Carlo method for numerical solution. In this paper we use the second approach.

For the sake of simplicity, let us assume that arrival flow is Poisson with rate $\lambda = B_T / \bar{\xi}$, where B_T – user traffic bitrate (see the bottom rows of Table. 1), and $\bar{\xi}$ – mean payload bit size.

In the numerical experiment we considered networks with up to ten stations; user traffic arrived either at the first station only, or at each station (cross-traffic). For each case a queueing network was constructed using PH distributions from sets S_1, S_2, S_3 defined above, and real wireless network properties were estimated using the simulation model developed for this experiment.

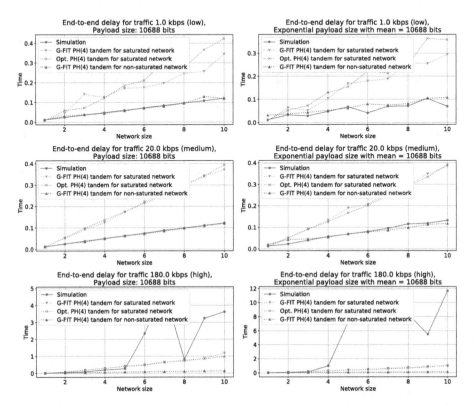

Fig. 5. End-to-end delays in wireless networks with up to 10 stations without cross-traffic. Results estimated with network simulation model and tandem queueing networks with PH distributions fitted for saturated and unsaturated channels.

5.1   Network with Single Traffic Source

Figure 5 shows end-to-end delays estimated with wireless network simulation model and queueing tandem network with PH distributions fitted for various traffic in unsaturated mode, and for saturated mode. It can be seen that the queueing network with PH distributions fitted for unsaturated mode provides very close to simulation results for low or medium traffic. However, in case of heavy user traffic all queueing models fail to provide accurate results for networks with more than four stations; for small networks with up to four stations, queueing networks with PH fitted from saturated channel model provide rather

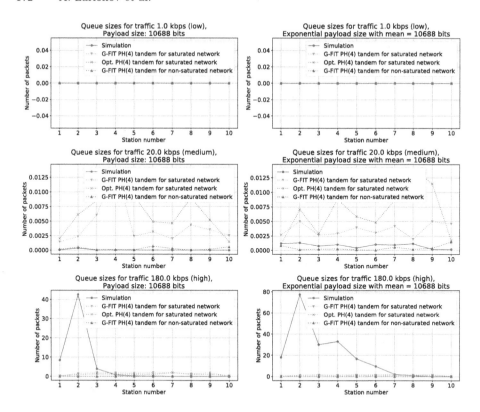

Fig. 6. Average queue sizes in wireless network with 10 stations and single traffic source.

close approximation. It should also be noticed, that queueing networks with PH distributions fitted with moments matching and EM procedure methods provide very close to each other results.

Figures 6 and 7 show the estimation results for queue sizes and utilization (busy) ratios mean values. Under low or medium user traffic tandem queueing networks with PH distributions fitted from unsaturated network provide very accurate results. However, under heavy user traffic the results are not very accurate. Moreover, one of the reasons for such simulation results may be the correlation of service times in the wireless network.

5.2 Network with Cross-Traffic

We also studied the network performance under cross-traffic, when sources at each station generate the same user traffic. Since this kind of network becomes saturated very easy (for instance, the gateway receives 200 kbps when each user generates only 20 kbps traffic), we did not consider heavy traffic case. The estimated end-to-end delays, queue sizes and utilization ratios are shown on Figs. 8, 9 and 10 respectively.

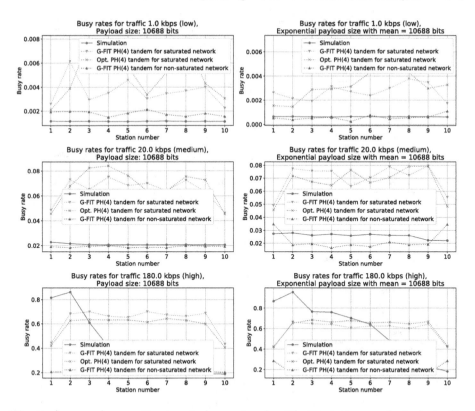

Fig. 7. Average busy ratios in wireless network with 10 stations and single traffic source.

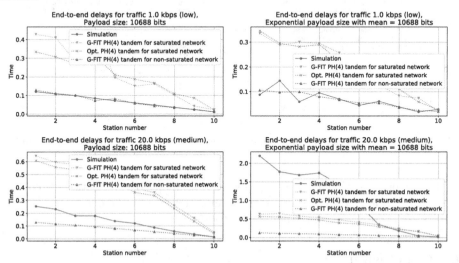

Fig. 8. Average end-to-end delays in wireless network with 10 stations and cross-traffic. Results estimated with network simulation model and tandem queueing networks with PH distributions fitted for saturated and unsaturated channels.

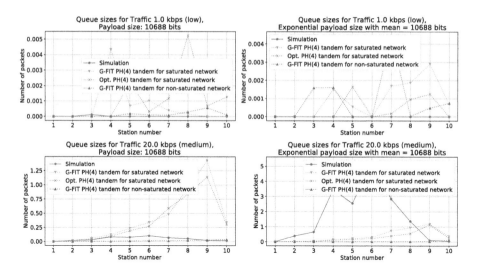

Fig. 9. Average queue sizes in wireless network with 10 stations and cross-traffic.

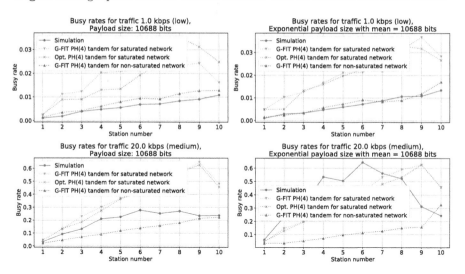

Fig. 10. Average utilization ratios in wireless network with 10 stations and cross-traffic.

It can be seen that considering low user traffic, PH-distribution fitted for unsaturated mode provides accurate results. However, for medium traffic the results are accurate for only the first 3–4 nodes.

6   Conclusion

The paper presented the results of the study of the queueing model $M/PH/1/N \rightarrow \bullet/PH/1/N \cdots \rightarrow \bullet/PH/1/N$ applicability for performance evaluation of a multi-hop wireless network with linear topology, when PH distributions used in the model are fitted from the transmission times in a single-hop collision domain network, operating under saturated or unsaturated mode. For saturated mode a semi-Markov absorbing process based on a model from [2] was used. For unsaturated mode PH distributions were fitted from transmission time samples collected from a simulation model execution.

The results provided in this paper show that queueing networks with PH distributions from a set $\mathcal{S}_3$ (see Sect. 4.3) provide very accurate estimations of end-to-end delays, average queue sizes and nodes busy rates when user traffic is rather low. In the same time, selecting PH distributions from sets $\mathcal{S}_1$ and $\mathcal{S}_2$ fitted from a semi-Markov process using moments matching or EM-procedure provide upper bounds with high error. However, these PH distributions can be used to estimate the network performance processing heavy user traffic, but for a small number of stations only. It should also be noticed that both fitting methods provide very close results, so choosing either fitting method is a question of time complexity: fitting using moments matching method took significantly less time then EM-procedure when only first two moments are considered. Boundaries definitions of various PH distributions applicability for wireless network properties estimation is a subject for future work, as well as more accurate protocol modelling.

All numerical experiments provided in this paper, as well as some additional data, including the fitted PH distributions subgenerators and initial probability distributions are available at GitHub[1].

References

1. Banchs, A., Serrano, P., Azcorra, A.: End-to-end delay analysis and admission control in 802.11 DCF WLANs. Comput. Commun. **29**(7), 842–854 (2006)
2. Bianchi, G.: Performance analysis of the IEEE 802.11 distributed coordination function. IEEE J. Sel. Areas Commun. **18**(3), 535–547 (2000)
3. Chatzimisios, P., Vitsas, V., Boucouvalas, A.: Throughput and delay analysis of IEEE 802.11 protocol. In: Proceedings 3rd IEEE International Workshop on System-on-Chip for Real-Time Applications, pp. 168–174. IEEE (2002)
4. Dai, L., Sun, X.: A unified analysis of IEEE 802.11 DCF networks: stability, throughput, and delay. IEEE Trans. Mob. Comput. **12**(8), 1558–1572 (2013)
5. Dong, L.F., Shu, Y.T., Chen, H.M., Ma, M.D.: Packet delay analysis on IEEE 802.11 DCF under finite load traffic in multi-hop ad hoc networks. Sci. China Ser. F Inf. Sci. **51**(4), 408–416 (2008)
6. Felemban, E., Ekici, E.: Single hop IEEE 802.11 DCF analysis revisited: accurate modeling of channel access delay and throughput for saturated and unsaturated traffic cases. IEEE Trans. Wirel. Commun. **10**(10), 3256–3266 (2011)
7. Haghani, E., Krishnan, M.N., Zakhor, A.: A method for estimating access delay distribution in IEEE 802.11 networks. In: 2011 IEEE Global Telecommunications Conference - GLOBECOM 2011, pp. 1–6. IEEE (December 2011)

---

[1] Experiment source code: https://github.com/larioandr/2019-itmm-paper-model.

8. Hung, F.Y., Marsic, I.: Access delay analysis of IEEE 802.11 DCF in the presence of hidden stations. In: IEEE GLOBECOM 2007–2007 IEEE Global Telecommunications Conference, pp. 2541–2545. IEEE (November 2007)
9. Issariyakul, T., Niyato, D., Hossain, E., Alfa, A.: Exact distribution of access delay in IEEE 802.11 DCF MAC. In: GLOBECOM 2005. IEEE Global Telecommunications Conference, 2005, p. 5. IEEE (2005). pp.-2538
10. Lauwens, B., Scheers, B., Van de Capelle, A.: Performance analysis of unslotted CSMA/CA in wireless networks. Telecommun. Syst. 44(1–2), 109–123 (2010)
11. Sakurai, T., Vu, H.: MAC access delay of IEEE 802.11 DCF. IEEE Trans. Wirel. Commun. 6(5), 1702–1710 (2007)
12. Thummler, A., Buchholz, P., Telek, M.: A novel approach for fitting probability distributions to real trace data with the EM algorithm. In: 2005 International Conference on Dependable Systems and Networks (DSN 2005), pp. 712–721. IEEE (2005)
13. Tickoo, O., Sikdar, B.: Modeling queueing and channel access delay in unsaturated IEEE 802.11 random access MAC based wireless networks. IEEE/ACM Trans. Netw. 16(4), 878–891 (2008)
14. Vardakas, J., Papapanagiotou, I., Logothetis, M., Kotsopoulos, S.: On the end-to-end delay analysis of the IEEE 802.11 distributed coordination function. In: Second International Conference on Internet Monitoring and Protection (ICIMP 2007), pp. 16–16. IEEE (July 2007)
15. Vishnevsky, V., Dudin, A., Kozyrev, D., Larionov, A.: Methods of performance evaluation of broadband wireless networks along the long transport routes. In: Vishnevsky, V., Kozyrev, D. (eds.) DCCN 2015. CCIS, vol. 601, pp. 72–85. Springer, Cham (2016). https://doi.org/10.1007/978-3-319-30843-2_8
16. Vishnevsky, V., Larionov, A., Semenova, O., Ivanov, R.: State reduction in analysis of a tandem queueing system with correlated arrivals. In: Dudin, A., Nazarov, A., Kirpichnikov, A. (eds.) ITMM 2017. CCIS, vol. 800, pp. 215–230. Springer, Cham (2017). https://doi.org/10.1007/978-3-319-68069-9_18

Single Server Queues with Batch Poisson Input and Multiple Types of Outgoing Calls

Anatoly Nazarov[1], Tuan Phung-Duc[2], Svetlana Paul[1(✉)],
and Olga Lizura[1]

[1] Institute of Applied Mathematics and Computer Science, National Research
Tomsk State University, 36 Lenina Avenue, Tomsk 634050, Russia
`nazarov.tsu@gmail.com, paulsv82@mail.ru, oliztsu@mail.ru`
[2] Faculty of Engineering Information and Systems, University of Tsukuba,
1-1-1 Tennodai, Tsukuba, Ibaraki 305-8573, Japan
`tuan@sk.tsukuba.ac.jp`

Abstract. In this paper, we consider a single server queueing model $M^{[n]}|GI|GI|1|L$ with Batch Poisson input flow. Upon arrival, an incoming call from the batch occupies the server, if the server is idle. Other calls from the batch join the orbit and try to occupy the server after an exponentially distributed time. If the server is busy all incoming calls from the batch join the orbit and make a delay for an exponentially distributed time then repeat their request for service. The server makes an outgoing call in its idle time. Our contribution is to derive the stationary probability distribution of the number of incoming calls in the system.

Keywords: Retrial queueing system · Incoming and outgoing calls · Batch Poisson input flow

1 Introduction

Retrial queues arise naturally in many situations in both communication and service systems. They reflect the fact that customers who are blocked upon arrivals might not wait for service but retry for the service in a later time [1,2,8]. In service systems such as call centers, it is common that customers will call again after some random time if they are not served immediately upon arrivals. In random access systems in communication networks, multiple users share a channel and only one user can transmit data at a time. In this case, blocked users will retransmit their data in a random time. The retrial time interval is increased with the number of blockings [11].

On the other hand, the server might utilize the idle time for a secondary service. This also fits the situations in blended call centers where the operators

---

The reported study was funded by RFBR according to the research project 18-01-00277.

A. Dudin et al. (Eds.): ITMM 2019, CCIS 1109, pp. 177–187, 2019.
https://doi.org/10.1007/978-3-030-33388-1_15

serve inbound and outbound calls [5,7]. In particular, the server may make outgoing calls in its idle time. These outgoing calls are not for those in the orbit. In this paper, we consider a model where the server makes outgoing calls to multiple type of customers with distinct service time distributions.

Retrial queues with outgoing calls are also referred to as the models with two-way communications for which a number of works were devoted [3,4,6,9,10, 12,13]. In all of these previous works, the input flow is assumed to be Poisson process with individual arrival. In this paper, we extend the arrival process to batch Poisson process which may be more realistic in most communication and service systems. In the former, users seem to send data in a file which is derived into multiple packets while in service systems, customers often arrive the system in groups. A special case of our model is presented in [13] where a model with multiple type of outgoing calls and Poisson arrival process was considered.

The rest of our paper is organized as follows. In Sect. 2, we present the model and define some notations. Section 3 provides the analysis of the model, deriving the generating functions of the number of jobs in the system. In Sect. 4, we present the concluding remarks.

2   Model Description and Problem Definition

In this paper, we consider a single server queueing model $M^{[n]}|GI|GI|1|L$ with Batch Poisson input flow with the arrival rate λ. An event of the flow causes an arrival of the batch of demands of volume n with probability r_n, $n \geqslant 0$. We will call the demands from the flow as incoming calls.

Upon arrival, an incoming call from the batch occupies the server for an arbitrary distributed time with probability function $B(x)$, if the server is idle. Other calls from the batch join the orbit and try to occupy the server after an exponentially distributed time with rate σ independently. If the server is busy all incoming calls from the batch join the orbit and make a delay for an exponentially distributed time with rate σ then repeat their request for service.

On the other hand, the server makes an outgoing calls after an exponentially distributed idle time. The server makes an outgoing call of type l with rate α_l, $l = \overline{2, L}$ and serves it for an arbitrary distributed time with probability function $V_l(x)$, $l = \overline{2, L}$.

We denote random process $i(t)$ as the number of incoming calls in the system at the moment t. The aim of the current research is to derive the stationary probability distribution of this process.

3   Probability Distribution of the Number of Incoming Calls in Retrial Queue

Random process $k(t)$ is the state of the server at the moment t. This process has the following set of states: 0 if the server is idle, 1 if an incoming call is in service, l if an outgoing call of type l is in service, $l = \overline{2, L}$.

Let $y(t)$ is an elapsed service time at time t.

We denote

$$\mu(x) = \frac{B'(x)}{1 - B(x)}, \quad v_l(x) = \frac{V'(x)}{1 - V(x)}$$

conditional rates of service for incoming and outgoing call under the condition that elapsed service time is x, respectively. Under the current setting the process $\{k(t), i(t)\}$, $k = 0$, $\{k(t), i(t), y(t)\}$, $k = \overline{1, L}$ with variable number of dimensions is a Markov process. Let

$$P_0(i, t) = P\{k(t) = 0, i(t) = i\},$$

$$P_k(i, y, t) = \frac{\partial P\{k(t) = k, i(t) = i, y(t) \leq y\}}{\partial y}, \quad k = \overline{1, L},$$

denote the probability distribution of the system state which is the unique solution of Kolmogorov system of equations. To derive the system we write the following equations

$$P_0(i, t + \Delta t) = P_0(i, t)(1 - \lambda \Delta t)(1 - i\sigma \Delta t)\left(1 - \sum_{l=2}^{L} \alpha_l \Delta t\right)$$

$$+ \int_0^\infty P_1(i + 1, y, t)\mu(y)\Delta t dy + \sum_{l=2}^{L} \int_0^\infty P_l(i, y, t)v_l(y)\Delta t dy + o(\Delta t),$$

$$P_1(i, y + \Delta t, t + \Delta t) = P_1(i, y, t)(1 - \lambda \Delta t)(1 - \mu(y)\Delta t)$$

$$+ \sum_{n=0}^{i} P_1(i - n, y, t)\lambda r_n \Delta t + o(\Delta t),$$

$$\int_0^{\Delta t} P_1(i, x, t)dx = \sum_{n=0}^{i} P_0(i - n, t)\lambda r_n \Delta t + P_0(i, t)i\sigma \Delta t + o(\Delta t),$$

$$P_l(i, y + \Delta t, t + \Delta t) = P_l(i, y, t)(1 - \lambda \Delta t)(1 - v_l(y)\Delta t) +$$

$$\sum_{n=0}^{i} P_l(i - n, y, t)\lambda r_n \Delta t + o(\Delta t), \quad l = \overline{2, L},$$

$$\int_0^{\Delta t} P_l(i, x, t)dx = P_0(i, t)\alpha_l \Delta t + o(\Delta t), \quad l = \overline{2, L}.$$

From these equations we derive the Kolmogorov system of equations for the probability distribution of the system state in stationary regime

$$-\left(\lambda + i\sigma + \sum_{l=2}^{L} \alpha_l\right) P_0(i) + \int_0^\infty P_1(i + 1, y)\mu(y)dy +$$

$$\sum_{l=2}^{L} \int_0^\infty P_l(i, y)v_l(y)dy = 0,$$

$$\frac{\partial P_1(i,y)}{\partial y} = -(\lambda + \mu(y))P_1(i,y) + \lambda \sum_{n=0}^{i} P_1(i-n,y)r_n,$$

$$P_1(i,0) = \lambda \sum_{n=0}^{i} P_0(i-n)r_n + i\sigma P_0(i),$$

$$\frac{\partial P_l(i,y)}{\partial y} = -(\lambda + v_l(y))P_l(i,y) + \lambda \sum_{n=0}^{i} P_l(i-n,y)r_n, \ l = \overline{2,L},$$

$$P_l(i,0) = P_0(i)\alpha_l, \ l = \overline{2,L}. \tag{1}$$

We denote the partial characteristic functions

$$H_0(u) = \sum_{i=0}^{\infty} e^{jui} P_0(i), \ H_1(u,y) = \sum_{i=0}^{\infty} e^{jui} P_1(i,y),$$

$$H_l(u,y) = \sum_{i=0}^{\infty} e^{jui} P_l(i,y), \ r(u) = \sum_{n=1}^{\infty} e^{jun} r_n,$$

where

$$P_1(0,y) \equiv 0,$$
$$r_0 = 0.$$

We rewrite system of Eq. (1) in the following form

$$-\left(\lambda + \sum_{l=2}^{L} \alpha_l\right) H_0(u) + e^{-ju} \int_0^{\infty} H_1(u,y)\mu(y)dy$$

$$+ \sum_{l=2}^{L} \int_0^{\infty} H_l(u,y)v_l(y)dy + j\sigma H_0'(u) = 0,$$

$$\frac{\partial H_1(u,y)}{\partial y} = ((r(u)-1)\lambda - \mu(y))H_1(u,y),$$

$$H_1(u,0) = \lambda r(u)H_0(u) - j\sigma H_0'(u),$$

$$\frac{\partial H_l(u,y)}{\partial y} = ((r(u)-1)\lambda - v_l(y))H_l(u,y), \ l = \overline{2,L},$$

$$H_l(u,0) = \alpha_l H_0(u), \ l = \overline{2,L}. \tag{2}$$

Theorem 1. *The characteristic function of the number of incoming calls in* $M^{[n]}|GI|GI|1|L$ *retrial queue with batch Poisson input and multiple types of outgoing calls has the following form:*

$$H(u)$$
$$= H_0(u)\left(1 + (\lambda r(u) - j\sigma h(u))\frac{1 - B^*(\lambda - \lambda r(u))}{\lambda - \lambda r(u)} + \sum_{l=2}^{L} \alpha_l \frac{1 - V_l^*(\lambda - \lambda r(u))}{\lambda - \lambda r(u)}\right).$$

Here $B^*(u)$ and $V_l^*(u)$ are the Laplas-Stieltjes transforms of the functions $B(x)$ and $V_l(x)$, $l = \overline{2, L}$ respectively. Function $H_0(u)$ is defined by the equations

$$H_0(u) = P_0 \exp \left\{ \int_0^u h(x)dx \right\},$$

$$h(u) = j \frac{\lambda \left(e^{-ju} r(u) B^*(\lambda - \lambda r(u)) - 1 \right) + \sum_{l=2}^{L} \alpha_l \left(V_l^*(\lambda - \lambda r(u)) - 1 \right)}{\sigma \{ 1 - e^{-ju} B^*(\lambda - \lambda r(u)) \}},$$

$$h(0) = j \frac{\lambda}{\sigma} \cdot \frac{r \left(1 + \sum_{l=2}^{L} \alpha_l v_l \right) - (1 - r\lambda b)}{1 - r\lambda b},$$

$$P_0 = \frac{1 - r\lambda b}{1 + \sum_{l=2}^{L} \alpha_l v_l},$$

where b and v_l are the mean values of probability distributions $B(x)$ and $V_l(x)$, $l = \overline{2, L}$; r is the mean number of calls in batch.

Proof. We introduce the notations in the system (2)

$$\begin{aligned} H_1(u, y) &= (1 - B(y)) h_1(u, y), \\ H_l(u, y) &= (1 - V_l(y)) h_l(u, y), \quad l = \overline{2, L}, \end{aligned} \tag{3}$$

in order to obtain the following system of equations

$$- \left(\lambda + \sum_{l=2}^{L} \alpha_l \right) H_0(u) + e^{-ju} \int_0^\infty h_1(u, y) dB(y)$$

$$+ \sum_{l=2}^{L} \int_0^\infty h_l(u, y) dV_l(y) + j\sigma H_0'(u) = 0,$$

$$-B'(y) h_1(u, y) + (1 + B(y)) \frac{\partial h_1(u, y)}{\partial y} =$$

$$\left((r(u) - 1)\lambda - \frac{B'(y)}{1 - B(y)} \right) (1 - B(y)) h_1(u, y), \tag{4}$$

$$h_1(u, 0) = \lambda r(u) H_0(u) - j\sigma H_0'(u),$$

$$-V_l'(y) h_1(u, y) + (1 - V_l(y)) \frac{\partial h_1(u, y)}{\partial y}$$

$$= \left((r(u) - 1)\lambda - \frac{V_l'(y)}{1 - V_l(y)} \right) (1 - V_l(y)) h_l(u, y), \quad l = \overline{2, L},$$

$$h_l(u, 0) = \alpha_l H_0(u), \quad l = \overline{2, L}.$$

From (4) we have

$$-\left(\lambda + \sum_{l=2}^{L} \alpha_l\right) H_0(u) + e^{-ju} \int_0^{\infty} h_1(u,y) dB(y) +$$

$$\sum_{l=2}^{L} \int_0^{\infty} h_l(u,y) dV_l(y) + j\sigma H_0'(u) = 0,$$

$$\frac{\partial h_1(u,y)}{\partial y} = \lambda(r(u)-1) h_1(u,y),$$

$$h_1(u,0) = \lambda r(u) H_0(u) - j\sigma H_0'(u),$$

$$\frac{\partial h_l(u,y)}{\partial y} = \lambda(r(u)-1) h_l(u,y), \ l = \overline{2,L},$$

$$h_l(u,0) = \alpha_l H_0(u), \ l = \overline{2,L}.$$

$$(5)$$

The solutions of the second and fourth equations were obtained as follows.

$$h_1(u,y) = h_1(u,0) \exp\{\lambda(r(u)-1)y\},$$
$$h_l(u,y) = h_l(u,0) \exp\{\lambda(r(u)-1)y\}.$$

$$(6)$$

Substituting (6) into the first equation of the system (5) and denoting the Laplas-Stieltjes transforms as $B^*(u)$ and $V_l^*(u)$ we obtain

$$-\left(\lambda + \sum_{l=2}^{L} \alpha_l\right) H_0(u) + e^{-ju} h_1(u,0) B^*(\lambda - \lambda r(u))$$

$$+ \sum_{l=2}^{L} h_l(u,0) V_l^*(\lambda - \lambda r(u)) + j\sigma H_0'(u) = 0.$$

We transform the last equation

$$-\left(\lambda + \sum_{l=2}^{L} \alpha_l\right) H_0(u) + e^{-ju}\{\lambda r(u) H_0(u) - j\sigma H_0'(u)\} B^*(\lambda - \lambda r(u))$$

$$+ \sum_{l=2}^{L} \alpha_l H_0(u) V_l^*(\lambda - \lambda r(u)) + j\sigma H_0'(u) = 0.$$

$$\left(-\lambda - \sum_{l=2}^{L} \alpha_l + e^{-ju} \lambda r(u) B^*(\lambda - \lambda r(u)) + \sum_{l=2}^{L} \alpha_l V_l^*(\lambda - \lambda r(u))\right) H_0(u) \quad (7)$$

$$+ j\sigma H_0'(u)\{1 - e^{-ju} B^*(\lambda - \lambda r(u))\} = 0.$$

$$j H_0(u) \left(\lambda(e^{-ju} r(u) B^*(\lambda - \lambda r(u)) - 1) + \sum_{l=2}^{L} \alpha_l (V_l^*(\lambda - \lambda r(u)) - 1))\right) =$$

$$\sigma H_0'(u)\{1 - e^{-ju} B^*(\lambda - \lambda r(u))\}.$$

We denote

$$h(u) = j \frac{\lambda(e^{-ju} r(u) B^*(\lambda - \lambda r(u)) - 1) + \sum_{l=2}^{L} \alpha_l (V_l^*(\lambda - \lambda r(u)) - 1)}{\sigma\{1 - e^{-ju} B^*(\lambda - \lambda r(u))\}}, \quad (8)$$

Then we rewrite (7) as follows

$$H_0'(u) = H_0(u)h(u). \tag{9}$$

The solution of Eq. (9) has the following form:

$$H_0(u) = P_0 \exp\left\{ \int_0^u h(x)dx \right\},$$

$$h(u) = j\frac{\lambda(e^{-ju}r(u)B^*(\lambda - \lambda r(u)) - 1) + \sum_{l=2}^{L} \alpha_l(V_l^*(\lambda - \lambda r(u)) - 1)}{\sigma\{1 - e^{-ju}B^*(\lambda - \lambda r(u))\}}. \tag{10}$$

Let us consider Eqs. (3) and (6)

$$H_1(u, y) = (1 - B(y))h_1(u, y) = h_1(u, 0)\exp\{\lambda(r(u) - 1)y\}(1 - B(y)),$$
$$H_l(u, y) = (1 - V_l(y))h_l(u, y) =$$
$$h_l(u, 0)\exp\{\lambda(r(u) - 1)y\}(1 - V_l(y)), \quad l = \overline{2, L}. \tag{11}$$

Integrating Eq. (11) by y we have

$$H_1(u) = \int_0^\infty H_1(u, y)dy = h_1(u, 0)\frac{1 - B^*(\lambda - \lambda r(u))}{\lambda - \lambda r(u)},$$

$$H_l(u) = \int_0^\infty H_l(u, y)dy = h_l(u, 0)\frac{1 - V_l^*(\lambda - \lambda r(u))}{\lambda - \lambda r(u)}, \quad l = \overline{2, L}.$$

Taking (4) into account we obtain

$$H_1(u) = H_0(u)(\lambda r(u) - j\sigma h(u))\frac{1 - B^*(\lambda - \lambda r(u))}{\lambda - \lambda r(u)},$$

$$H_l(u) = \alpha_l H_0(u)\frac{1 - V_l^*(\lambda - \lambda r(u))}{\lambda - \lambda r(u)}, \quad l = \overline{2, L}.$$

The characteristic function of the number of incoming calls in the system can be obtained as follows

$$H(u) = H_0(u) + H_1(u) + \sum_{l=2}^{L} H_l(u)$$

$$= H_0(u)\left(1 + (\lambda r(u) - j\sigma h(u))\frac{1 - B^*(\lambda - \lambda r(u))}{\lambda - \lambda r(u)} + \sum_{l=2}^{L} \alpha_l \frac{1 - V_l^*(\lambda - \lambda r(u))}{\lambda - \lambda r(u)}\right),$$

where function $H_0(u)$ is defined by (10). Let us consider the normalization condition

$$\frac{1 - B^*(\lambda - \lambda r(u))}{\lambda - \lambda r(u)}\bigg|_{u=0} = b = \int_0^\infty x dB(x),$$

$$\frac{1 - V_l^*(\lambda - \lambda r(u))}{\lambda - \lambda r(u)}\bigg|_{u=0} = v_l = \int_0^\infty x dV_l(x),$$

where b and v_l are the mean values of probability distributions $B(x)$ and $V_l(x)$, $l = \overline{2, L}$. On the other hand we have

$$1 = P_0 \left(1 + (\lambda - j\sigma h(0))b + \sum_{l=2}^{L} \alpha_l v_l \right). \tag{12}$$

We consider (10) in order to obtain the function $h(0)$

$$\sigma \left\{ 1 - e^{-ju} B^*(\lambda - \lambda r(u)) \right\} h(u)$$
$$= j\lambda(e^{-ju} r(u) B^*(\lambda - \lambda r(u)) - 1) + j \sum_{l=2}^{L} \alpha_l (V_l^*(\lambda - \lambda r(u)) - 1).$$

Differentiating left and right parts of the equation by u we obtain

$$\left\{ j\sigma e^{-ju} B^*(\lambda - \lambda r(u)) + \sigma e^{-ju} r'(u) \lambda B^{*'}(\lambda - \lambda r(u)) \right\} h(u)$$
$$= j\lambda \left(-je^{-ju} r(u) B^*(\lambda - \lambda r(u)) + e^{-ju} r'(u) B^*(\lambda - \lambda r(u)) \right.$$
$$\left. - \lambda e^{-ju} r(u) r'(u) B^{*'}(\lambda - \lambda r(u)) \right) - j\lambda \sum_{l=2}^{L} \alpha_l r'(u) V_l^{*'}(\lambda - \lambda r(u)). \tag{13}$$

Denoting $r'(u)|_{u=0} = jr$, where r is mean number of calls in the batch, we substitute $u = 0$ in equality (13)

$$\sigma\{1 - r\lambda b\} h(0) = j\lambda \left(r - 1 + r\lambda b + r \sum_{l=2}^{L} \alpha_l v_l \right)$$
$$= j\lambda \left(r\lambda b - 1 + r \left(1 + \sum_{l=2}^{L} \alpha_l v_l \right) \right). \tag{14}$$

Thus,

$$h(0) = j\frac{\lambda}{\sigma} \cdot \frac{r \left(1 + \sum_{l=2}^{L} \alpha_l v_l \right) - (1 - r\lambda b)}{1 - r\lambda b}. \tag{15}$$

From the last equation we obtain

$$P_0 = \frac{1 - r\lambda b}{1 + \sum_{l=2}^{L} \alpha_l v_l}.$$

4   Numerical Example

We fix probability functions $B(x)$ as a gamma distribution with shape parameter s_1 and scale parameter γ_1, $s_1 = \gamma_1 = 2$.

The server makes an outgoing call of type l with rate α_l, $l = \overline{2, 4}$ and serves it for an arbitrary distributed time with probability function $V_l(x)$, $l = \overline{2, 4}$ as a gamma distribution with shape parameter s_l, $l = \overline{2, 4}$ and scale parameter γ_l, $l = \overline{2, 4}$ and $s_2 = \gamma_2 = 0.5$, $s_3 = \gamma_3 = 1.5$, $s_4 = \gamma_4 = 3$.

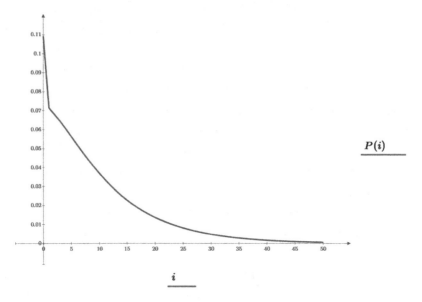

Fig. 1. Probability distribution of the number of incoming calls in the system, $\sigma = 5$

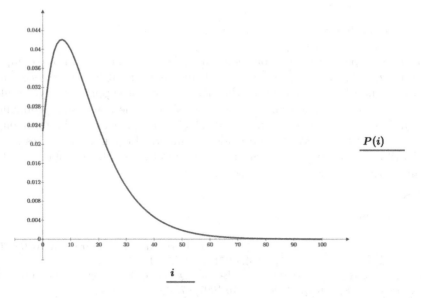

Fig. 2. Probability distribution of the number of incoming calls in the system, $\sigma = 1$.

The arrival rate of the batch Poisson input flow is $\lambda = 0.4$ and probability distribution of the number of customers in the batch is shifted geometric distribution with parameter $q = 0.5$.

Figures 1, 2 and 3 show the probability distribution of the number of incoming calls in the system in cases of $\sigma = 5$, $\sigma = 1$ and $\sigma = 0.2$.

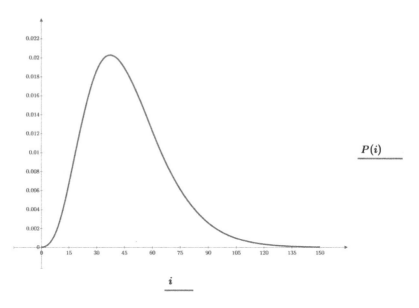

Fig. 3. Probability distribution of the number of incoming calls in the system, $\sigma = 0.2$.

5 Conclusions

In the current paper we have considered retrial queue with batch Poisson input, multiple types of outgoing calls and arbitrary distributions of the service durations and the number of calls in the batch. Using a supplementary variable method we have obtained an explicit expression of the stationary probability distribution of the number of incoming calls in the system and the probability that the server is idle. We have provided a numerical examples for cases where the distribution of service durations is gamma and the distribution of the number of calls in the batch is shifted geometric.

References

1. Artalejo, J.R.: Accessible bibliography on retrial queues: progress in 2000–2009. Math. Comput. Model. **51**(9–10), 1071–1081 (2010)
2. Artalejo, J.R., Gómez-Corral, A.: Retrial Queueing Systems: A Computational Approach (2008). https://doi.org/10.1007/978-3-540-78725-9
3. Artalejo, J.R., Phung-Duc, T.: Markovian retrial queues with two way communication. J. Ind. Manag. Optim. **8**(4), 781–806 (2012)
4. Artalejo, J.R., Phung-Duc, T.: Single server retrial queues with two way communication. Appl. Math. Model. **37**(4), 1811–1822 (2013)
5. Bhulai, S., Koole, G.: A queueing model for call blending in call centers. IEEE Trans. Autom. Control **48**(8), 1434–1438 (2003)
6. Choi, B.D., Choi, K.B., Lee, Y.W.: M/g/1 retrial queueing systems with two types of calls and finite capacity. Queueing Syst. **19**(1–2), 215–229 (1995)

7. Deslauriers, A., L'Ecuyer, P., Pichitlamken, J., Ingolfsson, A., Avramidis, A.N.: Markov chain models of a telephone call center with call blending. Comput. Oper. Res. **34**(6), 1616–1645 (2007)
8. Falin, G., Templeton, J.G.: Retrial Queues, vol. 75. CRC Press, Boca Raton (1997)
9. Falin, G.: Model of coupled switching in presence of recurrent calls. Eng. Cybern. **17**(1), 53–59 (1979)
10. Falin, G., Artalejo, J.R., Martin, M.: On the single server retrial queue with priority customers. Queueing Syst. **14**(3–4), 439–455 (1993)
11. Fiems, D., Phung-Duc, T.: Light-traffic analysis of random access systems without collisions. Ann. Oper. Res. **277**(3), 311–327 (2019)
12. Phung-Duc, T., Rogiest, W., Takahashi, Y., Bruneel, H.: Retrial queues with balanced call blending: analysis of single-server and multiserver model. Ann. Oper. Res. **239**(2), 429–449 (2016)
13. Sakurai, H., Phung-Duc, T.: Two-way communication retrial queues with multiple types of outgoing calls. Top **23**(2), 466–492 (2015). https://doi.org/10.1007/s11750-014-0349-5

Refined Approximate Algorithm for Steady-State Probabilities of the Large Scale Queuing Systems with Instantaneous and Delayed Feedback

A. Melikov[1]($\boxtimes$) and S. Aliyeva[2]

[1] Institute of Control Systems, National Academy of Science of Azerbaijan,
Baku, Azerbaijan
agassi.melikov@gmail.com
[2] Baku State University, Baku, Azerbaijan
s@aliyeva.info

Abstract. Mathematical models of the multichannel queueing systems with both instantaneous and delayed feedbacks are proposed. Probabilities of leaving the system, instantaneous returning to channels and entering to orbit depend on the number of calls in orbit. Both models with finite and infinite size of orbit are investigated. Refined approximate algorithms to calculate the steady-state probabilities as well as quality of service (QoS) metrics of the system are developed. Execution time of the proposed algorithms is negligible. Results of numerical experiments are given.

Keywords: Queuing system · Feedback calls · Numerical analysis

1 Introduction

Queuing systems with feedback are adequate mathematical models of many real situations in which part of already serviced calls return to the system to get additional service. Among models of queuing systems with feedback two kinds of models should be distinguished: (1) models with instantaneous feedback (i.e. models without orbit) and (2) models with delayed feedback (i.e. models with orbit). In the literature both kinds of models have been investigated separately. Note that first papers devoted to both kinds of feedback were published by Takacs in [1] and [2]. In future the models of queuing systems with single node and instantaneous and delayed feedback were investigated in [3–9] and [10–17] correspondingly. In [18–22] tandem open queueing networks with instantaneous feedback are used for the analysis of integrated cellular communication network and wireless local area network. A similar queuing network model with delayed feedback is presented in [23] to analyze the call center, where it is assumed that there is no place for waiting before first node of the network and the size of intermediate buffer between nodes is limited. Models of queuing systems with

© Springer Nature Switzerland AG 2019
A. Dudin et al. (Eds.): ITMM 2019, CCIS 1109, pp. 188–201, 2019.
https://doi.org/10.1007/978-3-030-33388-1_16

simultaneously instantaneous and delayed feedback have been investigated in recent papers [24–26]. In last papers detailed review of indicated above work might be found. In this paper, the Markov models of multi-channel queuing systems with instantaneous and delayed feedback are examined. Following the above mentioned works it is assumed that primary and feedback calls have the same channel holding times. However unlike the known works here it is assumed that both departure and feedback probabilities are state-dependent. Moreover, we consider models with linear but not constant retrial rate from the orbit which has either finite or infinite size. These assumptions essentially broaden applicability of their results in real systems.

The main task in the analysis of the investigated queuing systems with feedback is to find steady-state probabilities of the appropriate two-dimensional Markov Chains (2D MC) with large number of states. In known works to solve this problem mainly matrix-geometric method (MGM) [27] and spectral expansion method (SEM) [28], are used. However, in order to apply these methods some unrealistic assumptions should be accepted. For example, usually it is assumed that retrial rate from the orbit is constant, i.e. it is independent of the number of calls in the orbit. Unlike the known works investigated here 2D MC represent Level Dependent Quasi Birth-Death (LDQBD) process.

Note that an approach based on the system of equilibrium equations (SEE) for calculate the steady-state probabilities becomes inefficient for the large scale models. So, developing efficient methods for approximate analysis of models with a large number of channels and large size of orbit is highly desired. Below we develop space merging algorithms (SMA) to approximate calculate the steady-state probabilities of the queuing models with instantaneous and delayed feedback. Computation time which is needed by SMA is almost zero since it executed by explicit formulas many of which are even tabulated. Numerical results demonstrate high accuracy of the proposed method.

The rest of this paper is organized as follows. The description of the model with state-dependent feedback probabilities is presented in Sect. 2. Exact and approximate methods to calculate the QoS metrics are developed in Sect. 3. The results of numerical experiments performed by using the developed SMA are demonstrated in Sect. 4. Conclusion remarks are given in Sect. 5.

2 Description of the Model with Feedback

The system has $N > 1$ independent and identical channels and accepts Poisson flow of primary calls (p-calls) with λ intensity. After serving p-calls it either (1) leaving the system with probability $\sigma_1(x)$, either (2) instantaneously repeated service with probability $\sigma_2(x)$ (instantaneous feedback) or (3) it goes to the orbit with probability $\sigma_3(x) = 1 - \sigma_1(x) - \sigma_2(x)$ in order for repeated call after a random time (delayed feedback). These probabilities depend on the state x of a random environment, $x \in X$, where X is the set of possible states of the random environment. It is assumed that there is no limit on the number of repetition of calls, i.e. p-call and retrial call (r-call) can be repeated to be serviced arbitrary times.

Delayed r-calls forms orbit with the maximum size R, $0 < R \leq \infty$. The limitation on the size of the orbit (i.e. case $R < \infty$) means that the call can enter orbit if total number of retrial calls are less than R at the time of arrival; otherwise it leaves the system.

Delayed r-calls require repeated service at random times, which are subject to exponential distribution with mean $1/\eta$. If at the time of generation of delayed r-call has at least one free channel in the system, then it is accepted to service. Distribution functions of channel occupancy time of primary and repeated calls are assumed be independent and exponential with same parameter $1/\mu$.. It is expected that delayed r-calls are persistent, i.e. if all the channels are busy at the moment of arrival of delayed r-calls, then they either leave the system with probability α_j,or with probability $1 - \alpha_j$ return to the orbit, where j is the current number of calls in orbit, $j = 1, 2, ..., R$.

It is assumed that the state of a random environment is determined by number of r-call in orbit, i.e. set of all possible states of the random environment is defined as $X = \{0, 1, 2, ..., R\}$. Obviously, in case $R < \infty$ we have $\sigma_3(R) = 0$.

The problem is finding the joint probability distribution of the number of calls in system and number of repeated calls in the orbit. Determination of the indicated probability distribution allows calculate the desired QoS metrics as well.

3   Exact and Approximate Methods to Calculate the Steady-State Probabilities

First consider model with finite size of orbit, i.e. $R < \infty$. State of the system is defined by the two-dimensional (2D) vector (i, j), where i is the total number of calls (primary and repeated) in the channels, $i = 0, 1, ..., N$, and j indicates the number of repeated calls in the orbit $j = 0, 1, ..., R$. Based on the distribution function of the random variables involved in the formation of the model, we determine that the studied system is described by the two-dimensional Markov chain (2D MC). The set of all possible states of the system, i.e., state space of given 2D MC is defined as $S = \{0, 1, ..., N\} \times \{0, 1, ..., R\}$. The transition rate from the state $(i, j) \in S$ to the state $(m, n) \in S$ is denoted as $q((i, j), (m, n))$. The combination of these quantities involves Q-matrix of given 2D MC and are determined from the following relations:

For the case $0 \leq i \leq N - 1$:

$$q((i, j), (m, n)) = \begin{cases} \lambda, & \text{if } (m, n) = (i + 1, j), \\ i\mu\sigma_1(j), & \text{if } (m, n) = (i - 1, j), \\ i\mu\sigma_3(j), & \text{if } j < R, (m, n) = (i - 1, j + 1), \\ j\eta, & \text{if } (m, n) = (i + 1, j - 1), \\ 0 & \text{in other cases.} \end{cases} \qquad (1)$$

For the case $i = N$:

$$q\left((N,j),(m,n)\right) = \begin{cases} N\mu\sigma_1\left(j\right), & \text{if } (m,n) = (N-1,j), \\ N\mu\sigma_3\left(j\right), & \text{if } j < R, (m,n) = (N-1,j+1), \\ j\eta\alpha_j, & \text{if } (m,n) = (N,j-1), \\ 0 & \text{in other cases.} \end{cases} \quad (2)$$

As we seen from (1) and (2) the given finite 2D MC represent level dependent quasi birth-death (LDQBD) process and it is an irreducible. In other words, there exist its stationary steady-state probabilities. Let $p(i,j)$ means the stationary probability of state $(i,j) \in S$. These probabilities are determined by solving the system of equilibrium equations (SEE), which is compiled on the basis of (1) and (2). Due to the obviousness of the SEE it is not provided here. It is easy to derive desired QoS metrics via steady-state probabilities. There are several general QoS metrics, some of which are listed below:

• The loss probability of primary calls (P_p) is given by

$$P_p = \sum_{j=0}^{R} p\left(N,j\right) . \quad (3)$$

• The loss probability of repeated calls (P_r) is expressed as

$$P_r = \sum_{j=1}^{R} p\left(N,j\right)\alpha_j . \quad (4)$$

• The average number of busy channels (N_{av}) is given by

$$N_{av} = \sum_{i=1}^{N} i \sum_{j=0}^{R} p\left(i,j\right) . \quad (5)$$

• The average number of the repeated calls in orbit (L_0) is expressed as

$$L_o = \sum_{j=1}^{R} j \sum_{i=0}^{N} p\left(i,j\right) . \quad (6)$$

The dimension of SEE determined by the dimension of the state space S, which is defined as $(N+1)(R+1)$. Unfortunately, it's difficult to find the analytical solution of this system of equations. Therefore, it's required to use numerical methods of linear algebra for its solution. The known methods allow an accurate study the behavior QoS metrics (3)-(6) with respect to structural changes and load parameters of the models with moderate dimensions. With the growing dimension of the state space S, these methods are faced computational difficulties. To eliminate them, we use the SMA to calculate the stationary distribution of 2D MC (see [24–26]).

For correct use of this method, below we assume that $\sigma_3\left(j\right) \ll \sigma_1\left(j\right)+\sigma_2\left(j\right)$. It is important to note that the probabilities $\sigma_3\left(j\right)$ are not so small that they can be ignored. Otherwise we can just add the rate of r-calls to the p-calls rate and further investigate the model without r-calls.

Consider the following splitting of state space S:

$$S = \bigcup_{j=0}^{R} S_j \ , \ S_i \bigcap S_j = \emptyset \, , i \neq j, \tag{7}$$

where $S_j = \{(i, j) \in S : i = 0, 1, ..., N\} \, , \ j = 0, 1, ..., R.$

Merge function on the state space S is determined on the basis of the splitting (7) as follows:

$$U((i, j)) = < j >, \tag{8}$$

where $< j >$ is a merge state, which includes all the states of the S_j. Let $\Omega = \{< j >: j = 0, 1, ..., R\}$.

The approximate values of steady-state probabilities of the initial model are defined as follows:

$$\tilde{p}(i, j) \approx \rho_j(i) \pi(< j >) \, , \tag{9}$$

Where $\rho_j(i)$ denotes the state probability of (i, j) within the splitting model with state space S_j, and $\pi(< j >)$ is the probability of the merge state $< j > \in \Omega$.

From splitting scheme (7) it's clear that all the splitting models are one-dimensional birth and death processes (1D BDP), so that in the class of states S_j the second component is constant. Therefore, in the study of the splitting model with state space S_j microstate $(i, j) \in S$ can be representing by scalar$i, i = 0, 1, ..., N$. The transition intensity between states i and k in the splitting model with state space S_j is denoted by $q_j(i, k) \, , i, k = 0, 1, ..., N$. From (1) and (2) we get that these parameters are defined as follows:

$$q_j(i, k) = \begin{cases} \lambda, & \text{if } k = i + 1, \\ i\mu\sigma_1(j), & \text{if } k = i - 1, \\ 0 & \text{in other cases.} \end{cases} \tag{10}$$

From (10) we conclude that the state probabilities within the splitting model with state space S_j are defined as follows:

$$\rho_j(i) = \frac{\nu_j^i}{i!} \rho_j(0) \, , \ i = 1, 2, ..., N, \tag{11}$$

where $\nu_j = \lambda / \mu\sigma_1(j) \, , \rho_j(0)$ is derived from normalizing condition, i.e. $\sum_{i=0}^{N} \rho_j(i) = 1$.

The transition intensity from the merge state $< i >$ to other merge state $< j >$ is denoted as $q(< i >, < j >) \, , < i >, < j > \in \Omega$. After certain algebras on the bases of (1), (2) and (11) we obtain:

$$q(< i >, < j >) = \begin{cases} \Lambda_i, & \text{if } j = i + 1, \\ i\Psi_i, & \text{if } j = i - 1, \\ 0 & \text{in other cases,} \end{cases} \tag{12}$$

where $\Lambda_i = \mu\sigma_3(i)\sum_{k=1}^{N} k\rho_i(k)$, $i = 0, 1, ..., R - 1$; $\Psi_i = \eta(1 - (1 - \alpha_i)\rho_i$ $(N))$, $i = 1, 2, .., R$. From (12) we conclude that the probabilities of the merging states $\pi(<j>)$, $<j>\in \Omega$, are calculated as the state probabilities of 1-D BDP. In other words,

$$\pi(<j>) = \frac{1}{j!} \prod_{i=1}^{j} \frac{\Lambda_{i-1}}{\Psi_i}\pi(<0>), \, j = 1, ..., R, \tag{13}$$

where $\pi(<0>)$ is derived from normalizing condition, i.e., $\sum_{j=0}^{R}\pi(<j>) = 1$. Therefore, taking into account the relations (11) and (13) from (9) we calculate the steady-state probabilities of the initial 2D MC. After certain algebras we obtain the following approximate formulas for calculating the desired QoS metrics of the system:

$$P_p \approx \sum_{i=0}^{R} \rho_i(N)\pi(<i>); \tag{14}$$

$$P_r \approx \sum_{i=1}^{R} \rho_i(N)\pi(<i>)\alpha_i; \tag{15}$$

$$N_{av} \approx \sum_{k=1}^{N} k \sum_{i=0}^{R} \rho_i(k)\pi(<i>); \tag{16}$$

$$L_o \approx \sum_{i=1}^{R} i\pi(<i>). \tag{17}$$

Special Case. Let the probabilities $\sigma_i(j)$, $i = 1, 2, 3$ and α_j are constants, i.e. they don't depend on the number of calls in orbit j, $j = 0, 1, ..., R$. In this case the above formulas (10)-(17) are getting more simplified. Thus, in this case the state probabilities within all splitting models coincide with state probabilities of classical Erlang's model $M/M/N/N$ with load$\nu = \lambda/\mu\sigma_1$, i.e.

$$\rho(i) = \frac{\nu^i}{i!} \bigg/ \sum_{j=0}^{N} \frac{\nu^j}{j!}, \, i = 1, 2, ..., N . \tag{18}$$

By taking into account (1), (2), (18) after certain algebras, we obtain:

$$q(<i>, <j>) = \begin{cases} \Lambda, & \text{if } j = i + 1, \\ i\Psi, & \text{if } j = i - 1, \\ 0 & \text{in other cases,} \end{cases} \tag{19}$$

where $\Lambda = \lambda\frac{\sigma_3}{\sigma_1}(1 - E_B(v, N))$; $\Psi = \eta(1 - (1 - \alpha)E_B(v, N))$. Here and below $E_B(v, N)$ indicate the Erlang's loss formula in $M/M/N/N$ with load ν, i.e. $E_B(\nu, N) = \rho(N)$.

Consequently, in this case, the probabilities of the merging states are defined as follows:

$$\pi(<j>) = \frac{(\Lambda/\Psi)^j}{j!}\pi(<0>), \, j = 0, 1, ..., R. \tag{20}$$

Then the QoS metrics are defined as follows:

$$P_p \approx E_B\left(\nu, N\right); \tag{21}$$

$$P_r \approx \alpha E_B\left(\nu, N\right)\left(1 - \pi\left(<0>\right)\right); \tag{22}$$

$$N_{av} \approx \nu\left(1 - E_B\left(\nu, N\right)\right); \tag{23}$$

$$L_o \approx \pi\left(<0>\right)\sum_{i=1}^{R}\frac{\left(\Lambda/\Psi\right)^i}{\left(i-1\right)!}. \tag{24}$$

From the formulas (21) and (23) we conclude that the loss probability of the primary calls and average number of busy channels doesn't depend on the rate of the retrial calls (see Eq. (20)). These metrics also does not depend on the size of the orbit. These facts are explained by the fact that the approximate formulas are based on the assumption that the probability of going to orbit essentially less than sum of other two probabilities, i.e. rate of primary calls significantly exceeds the rate of retrial calls from orbit. However, these QoS metrics depends on the probability of leaving of calls from the system; therefore, they depend on the probability of entering of the primary calls to the orbit after finishing the service. In other words, these metrics are indirectly dependent on the size of the orbit. It is important to note that in this case it is possible to obtain explicit formulas for the model with an infinite size of the orbit, i.e., if $R = \infty$ then from (20) we find that $\pi\left(<0>\right) = e^{-\Lambda/\Psi}$. Therefore, in the model with infinite size of the orbit and linear retrial rate the QoS metrics are determined by the following simple formulas:

$$P_r \approx \alpha E_B\left(\nu, N\right)\left(1 - e^{-\Lambda/\Psi}\right); \tag{25}$$

$$L_o \approx \Lambda/\Psi. \tag{26}$$

Let us now assume that in model with finite size of orbit retrial rate is constant, i.e. retrial rate is independent of the number of calls in the orbit, i.e below we assume that retrial rate is not linear as it was accepted above (see formulas (1) and (2)). Then the state probabilities within all splitting models are calculated by (18) and the transition rates between merged states are calculated similar to (19) but in (19) we should take into account that in the right site the coefficient of Ψ is equal to 1. So, in this case, for existence the stationary probabilities of the merging states the ergodicy condition $\Lambda < \Psi$ is required. Under satisfying ergodicy condition the state probabilities of merged model coincide with state probabilities of classical Erlang's model $M/M/1/\infty$ with load Λ/Ψ, i.e.

$$\pi\left(<j>\right) = \left(\frac{\Lambda}{\Psi}\right)^j\left(1 - \frac{\Lambda}{\Psi}\right), j = 0, 1, \ldots \tag{27}$$

Note 1. The ergodicy condition $\Lambda < \Psi$ might be substitute by following simple checkable but rough condition: $\lambda \frac{\sigma_3}{\sigma_1} < \eta$.

The QoS metrics P_p, P_r and N_{av} in this case are calculated by (21)-(23) while average number of calls in the orbit is defined as follows:

$$L_o \approx \frac{\Lambda}{\Psi - \Lambda}. \tag{28}$$

Under indicated above assumptions related to constancy of probabilities $\sigma_i(j)$, α_j application of the Spectral Expansion Method (SEM) to model with finite size of orbit looks like following.

First, we define matrix A_j for purely lateral transitions ($n = j$):

$$A_j(i,m) = \begin{cases} \lambda & \text{if } m = i+1, \ i \leq N-1, \\ i\mu\sigma_1 & \text{if } m = i-1, \\ 0 & \text{otherwise}. \end{cases} \tag{29}$$

Let's define matrix B_j for one-step upward transitions ($n = j+1$):

$$B_j(i,m) = \begin{cases} i\mu\sigma_3 & \text{if } m = i-1, \\ 0 & \text{otherwise}. \end{cases} \tag{30}$$

Let's define matrix C_j for one-step downward transitions ($n = j-1$):

$$C_j(i,m) = \begin{cases} \eta & \text{if } m = i+1, \ i \leq N-1, \\ \eta\alpha & \text{if } m = i = N, \\ 0 & \text{otherwise}. \end{cases} \tag{31}$$

According to (29)-(31) we have $A_j = A$, $B_j = B$, $C_j = C$ for any j. This means that we could choose SEM threshold parameter M arbitrary quantity such that $1 < M < R$.

We define matrices D^A, D^B, D^C with the element (m,m) as m^{th} row sum of the corresponding matrices A, B and C.

Let us introduce vectors $v_j = (p(0,j), p(1,j), ..., p(N,j))$, $j = 0, 1, ..., R$. Then steady-state balance equations will be as follows:

$$v_j(D^A + D^B + D^C) = v_{j-1}B + v_jA + v_{j+1}C, \tag{32}$$

$$\sum_{j=0}^{R} v_j e = 1, \tag{33}$$

where e is unit column vector of size $N+1$ and $v_{-1} = \mathbf{0}$.

From (32) matrix-difference equation of second order is derived:

$$v_jQ_0 + v_{j+1}Q_1 + v_{j+2}Q_2 = 0, \tag{34}$$

Where $Q_0 = B$, $Q_1 = A - D^A - D^B - D^C$, $Q_2 = C$. It is assumed that both matrices Q_0 and Q_2 has full rang and there exists their inverses. Based on (34) the following characteristic matrix polynomial of second order is obtained:

$$Q(\lambda) = Q_0 + Q_1\lambda + Q_2\lambda^2.$$

According to SEM we conclude that vectors v_j, $j = 0, 1, ..., R$, might be represents as follows:

$$v_j = \sum_{k=0}^{N} a_k \psi_k \lambda_k^{j+1}, \tag{35}$$

or equivalently

$$p(i,j) = \sum_{k=0}^{N} a_k \psi_k(i) \lambda_k^{j+1}, \ j = M - 1, ..., R,$$

Where (ψ_k, λ_k) be eigenvector and eigenvalue pairs of matrix $Q(\lambda)$. It is assumed that $|\lambda_k| < 1$ and unknown parameters a_k, $k = 0, 1, ..., N$, are calculated by using some recurrence procedure based on (32) and (35).

4   Numerical Results

This section has two-fold purpose. Firstly, we illustrate the high accuracy of developed algorithms and secondly, we show that execution time of these algorithms is less than appropriate algorithms based on original spectral expansion method. Consider the model with finite size of orbit. Below in numerical experiments, we choose the following parameters: $\sigma_1 = 0.5$, $\sigma_2 = 0.3$, $\alpha = 0.8$. The exact values (EV) of steady-state probabilities and performance measures are calculated by using SEE. The accuracy of the developed SMA to calculation of the steady-state probabilities is estimated by following norms: Maximum absolute difference:

$$\|N\|_1 = \max_{n \in E} |p(n) - \tilde{p}(n)|. \tag{36}$$

Cosine similarity:

$$\|N\|_2 = \frac{\sum_{n \in E} p(n)\tilde{p}(n)}{\left(\sum_{n \in E} (p(n))^2\right)^{\frac{1}{2}} \left(\sum_{n \in E} (\tilde{p}(n))^2\right)^{\frac{1}{2}}}. \tag{37}$$

Table 1. Estimation of accuracy of SMA to calculate the steady-state probabilities versus various norms.

(N, R)	(λ, η)	μ	Norms	
			(36)	(37)
(4,2)	(55,30)	15	0.0413	0.9939
	(60,40)	20	0.0380	0.9944
	(65,50)	25	0.0344	0.9951
(4,3)	(55,30)	15	0.0408	0.9939
	(60,40)	20	0.0375	0.9944
	(65,50)	25	0.0341	0.9951
(4,4)	(55,30)	15	0.0407	0.9939
	(60,40)	20	0.0375	0.9944
	(65,50)	25	0.0340	0.9951
(5,2)	(55,30)	15	0.0364	0.9932
	(60,40)	20	0.0312	0.9944
	(65,50)	25	0.0265	0.9957
(5,3)	(55,30)	15	0.0356	0.9932
	(60,40)	20	0.0306	0.9945
	(65,50)	25	0.0261	0.9957
(5,4)	(55,30)	15	0.0355	0.9932
	(60,40)	20	0.0305	0.9945
	(65,50)	25	0.0260	0.9957
(6,2)	(55,30)	15	0.0300	0.9933
	(60,40)	20	0.0235	0.9954
	(65,50)	25	0.0185	0.9969
(6,3)	(55,30)	15	0.0290	0.9934
	(60,40)	20	0.0229	0.9955
	(65,50)	25	0.0180	0.9970
(6,4)	(55,30)	15	0.0289	0.9934
	(60,40)	20	0.0228	0.9955
	(65,50)	25	0.0180	0.9970

The comparison results of the steady-state probabilities and performance measures are given in Tables 1 and 2 correspondingly. We conclude from these tables that the accuracy of the SMA is very high. We also compare the results obtained

Table 2. Estimation of accuracy of SMA to calculate performance measures, EV – Exact Value, AV – Approximate Value.

(N, R)	(λ, η)	μ	P_p		P_r		N_{av}		L_o	
			EV	AV	EV	AV	EV	AV	EV	AV
(4, 2)	(55, 30)	15	0.5017	0.5441	0.0886	0.1342	3.2648	3.3429	0.3393	0.3569
	(60, 40)	20	0.4321	0.4696	0.0690	0.1098	3.1071	3.1826	0.3199	0.3359
	(65, 50)	25	0.3808	0.4138	0.0562	0.0925	2.9769	3.0482	0.3046	0.3191
(4, 3)	(55, 30)	15	0.5009	0.5441	0.0892	0.1360	3.2625	3.3429	0.3501	0.3728
	(60, 40)	20	0.4315	0.4696	0.0695	0.1111	3.1052	3.1826	0.3289	0.3495
	(65, 50)	25	0.3804	0.4138	0.0566	0.0935	2.9752	3.0482	0.3123	0.3309
(4, 4)	(55, 30)	15	0.5009	0.5441	0.0893	0.1361	3.2623	3.3429	0.3511	0.3749
	(60, 40)	20	0.4315	0.4696	0.0695	0.1113	3.1050	3.1826	0.3296	0.3511
	(65, 50)	25	0.3803	0.4138	0.0566	0.0936	2.9751	3.0482	0.3129	0.3322
(5, 2)	(55, 30)	15	0.4053	0.4439	0.0831	0.1257	3.9906	4.0784	0.4022	0.4186
	(60, 40)	20	0.3291	0.3604	0.0608	0.0960	3.7572	3.8376	0.3760	0.3900
	(65, 50)	25	0.2753	0.3009	0.0469	0.0760	3.5633	3.6354	0.3552	0.3670
(5, 3)	(55, 30)	15	0.4043	0.4439	0.0839	0.1279	3.9867	4.0784	0.4203	0.4433
	(60, 40)	20	0.3285	0.3604	0.0614	0.0975	3.7540	3.8376	0.3909	0.4103
	(65, 50)	25	0.2749	0.3009	0.0473	0.0771	3.5607	3.6354	0.3679	0.3843
(5, 4)	(55, 30)	15	0.4042	0.4439	0.0840	0.1281	3.9863	4.0784	0.4224	0.4471
	(60, 40)	20	0.3284	0.3604	0.0614	0.0976	3.7537	3.8376	0.3925	0.4132
	(65, 50)	25	0.2749	0.3009	0.0473	0.0772	3.5605	3.6354	0.3691	0.3866
(6, 2)	(55, 30)	15	0.3192	0.3517	0.0731	0.1100	4.6622	4.7542	0.4566	0.4707
	(60, 40)	20	0.2412	0.2649	0.0495	0.0773	4.3320	4.4105	0.4226	0.4336
	(65, 50)	25	0.1891	0.2068	0.0357	0.0568	4.0588	4.1245	0.3954	0.4041
(6, 3)	(55, 30)	15	0.3182	0.3517	0.0741	0.1123	4.6568	4.7542	0.4831	0.5045
	(60, 40)	20	0.2406	0.2649	0.0502	0.0787	4.3278	4.4105	0.4441	0.4608
	(65, 50)	25	0.1888	0.2068	0.0361	0.0577	4.0556	4.1245	0.4135	0.4265
(6, 4)	(55, 30)	15	0.3180	0.3517	0.0742	0.1126	4.6560	4.7542	0.4868	0.5105
	(60, 40)	20	0.2405	0.2649	0.0503	0.0789	4.3273	4.4105	0.4468	0.4651
	(65, 50)	25	0.1888	0.2068	0.0362	0.0578	4.0552	4.1245	0.4155	0.4298

by SEM and SMA for the model with infinite orbit size (see Table 3). It is clear from Table 3 that results are very close. Additionally, the SMA is more computationally efficient than the SEM, as the SEM algorithm involves finding of eigen value/vectors and solving of the system of linear equation.

Table 3. The comparison of results obtained by SEM and SMA for the model with infinite orbit size.

(N, μ)	(λ, η)	P_p		P_r		N_{av}		L_o	
		SEM	SMA	SEM	SMA	SEM	SMA	SEM	SMA
(4, 25)	(20, 10)	0.0540	0.0565	0.0311	0.0345	1.4887	1.5097	2.9318	3.2274
	(30, 15)	0.1281	0.1387	0.0652	0.0786	2.0142	2.0671	2.0839	2.4328
	(40, 20)	0.2069	0.2281	0.0916	0.1181	2.3914	2.4699	1.5383	1.8329
(4, 30)	(20, 10)	0.0340	0.0351	0.0203	0.0218	1.2744	1.2865	3.2608	3.4916
	(30, 15)	0.0893	0.0952	0.0486	0.0562	1.7728	1.8095	2.4692	2.8148
	(40, 20)	0.1546	0.1687	0.0751	0.0929	2.1542	2.2168	1.8720	2.2078
(4, 40)	(20, 10)	0.0151	0.0154	0.0093	0.0097	0.9802	0.9846	3.6363	3.7647
	(30, 15)	0.0461	0.0480	0.0269	0.0295	1.4107	1.4281	3.0551	3.3292
	(40, 20)	0.0893	0.0952	0.0486	0.0562	1.7728	1.8095	2.4692	2.8148
(5, 25)	(20, 10)	0.0174	0.0177	0.0106	0.0112	1.5637	1.5716	3.5823	3.7304
	(30, 15)	0.0591	0.0624	0.0335	0.0379	2.2180	2.2502	2.8501	3.1588
	(40, 20)	0.1172	0.1274	0.0600	0.0730	2.7289	2.7923	2.1804	2.5252
(5, 30)	(20, 10)	0.0092	0.0093	0.0057	0.0059	1.3173	1.3210	3.7676	3.8555
	(30, 15)	0.0353	0.0367	0.0209	0.0228	1.9084	1.9266	3.2304	3.4711
	(40, 20)	0.0773	0.0825	0.0424	0.0493	2.4039	2.4465	2.6072	2.9414
(5, 40)	(20, 10)	0.0031	0.0031	0.0019	0.0020	0.9960	0.9969	3.9186	3.9514
	(30, 15)	0.0139	0.0142	0.0086	0.0090	1.4727	1.4787	3.6578	3.7823
	(40, 20)	0.0353	0.0367	0.0209	0.0228	1.9084	1.9266	3.2304	3.4711
(6, 25)	(20, 10)	0.0047	0.0047	0.0029	0.0030	1.5902	1.5925	3.8752	3.9257
	(30, 15)	0.0236	0.0244	0.0142	0.0153	2.3263	2.3415	3.4475	3.6368
	(40, 20)	0.0599	0.0636	0.0336	0.0386	2.9547	2.9964	2.8349	3.1451
(6, 30)	(20, 10)	0.0020	0.0021	0.0013	0.0013	1.3297	1.3306	3.9436	3.9673
	(30, 15)	0.0119	0.0121	0.0073	0.0077	1.9690	1.9758	3.7007	3.8134
	(40, 20)	0.0340	0.0354	0.0200	0.0220	2.5495	2.5723	3.2499	3.4881
(6, 40)	(20, 10)	0.0005	0.0005	0.0003	0.0003	0.9993	0.9995	3.9856	3.9918
	(30, 15)	0.0035	0.0035	0.0022	0.0023	1.4931	1.4947	3.9051	3.9441
	(40, 20)	0.0119	0.0121	0.0073	0.0077	1.9690	1.9758	3.7007	3.8134

5 Conclusion

In this paper, mathematical models of multi-channel queuing system with instantaneous and delayed feedbacks are proposed. The probabilities of the return of calls to immediate repeated service or going to orbit depends on the number of calls in orbit. Models with finite and infinite size of the orbit for repeated calls were investigated. Exact and approximate methods of calculating the steady-state probabilities as well as QoS metrics of the given model were developed. High accuracy of the developed approximate formulas is shown by numerical experiments. Comparisons of the developed method and spectral expansion method are shown.

References

1. Takacs, L.: A single-server queue with feedback. Bell Syst. Tech. J. **42**, 505–519 (1963)
2. Takacs, L.: A queueing model with feedback. Oper. Res. **11**(4), 345–354 (1977)
3. Wortman, M.A., Disney, R.L., Kiessler, P.C.: The M/GI/1 Bernoulli feedback queue with vacations. Queueing Syst. **9**(4), 353–363 (1991)
4. D'Avignon, G.R., Disney, R.L.: Queues with instantaneous feedback. Manag. Sci. **24**(2), 168–180 (1977)
5. Berg, J.L., Boxma, O.J.: The M/G/1 queue with processor sharing and its relation to feedback queue. Queueing Syst. **9**(4), 365–402 (1991)
6. Hunter, J.J.: Sojourn time problems in feedback queue. Queueing Syst. **5**(1–3), 55–76 (1989)
7. Dudin, A.N., Kazimirsky, A.V., Klimenok, V.I., Breuer, L., Krieger, U.: The queuing model MAP/PH/1/N with feedback operating in a Markovian random environment. Austrian J. Stat. **34**(2), 101–110 (2005)
8. Melikov, A.Z., Ponomarenko, L.A., Kuliyeva, K.N.: Calculation of the characteristics of multichannel queuing system with pure losses and feedback. J. Autom. Inf. Sci. **47**(5), 19–29 (2015)
9. Pekoz, E.A., Joglekar, N.: Poisson traffic flow in a general feedback. J. Appl. Probab. **39**(3), 630–636 (2002)
10. Lee, H.W., Seo, D.W.: Design of a production system with feedback buffer. Queueing Syst. **26**(1), 187–2002 (1997)
11. Lee, H.W., Ahn, B.Y.: Analysis of a production system with feedback buffer and general dispatching time. Math. Probl. Eng. **5**, 421–439 (2000)
12. Foley, R.D., Disney, R.L.: Queues with delayed feedback. Adv. Appl. Probab. **15**(1), 162–182 (1983)
13. Philippe, B., Saad, Y., Stewart, W.J.: Numerical methods in Markov chains modeling. Oper. Res. **40**(6), 1156–1179 (1992)
14. Melikov, A.Z., Ponomarenko, L.A., Kuliyeva, K.N.: Numerical analysis of the queuing system with feedback. Cybern. Syst. Anal. **51**(2), 566–573 (2015)
15. Kumar, B.K., Rukmani, R., Thangaraj, V.: On multi-server feedback retrial queue with finite buffer. Appl. Math. Model. **33**(4), 2062–2083 (2009)
16. Do, T.V.: An efficient computation algorithm for a multi-server feedback retrial queue with a large queuing capacity. Appl. Math. Model. **34**(8), 2272–2278 (2010)
17. Gemikonakli, O., Ever, E., Kocyigit, A.: Approximate solution for two stage open networks with Markov-modulated queues minimizing the state space explosion problem. J. Comput. Appl. Math. **223**(1), 519–533 (2009)
18. Kirsal, Y., Gemikonakli, E., Ever, E., Mapp, G., Gemikonakli, O.: An analytical approach for performance analysis of handoffs in the next generation integral cellular networks and WLANs. In: Proceedings of 19th IEEE International Conference on Computer Communications and Networks, pp. 1–6, Zurich, August 2–5, 2010. https://doi.org/10.1109/ICCCN.2010.5560093
19. Ever, E., Gemikonakli, O., Kocyigit, A., Gemikonakli, E.: A hybrid approach to minimize state explosion problem for the solution of two stage tandem queues. J. Netw. Comput. Appl. **36**, 908–926 (2013)

20. Kirsal, Y., Ever, E., Kocyigit, A., Gemikonakli, O., Mapp, G.: A generic analytical modeling approach for performance evaluation of the handover schemes in heterogeneous environments. Wirel. Pers. Commun. **79**, 1247–1276 (2014)
21. Kirsal, Y., Ever, E., Kocyigit, A., Gemikonakli, O., Mapp, G.: Modeling and analysis of vertical handover in highly mobile environments. J. Supercomputing. **71**, 4352–4380 (2015)
22. Kim, C., Klimenok, V.I., Dudin, A.N.: Priority tandem queueing system with retrials and reservation of channels as a model of call center. Comput. Ind. Eng. **96**, 61–71 (2016)
23. Melikov, A.Z., Ponomarenko, L.A., Rustamov, A.İ.: Methods for the analysis of queueing models with instantaneous and delayed feedbacks. Commun. Comput. Inf. Sci. **564**, 185–199 (2015)
24. Koroliuk, V.S., Melikov, A.Z., Ponomarenko, L.A., Rustamov, A.M.: Methods for the analysis of multi-channel queueing models with instantaneous and delayed feedbacks. Cybern. Syst. Anal. **52**(1), 58–70 (2016)
25. Melikov, A.Z., Ponomarenko, L.A., Rustamov, A.M.: Hierarchical space merging algorithm to analysis of open tandem queuing networks. Cybern. Syst. Anal. **52**(6), 867–877 (2016)
26. Neuts, M.F.: Matrix-Geometric Solutions in Stochastic Models: An Algorithmic Approach, p. 332. John Hopkins University Press, Baltimore (1981)
27. Mitrani, I., Chakka, R.: Spectral expansion solution for a class of Markov models: application and comparison with the matrix-geometric method. Perform. Eval. **23**, 241–260 (1995)
28. Chakka, R.: Spectral expansion solution for some finite capacity queues. Ann. Oper. Res. **79**, 27–44 (1998)

Estimation of the Probability Density Parameters of the Interval Duration Between Events in Correlated Synchronous Generalized Flow of the Second Order

Lyudmila Nezhelskaya[1], Michele Pagano[2], and Ekaterina Sidorova[1(✉)]

[1] National Research Tomsk State University, 36 Lenina Ave., Tomsk 634050, Russia
{ludne,katusha_sidorova}@mail.ru
[2] University of Pisa, 16 Via Caruso, 56122 Pisa, Italy
m.pagano@iet.unipi.it

Abstract. We consider the problem of estimating the probability density parameters of the interval duration values between adjacent moments of occurrence of events in correlated synchronous generalized doubly stochastic event flow of the second order under conditions of its complete observability. The explicit form of parameter estimates is found by the method of moments and the quality of estimates within the selected criteria is established through the work of the flow simulation model. Finally, numerical results of statistical experiments obtained using computational analytical formulas and simulation modeling are given.

Keywords: Synchronous generalized event flow of the second order · Doubly stochastic flows · Probability density · Parameters estimation · Method of moments

1 Introduction

The intensive development of computer appliance and innovations in the field of information technologies have stimulated the development of queueing theory and the improvement of the mathematical apparatus used within it. The most relevant research related to the design, subsequent implementation and maintenance of information and computing systems and networks of various configurations, whose mathematical models are queuing systems (QS) and queuing networks (QN), arises primarily in the framework of incoming streams of events (messages, requests) [1–3].

The complication of the structure of telecommunication systems, global computer networks, satellite communication networks, various software and hardware, integration of various communication systems have revealed the need to construct new mathematical models of incoming streams in the form of doubly stochastic flows, the studies of which are described in [4–9]. The intensity of

A. Dudin et al. (Eds.): ITMM 2019, CCIS 1109, pp. 202–216, 2019.
https://doi.org/10.1007/978-3-030-33388-1_17

such flows can be represented by both continuous and piecewise constant random processes with a finite number of states.

Flows with a step intensity function are most characteristic of real telecommunication networks and called MC (Markov chain) or MAP (Markovian Arrival Process) flows [6–8]. The work [10] is devoted to the generalization of the MAP-flows of events, while [11] is about establishing the relationship between MC-flows and MAP-flows. This class of flows, in particular, includes synchronous generalized event flow of the second order studied within this work [12,13]. It should be noted separately that works on the study of QS and QN with incoming MAP-flow of requests (as well as MMPP (Markov Modulated Poisson Process) as its particular case) are of particular scientific and practical interest due to the adequacy of such mathematical models to real processes and systems [14–16].

In the direct study of doubly stochastic flows, two main classes of problems are addressed: estimating flow states [12,13,17–21] and estimating their parameters [22–26] from the observed moments of occurrence of events.

In papers [12,13], the problem of optimal estimation of the states of synchronous generalized flow of the second order was solved: under conditions of accessibility to observation of all its events in [12], in the presence of an unextandable dead time in [13]. In this paper, we are solving the problem of estimating the parameters of the probability density function of the interval duration values between events in correlated flow using the method of moments.

2   Problem Statement

We consider a synchronous generalized doubly stochastic event flow of the second order (flow) in stationary conditions and assume that the accompanying process $\lambda(t)$ is a piecewise constant random process with two states S_1 and S_2. Hereinafter, S_i is understood as ith state of $\lambda(t)$, $i = 1, 2$.

The interval duration between flow events at the ith state is determined by the random variable $\eta_i = \min(\xi_i^{(1)}, \xi_i^{(2)})$, where $\xi_i^{(1)}$, $\xi_i^{(2)}$ are mutually independent random variables with distribution $F_i^{(1)}(t) = 1 - e^{-\lambda_i t}$, $F_i^{(2)}(t) = 1 - e^{-\alpha_i t}$, $i = 1, 2$, respectively. At the moment when a flow event occurs, depending on the value η_i, $i = 1, 2$, the process $\lambda(t)$ either transits from the ith state to the jth state, $i \neq j$, or remains at the ith state, $i = j$, with probability $P_1^{(1)}(\lambda_j|\lambda_i)$ or $P_1^{(2)}(\lambda_j|\lambda_i)$, $i, j = 1, 2$. Here $P_1^{(1)}(\lambda_j|\lambda_i) + P_1^{(1)}(\lambda_i|\lambda_i) = 1$, $P_1^{(2)}(\lambda_j|\lambda_i) + P_1^{(2)}(\lambda_i|\lambda_i) = 1$, $i, j = 1, 2$, $i \neq j$. Thus, the flow inter-event interval duration at the ith state of the process $\lambda(t)$ is a random variable with exponential distribution function $F_i(t) = 1 - e^{-(\lambda_i + \alpha_i)t}$, $i = 1, 2$. In the sequel it is assumed that the state S_i takes place if $\lambda(t) = \lambda_i$, $i = 1, 2$ ($\lambda_1 > \lambda_2 \geq 0$).

Under the above assumptions, $\lambda(t)$ for the considered event flow is a hidden Markov process [12] with infinitesimal characteristics matrices of the form

$$\mathbf{D}_0 = \left\| \begin{matrix} -(\lambda_1 + \alpha_1) & 0 \\ 0 & -(\lambda_2 + \alpha_2) \end{matrix} \right\|,$$

$$\mathbf{D}_1 = \left\| \begin{matrix} \lambda_1 P_1^{(1)}(\lambda_1|\lambda_1) + \alpha_1 P_1^{(2)}(\lambda_1|\lambda_1) & \lambda_1 P_1^{(1)}(\lambda_2|\lambda_1) + \alpha_1 P_1^{(2)}(\lambda_2|\lambda_1) \\ \lambda_2 P_1^{(1)}(\lambda_1|\lambda_2) + \alpha_2 P_1^{(2)}(\lambda_1|\lambda_2) & \lambda_2 P_1^{(1)}(\lambda_2|\lambda_2) + \alpha_2 P_1^{(2)}(\lambda_2|\lambda_2) \end{matrix} \right\| .$$

Elements of $\mathbf{D}_1$ are the intensities of the process $\lambda(t)$ transitions from state to state with a flow event occurrence. Off-diagonal elements of $\mathbf{D}_0$ represent the transition intensities without an event occurrence; diagonal elements are the intensities of the $\lambda(t)$ output from its states, taken with the opposite sign [7].

A variant of the arising situation is shown in Fig. 1, where S_1, S_2 are the states of the fundamentally unobservable random process $\lambda(t)$; $t_1, t_2, ..., t_k, ...$ are observable moments of occurrence of events in the flow under consideration.

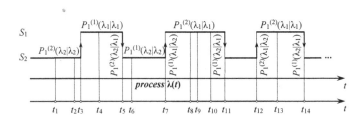

Fig. 1. Formation of the observed event flow

Let t_0 and t represent the start and the end of observations, respectively; then, the sequence of observable moments of occurrence of events on the interval (t_0, t) generates an embedded Markov chain $\{\lambda(t_k)\}$ due to the formulated prerequisites. In other words, the observed event flow has the Markov property, if its evolution is considered from the moment of an event occurrence t_k, $k = 1, 2, ...$

The study aim is to determine the explicit form of the probability density of the interval duration between instants of occurrence of events in a synchronous generalized flow of the second order, which is a correlated event flow in the general case, and to estimate its parameters using the method of moments.

3   Derivation of the Probability Density

Let us denote the probability density of the duration values of the kth interval between neighboring events t_k and t_{k+1}, $k = 1, 2, ...$, in the studied flow by $p(\tau)$. Since we consider the steady-state operation mode of the flow, for any $k \geq 1$ the equality $p(\tau_k) = p(\tau)$, $\tau \geq 0$, is valid. As a consequence, the moment of an event occurrence without any loss of generality can be set equal to zero, or, equivalently, the moment when a flow event occurs is $\tau = 0$.

Let $p_{ij}(\tau)$ be the conditional probability that there are no flow events on the interval $(0, \tau)$ and the value of the process at the moment τ is $\lambda(\tau) = \lambda_j$, provided that $\lambda(0) = \lambda_i$, i.e. the conditional probability that during the time interval $(0, \tau)$ a non-conjugate with an event occurrence transition of the process $\lambda(\tau)$ from the state S_i to the state S_j, $i, j = 1, 2$, takes place.

Lemma 1. *The conditional probabilities $p_{ij}(\tau)$, $i,j = 1,2$, in a correlated synchronous generalized event flow of the second order have the form*

$$p_{11}(\tau) = e^{-(\lambda_1+\alpha_1)\tau}, p_{12}(\tau) = 0, p_{21}(\tau) = 0, p_{22}(\tau) = e^{-(\lambda_2+\alpha_2)\tau}, \tau \geq 0. \quad (1)$$

Proof. In accordance with the flow definition, the probabilities $p_{12}(\tau)$, $p_{21}(\tau)$ are identically equal to zero, since transitions from state to state are accompanied by an event occurrence. For $p_{11}(\tau)$, $p_{22}(\tau)$, the differential equations

$$p_{11}'(\tau) = -(\lambda_1 + \alpha_1)p_{11}(\tau), p_{22}'(\tau) = -(\lambda_2 + \alpha_2)p_{22}(\tau) \quad (2)$$

with the initial conditions $p_{11}(0) = 1$, $p_{22}(0) = 1$ are valid. Integrating equations with separable variables (2) between $\tau = 0$ and τ, we get (1).

Lemma 2. *The probability densities $\tilde{p}_{ij}(\tau)$, $i,j = 1,2$, in a correlated synchronous generalized flow of the second order are determined by the formulas*

$$
\begin{aligned}
\tilde{p}_{11}(\tau) &= (\lambda_1 P_1^{(1)}(\lambda_1|\lambda_1) + \alpha_1 P_1^{(2)}(\lambda_1|\lambda_1))e^{-(\lambda_1+\alpha_1)\tau}, \\
\tilde{p}_{12}(\tau) &= (\lambda_1 P_1^{(1)}(\lambda_2|\lambda_1) + \alpha_1 P_1^{(2)}(\lambda_2|\lambda_1))e^{-(\lambda_1+\alpha_1)\tau}, \\
\tilde{p}_{21}(\tau) &= (\lambda_2 P_1^{(1)}(\lambda_1|\lambda_2) + \alpha_2 P_1^{(2)}(\lambda_1|\lambda_2))e^{-(\lambda_2+\alpha_2)\tau}, \\
\tilde{p}_{22}(\tau) &= (\lambda_2 P_1^{(1)}(\lambda_2|\lambda_2) + \alpha_2 P_1^{(2)}(\lambda_2|\lambda_2))e^{-(\lambda_2+\alpha_2)\tau}.
\end{aligned}
\quad (3)
$$

Proof. The joint probabilities that, without an event occurrence on the interval $(0,\tau)$, the process $\lambda(\tau)$ transits on this interval from the ith to the jth state, $i,j = 1,2$, the state S_j of the process $\lambda(\tau)$ ends on the half-closed interval $[\tau, \tau + \Delta\tau)$ and at the moment when a flow event occurs, the process $\lambda(\tau)$ transits from the jth state to the ith, $i,j = 1,2$, will be written as

$$
\begin{aligned}
P_{11}(\tau) &= p_{11}(\tau)(\lambda_1 P_1^{(1)}(\lambda_1|\lambda_1) + \alpha_1 P_1^{(2)}(\lambda_1|\lambda_1))\Delta\tau + o(\Delta\tau), \\
P_{12}(\tau) &= p_{11}(\tau)(\lambda_1 P_1^{(1)}(\lambda_2|\lambda_1) + \alpha_1 P_1^{(2)}(\lambda_2|\lambda_1))\Delta\tau + o(\Delta\tau), \\
P_{21}(\tau) &= p_{22}(\tau)(\lambda_2 P_1^{(1)}(\lambda_1|\lambda_2) + \alpha_2 P_1^{(2)}(\lambda_1|\lambda_2))\Delta\tau + o(\Delta\tau), \\
P_{22}(\tau) &= p_{22}(\tau)(\lambda_2 P_1^{(1)}(\lambda_2|\lambda_2) + \alpha_2 P_1^{(2)}(\lambda_2|\lambda_2))\Delta\tau + o(\Delta\tau).
\end{aligned}
\quad (4)
$$

Note that each of the considered joint probabilities (4) can be represented as $p_{ii}(\tau)(\lambda_i P_1^{(1)}(\lambda_j|\lambda_i) + \alpha_i P_1^{(2)}(\lambda_j|\lambda_i))\Delta\tau + o(\Delta\tau) = \int_{\tau}^{\tau+\Delta\tau} \tilde{p}_{ij}(u)du = \tilde{p}_{ij}(\tau)\Delta\tau + o(\Delta\tau)$, $\tilde{p}_{ij}(\tau)$ is the probability density corresponding to $P_{ij}(\tau)$, $i,j = 1,2$.

Let us rewrite the last equality as $p_{ii}(\tau)(\lambda_i P_1^{(1)}(\lambda_j|\lambda_i) + \alpha_i P_1^{(2)}(\lambda_j|\lambda_i)) + \frac{o(\Delta\tau)}{\Delta\tau} = \tilde{p}_{ij}(\tau) + \frac{o(\Delta\tau)}{\Delta\tau}$, $i,j = 1,2$, and let $\Delta\tau$ goes to zero; as a result we find

$$\tilde{p}_{ij}(\tau) = p_{ii}(\tau)(\lambda_i P_1^{(1)}(\lambda_j|\lambda_i) + \alpha_i P_1^{(2)}(\lambda_j|\lambda_i)), i,j = 1,2. \quad (5)$$

Substituting in (5) $p_{ii}(\tau)$ from (1), we obtain (3).

Lemma 3. *For a correlated synchronous generalized event flow of the second order, the transition probabilities p_{ij}, $i, j = 1, 2$, are determined by the formulas*

$$
\begin{aligned}
p_{11} &= (\lambda_1 P_1^{(1)}(\lambda_1|\lambda_1) + \alpha_1 P_1^{(2)}(\lambda_1|\lambda_1))(\lambda_1 + \alpha_1)^{-1}, \\
p_{12} &= (\lambda_1 P_1^{(1)}(\lambda_2|\lambda_1) + \alpha_1 P_1^{(2)}(\lambda_2|\lambda_1))(\lambda_1 + \alpha_1)^{-1}, \\
p_{21} &= (\lambda_2 P_1^{(1)}(\lambda_1|\lambda_2) + \alpha_2 P_1^{(2)}(\lambda_1|\lambda_2))(\lambda_2 + \alpha_2)^{-1}, \\
p_{22} &= (\lambda_2 P_1^{(1)}(\lambda_2|\lambda_2) + \alpha_2 P_1^{(2)}(\lambda_2|\lambda_2))(\lambda_2 + \alpha_2)^{-1}.
\end{aligned}
\tag{6}
$$

Proof. Since τ is an arbitrary time moment, then p_{ij} (the transition probabilities of the process $\lambda(\tau)$ from the state S_i to the state S_j, $i, j = 1, 2$, in the time from $\tau = 0$ to the moment of occurrence of the next flow event) are given by

$$
p_{ij} = \int_0^\infty \tilde{p}_{ij}(\tau)d\tau, i, j = 1, 2.
\tag{7}
$$

Substituting (3) into (7), we find (6).

We denote by $\pi_i(0)$ the conditional stationary probability that the process $\lambda(\tau)$ at the time moment $\tau = 0$ is in the ith state provided that $\tau = 0$ is the moment of a flow event occurrence, $i = 1, 2$; $\pi_1(0) + \pi_2(0) = 1$.

Lemma 4. *The conditional final probabilities $\pi_i(0)$, $i = 1, 2$, in a correlated synchronous generalized flow of the second order are given by the expressions*

$$
\pi_1(0) = \frac{(\lambda_1 + \alpha_1)\phi_2}{(\lambda_1 + \alpha_1)\phi_2 + (\lambda_2 + \alpha_2)\phi_1}, \pi_2(0) = \frac{(\lambda_2 + \alpha_2)\phi_1}{(\lambda_1 + \alpha_1)\phi_2 + (\lambda_2 + \alpha_2)\phi_1},
\tag{8}
$$

where $\phi_1 = \lambda_1 P_1^{(1)}(\lambda_2|\lambda_1) + \alpha_1 P_1^{(2)}(\lambda_2|\lambda_1)$, $\phi_2 = \lambda_2 P_1^{(1)}(\lambda_1|\lambda_2) + \alpha_2 P_1^{(2)}(\lambda_1|\lambda_2)$.

Proof. Due to the fact that $\{\lambda(t_k)\}$ is an embedded Markov chain, the following equations are valid for the probabilities $\pi_i(0)$, $i = 1, 2$,

$$
\pi_1(0) = p_{11}\pi_1(0) + p_{21}\pi_2(0), \pi_2(0) = p_{12}\pi_1(0) + p_{22}\pi_2(0),
\tag{9}
$$

p_{ij}, $i, j = 1, 2$, are determined by Lemma 3 and have the form (6). Substituting (6) into (9), taking into account the normalization condition, we arrive at (8).

Lemmas 2 and 4 allow us to formulate the following theorem.

Theorem 1. *In a correlated synchronous generalized flow of the second order, the probability density of the interval duration values between neighboring flow events takes the form*

$$
p(\tau) = \gamma(\lambda_1 + \alpha_1)e^{-(\lambda_1+\alpha_1)\tau} + (1 - \gamma)(\lambda_2 + \alpha_2)e^{-(\lambda_2+\alpha_2)\tau}, \tau \geq 0,
\tag{10}
$$

$\gamma = \pi_1(0)$, *where $\pi_1(0)$ is defined in (8).*

Proof. Since the sequence of instants of occurrence of events generates an embedded Markov chain, the density $p(\tau)$ in the correlated flow has the form

$$p(\tau) = \sum_{i=1}^{2} \pi_i(0) \sum_{j=1}^{2} \tilde{p}_{ij}(\tau), \tau \geq 0. \tag{11}$$

Substituting into formula (11) first (3), and then explicit expressions (9) for the probabilities $\pi_i(0)$, $i = 1, 2$, after the necessary manipulations, we obtain (10).

Note that (10) is the probability density $p(\tau)$ of a random variable distributed according to the hyperexponential law with the corresponding parameters. The last statement leads to the following remark.

Remark 1. The probability density of the inter-event interval duration of the flow under study has the form (10) if and only if there is an additional condition for setting the distribution parameters of the random variables $\xi_i^{(1)}$ and $\xi_i^{(2)}$, $i = 1, 2$, namely, $(\lambda_1 + \alpha_1) \neq (\lambda_2 + \alpha_2)$, which may well be violated in the original problem formulation due to the fact that λ_1 and λ_2 satisfy the condition $\lambda_1 > \lambda_2 \geq 0$, but, at the same time, α_1 and α_2, generally speaking, are arbitrary non-negative numbers. The equality $(\lambda_1 + \alpha_1) = (\lambda_2 + \alpha_2)$ leads to the probability density of a Poisson flow of events with parameter $(\lambda_1 + \alpha_1)$.

In the following, it is assumed that $(\lambda_1 + \alpha_1) \neq (\lambda_2 + \alpha_2)$. Putting $\alpha_1 = \alpha_2 = 0$ in (11), we obtain the probability density $p(\tau)$ for a synchronous event flow [27].

4 Estimation of the Probability Density Parameters

It is not possible to estimate by the method of moments the twelve unknown parameters of a synchronous generalized flow of events of the second order λ_i, α_i, $P_1^{(1)}(\lambda_j|\lambda_i)$, $P_1^{(2)}(\lambda_j|\lambda_i)$, $i, j = 1, 2$, or eight parameters with regard to $P_1^{(1)}(\lambda_i|\lambda_i) = 1 - P_1^{(1)}(\lambda_j|\lambda_i)$, $P_1^{(2)}(\lambda_i|\lambda_i) = 1 - P_1^{(2)}(\lambda_j|\lambda_i)$, $i, j = 1, 2$, $i \neq j$, having only information on the form of $p(\tau)$, as will be seen below. We will estimate the unknown probability density parameters $(\lambda_1 + \alpha_1)$, $(\lambda_2 + \alpha_2)$, γ.

Remark 2. The considered event flow is, in general, correlated, i.e. the interval durations between the instants of occurrence of events in the flow are dependent random variables. This dependency does not allow us to speak about the consistency of the estimates obtained by the method of moments, and we can state their quality only on the basis of the results of simulation in sense of one or another estimation quality criterion.

Let us introduce statistics $C_l = \frac{1}{n} \sum_{k=1}^{n} \tau_k{}^l$, where $\tau_k = t_{k+1} - t_k$ (here, we observe $n + 1$ flow events). Let $\tau_1, \tau_2, ..., \tau_n$ be a sample from the distribution $p(\tau|z_1, z_2, \gamma) = \gamma z_1 e^{-z_1 \tau} + (1 - \gamma) z_2 e^{-z_2 \tau}$, depending on the three unknown parameters $z_1 = \lambda_1 + \alpha_1$, $z_2 = \lambda_2 + \alpha_2$ (without any loss of generality, we assume $z_1 > z_2$), $\gamma = \pi_1(0)$, which must be estimated. The theoretical initial moment of the lth order $E[\tau^l] = \int_0^\infty \tau^l p(\tau|z_1, z_2, \gamma)d\tau$ is a function of the unknown parameters and, due to the closeness of the theoretical and empirical distribution functions for a sufficiently large n, the theoretical moment is close to the corresponding sample moment (the statistics of the same order). Thus, for estimating z_1, z_2, γ it is necessary to have three moment equations, i.e. $E[\tau^l] = C_l$, $l = \overline{1,3}$. The theoretical initial moment of the lth order is determined by the formula $E[\tau^l] = l!\gamma z_1^{-l} + l!(1-\gamma)z_2^{-l}$, $l = \overline{1,3}$, on the basis of which we write the system of equations for the unknowns z_1, z_2, γ:

$$\frac{\gamma}{z_1} + \frac{(1-\gamma)}{z_2} = C_1, 2\frac{\gamma}{z_1^2} + 2\frac{(1-\gamma)}{z_2^2} = C_2, 6\frac{\gamma}{z_1^3} + 6\frac{(1-\gamma)}{z_2^3} = C_3. \tag{12}$$

As a result of simple manipulations, we bring the system (12) to the form

$$\gamma z_2 + (1-\gamma)z_1 - z_1 z_2 C_1 = 0, (z_1 + z_2)C_1 - z_1 z_2 C_2/2 = 1,$$
$$(z_1 + z_2)C_2 - z_1 z_2 C_3/3 = 2C_1. \tag{13}$$

From the system (13), we find $\hat{z}_1 + \hat{z}_2 = \frac{2(3C_1C_2 - C_3)}{3C_2^2 - 2C_1C_3}$, $\hat{z}_1\hat{z}_2 = \frac{6(2C_1^2 - C_2)}{3C_2^2 - 2C_1C_3}$. Then the estimates $\hat{z}_1$, $\hat{z}_2$ of parameters z_1, z_2 are roots of the quadratic equation

$$z^2 - x_1 z + x_2 = 0 \tag{14}$$

with known coefficients $x_1 = \hat{z}_1 + \hat{z}_2 = \frac{2(3C_1C_2 - C_3)}{3C_2^2 - 2C_1C_3}$ and $x_2 = \hat{z}_1\hat{z}_2 = \frac{6(2C_1^2 - C_2)}{3C_2^2 - 2C_1C_3}$.
According to the condition $z_1 > z_2$, we have

$$\hat{z}_{1,2} = \frac{3C_1C_2 - C_3}{3C_2^2 - 2C_1C_3} \pm \frac{1}{2}\sqrt{\left(\frac{2(3C_1C_2 - C_3)}{3C_2^2 - 2C_1C_3}\right)^2 - 4\frac{6(2C_1^2 - C_2)}{3C_2^2 - 2C_1C_3}}. \tag{15}$$

Remark 3. $\hat{z}_1$, $\hat{z}_2$ as real positive roots of (14) exist if and only if all of conditions are met: $\left(\frac{2(3C_1C_2 - C_3)}{3C_2^2 - 2C_1C_3}\right)^2 - 4\frac{6(2C_1^2 - C_2)}{3C_2^2 - 2C_1C_3} > 0$, $\frac{3C_1C_2 - C_3}{3C_2^2 - 2C_1C_3} > 0$, $\frac{6(2C_1^2 - C_2)}{3C_2^2 - 2C_1C_3} > 0$.

Estimate $\hat{\gamma}$ is determined uniquely from the first equation of the system (13)

$$\hat{\gamma} = \frac{\hat{z}_1(1 - C_1\hat{z}_2)}{\hat{z}_1 - \hat{z}_2}. \tag{16}$$

Thus, system (13) has the unique solution $\hat{z}_1$, $\hat{z}_2$, $\hat{\gamma}$.

Let us consider the representation of the parameter γ as $\gamma = \gamma_1(\gamma_1 + \gamma_2)^{-1}$, $1 - \gamma = \gamma_2(\gamma_1 + \gamma_2)^{-1}$, where $\gamma_1 = (\lambda_1 + \alpha_1)[\lambda_2 P_1^{(1)}(\lambda_1|\lambda_2) + \alpha_2 P_1^{(2)}(\lambda_1|\lambda_2)]$, $\gamma_2 = (\lambda_2 + \alpha_2)[\lambda_1 P_1^{(1)}(\lambda_2|\lambda_1) + \alpha_1 P_1^{(2)}(\lambda_2|\lambda_1)]$. As a result, $\tau_1, \tau_2, ..., \tau_n$ is a sample drawn from $p(\tau|z_1, z_2, \gamma_1, \gamma_2) = \gamma_1(\gamma_1 + \gamma_2)^{-1}z_1 e^{-z_1\tau} + \gamma_2(\gamma_1 + \gamma_2)^{-1}z_2 e^{-z_2\tau}$, depending on the four unknown parameters z_1, z_2 ($z_1 > z_2$), γ_1, γ_2. For the first four initial moments, as before, we write the exact equalities, in which, instead of the true parameter values, we substitute their estimates: $E[\tau^l] = C_l$, $l = \overline{1,4}$. Thus, the analogue of (13) is written as

$$\gamma_1 z_2 + \gamma_2 z_1 - (\gamma_1 + \gamma_2)z_1 z_2 C_1 = 0, (z_1 + z_2)C_1 - z_1 z_2 C_2/2 = 1,$$
$$(z_1 + z_2)C_2 - z_1 z_2 C_3/3 = 2C_1, (z_1 + z_2)C_3 - z_1 z_2 C_4/4 = 3C_2. \tag{17}$$

Theorem 2. *The system of the four moment equations* (17) *with respect to the unknown parameters* z_1, z_2, γ_1, γ_2 *of the probability density* $p(\tau)$ *is incompatible.*

Proof. We perform the variables change in system (17) $z_1 + z_2 = x_1$, $z_1 z_2 = x_2$, $\gamma_1 z_2 + \gamma_2 z_1 = x_3$ and $(\gamma_1 + \gamma_2) z_1 z_2 = x_4$, which leads to a linear form

$$x_3 - x_4 C_1 = 0, x_1 C_1 - x_2 C_2/2 = 1,$$
$$x_1 C_2 - x_2 C_3/3 = 2C_1, x_1 C_3 - x_2 C_4/4 = 3C_2. \tag{18}$$

The ranks of the matrix and extended matrix of the resulted system of four linear inhomogeneous equations in four unknowns x_l, $l = \overline{1,4}$, are respectively 3 and 4. Consequently, according to the corollary of the Kronecker–Capelli theorem, the system is incompatible, i.e. the solution of (18) does not exist.

Thus, Theorem 2 determines that the knowledge of only the probability density form does not allow one to estimate z_1, z_2, γ_1, γ_2, and, all the more, all the correlated flow parameters, i.e. the information contained in $p(\tau)$ is not enough to estimate more than three parameters. Indeed, the density (10) makes it possible to estimate by the method of moments $z_1 = \lambda_1 + \alpha_1$, $z_2 = \lambda_2 + \alpha_2$, $\gamma = \pi_1(0)$.

5   Numerical Results of Parameter Estimation

In order to obtain numerical results, a two-stage algorithm for computing $\hat{z}_1$, $\hat{z}_2$, $\hat{\gamma}$ has been developed. The first stage involves simulation [28] of the flow to obtain statistics, the second is the calculation of estimates by (15), (16).

Table 1. Model parameters

$\lambda_1 = 6,1$	$P_1^{(1)}(\lambda_1|\lambda_1) = 0,6$	$P_1^{(1)}(\lambda_2|\lambda_1) = 0,4$
$\lambda_2 = 1,3$	$P_1^{(1)}(\lambda_2|\lambda_2) = 0,5$	$P_1^{(1)}(\lambda_1|\lambda_2) = 0,5$
$\alpha_1 = 5,1$	$P_1^{(2)}(\lambda_1|\lambda_1) = 0,3$	$P_1^{(2)}(\lambda_2|\lambda_1) = 0,7$
$\alpha_2 = 0,1$	$P_1^{(2)}(\lambda_2|\lambda_2) = 0,8$	$P_1^{(2)}(\lambda_1|\lambda_2) = 0,2$

Table 2. Results of the first statistical experiment

θ	T_m	400	500	600	700	800	900	1000
$z_1 = 11,2$	$\hat{E}[\hat{z}_1]$	12,6151	10,8123	12,2031	11,1435	11,8946	11,7036	11,2358
	$\sqrt{\hat{V}[\hat{z}_1]}$	7,2285	9,2995	5,6807	4,8109	2,7360	2,2163	1,7365
$z_2 = 1,4$	$\hat{E}[\hat{z}_2]$	1,4029	1,2701	1,4159	1,3762	1,4180	1,4158	1,3958
	$\sqrt{\hat{V}[\hat{z}_2]}$	0,7827	1,0338	0,5207	0,4917	0,0685	0,0774	0,0371
$\gamma = 0,4714$	$\hat{E}[\hat{\gamma}]$	0,4799	0,5227	0,4757	0,4878	0,4718	0,4742	0,4714
	$\sqrt{\hat{V}[\hat{\gamma}]}$	0,2675	0,3975	0,1790	0,1389	0,0365	0,0361	0,0259

To establish the quality of the obtained estimates, statistical experiments were carried out. In the first of them, the time interval for reaching the stationary operation mode is monitored: for each modeling time value (duration T_m of the flow observation) with fixed event flow parameters, $N = 100$ independent realizations of the correlated event flow are simulated; for each of them estimates $\hat{\theta}^k$, $k = \overline{1, N}$, of the corresponding parameters θ are determined according to formulas (15), (16), after which the sample mean estimate value $\hat{E}[\hat{\theta}] = \frac{1}{N} \sum_{k=1}^{N} \hat{\theta}^{(k)}$ and sample square error $\sqrt{\hat{V}[\hat{\theta}]}$ as square root of $\hat{V}[\hat{\theta}] = \frac{1}{N} \sum_{k=1}^{N} (\hat{\theta}^k - \theta)^2$, $\theta \in \{z_1, z_2, \gamma\}$, $\hat{\theta} \in \{\hat{z}_1, \hat{z}_2, \hat{\gamma}\}$, are calculated. The corresponding results for the parameter set of Table 1 are presented in Table 2.

Based on the analysis of numerous experiment results, including the ones presented in Table 2, it can be argued that the quality of the estimates $\hat{\theta} \in \{\hat{z}_1, \hat{z}_2, \hat{\gamma}\}$ significantly depends on the modeling time of the flow implementations: namely, with an increase of T_m, the quality of the estimates improves in sense of decreasing the sample quadratic error. The latter is quite natural and is explained first of all by the very concept of the method of moments, which is based on the statistics values required for the calculations and taken from the implementations of a synchronous generalized flow of the second order. The statistics C_l, $l = \overline{1, 3}$ contain information about the intervals between the observed instants of occurrence of events, consequently, the greater number of intervals in the implementation, the more information will be collected in the statistics and, therefore, the accuracy of density parameters estimation will be higher. We also note the oscillatory behavior of $\hat{E}[\hat{\theta}]$, $\sqrt{\hat{V}[\hat{\theta}]}$, $\hat{\theta} \in \{\hat{z}_1, \hat{z}_2, \hat{\gamma}\}$, which, however, with a significant increase in T_m is less significant. As a consequence, for further experiments the modeling time value was chosen equal to $T_m = 1000$ time units.

The subject of research in the second statistical experiment is the dependence of $\hat{E}[\hat{\theta}]$ and $\sqrt{\hat{V}[\hat{\theta}]}$, $\hat{\theta} \in \{\hat{z}_1, \hat{z}_2, \hat{\gamma}\}$, on changes in λ_i, α_i, $i = 1, 2$, with fixed $T_m = 1000$, $N = 100$ and the initial data of Table 3. As an illustration the following tables show the results of the current experiment: for $\lambda_1 = 4; 6; 8; 10$ in Table 4, for $\alpha_1 = 4; 6; 8; 10$ in Table 5, for $\lambda_2 = 1, 125; 1, 25; 1, 375; 1, 5$ in Table 6, for $\alpha_2 = 41, 125; 1, 25; 1, 375; 1, 5$ in Table 7.

Table 3. Model parameters

$P_1^{(1)}(\lambda_1|\lambda_1) = 0,2$	$P_1^{(1)}(\lambda_2|\lambda_2) = 0,1$	$P_1^{(2)}(\lambda_1|\lambda_1) = 0,1$	$P_1^{(2)}(\lambda_2|\lambda_2) = 0,2$
$P_1^{(1)}(\lambda_2|\lambda_1) = 0,8$	$P_1^{(1)}(\lambda_1|\lambda_2) = 0,9$	$P_1^{(2)}(\lambda_2|\lambda_1) = 0,9$	$P_1^{(2)}(\lambda_1|\lambda_2) = 0,8$

Table 4. Results of the experiment ($\lambda_2 = 1,125$, $\alpha_1 = 4$, $\alpha_2 = 1,125$)

λ_1	4	6	8	10
z_1	8	10	12	14
$\hat{E}[\hat{z}_1]$	8,1755	10,0780	11,8932	14,1534
$\sqrt{\hat{V}[\hat{z}_1]}$	1,0134	0,8796	0,9852	0,6873
z_2	2,25	2,25	2,25	2,25
$\hat{E}[\hat{z}_2]$	2,2686	2,2644	2,2598	2,2601
$\sqrt{\hat{V}[\hat{z}_2]}$	0,0861	0,0872	0,0667	0,0658
γ	0,5000	0,5030	0,5050	0,5064
$\hat{E}[\hat{\gamma}]$	0,5071	0,5126	0,5102	0,5108
$\sqrt{\hat{V}[\hat{\gamma}]}$	0,0361	0,0341	0,0254	0,0272

Table 5. Results of the experiment ($\lambda_1 = 4$, $\lambda_2 = 1,125$, $\alpha_2 = 1,125$)

α_1	4	6	8	10
z_1	8	10	12	14
$\hat{E}[\hat{z}_1]$	8,0959	10,0505	12,1694	14,1215
$\sqrt{\hat{V}[\hat{z}_1]}$	1,1017	0,8603	0,9396	0,6383
z_2	2,25	2,25	2,25	2,25
$\hat{E}[\hat{z}_2]$	2,2637	2,2615	2,2521	2,2722
$\sqrt{\hat{V}[\hat{z}_2]}$	0,0832	0,0694	0,0669	0,0572
γ	0,5000	0,4971	0,4951	0,4938
$\hat{E}[\hat{\gamma}]$	0,5098	0,5065	0,4927	0,4979
$\sqrt{\hat{V}[\hat{\gamma}]}$	0,0411	0,0283	0,0248	0,0231

Table 6. Results of the experiment ($\lambda_1 = 4$, $\alpha_1 = 4$, $\alpha_2 = 1,125$)

λ_2	1,125	1,25	1,375	1,5
z_1	8	8	8	8
$\hat{E}[\hat{z}_1]$	8,1549	8,2074	7,9946	8,1019
$\sqrt{\hat{V}[\hat{z}_1]}$	0,9934	0,9525	0,9614	0,9141
z_2	2,25	2,375	2,5	2,625
$\hat{E}[\hat{z}_2]$	2,2653	2,3993	2,5012	2,6300
$\sqrt{\hat{V}[\hat{z}_2]}$	0,0897	0,0804	0,0802	0,0807
γ	0,5000	0,5001	0,5015	0,5021
$\hat{E}[\hat{\gamma}]$	0,5100	0,5052	0,5109	0,5116
$\sqrt{\hat{V}[\hat{\gamma}]}$	0,0402	0,0383	0,0307	0,0356

Table 7. Results of the experiment ($\lambda_1 = 4$, $\lambda_2 = 1,125$, $\alpha_1 = 4$)

α_2	1,125	1,25	1,375	1,5
z_1	8	8	8	8
$\hat{E}[\hat{z}_1]$	8,1756	8,1253	8,2586	8,1104
$\sqrt{\hat{V}[\hat{z}_1]}$	1,0760	1,0578	0,9411	0,8680
z_2	2,25	2,375	2,5	2,625
$\hat{E}[\hat{z}_2]$	2,2620	2,3853	2,5364	2,6587
$\sqrt{\hat{V}[\hat{z}_2]}$	0,0900	0,0826	0,0887	0,0809
γ	0,5000	0,4992	0,4985	0,4979
$\hat{E}[\hat{\gamma}]$	0,5048	0,5044	0,4966	0,4958
$\sqrt{\hat{V}[\hat{\gamma}]}$	0,0443	0,0349	0,0353	0,0313

The numerical results reported in Tables 4, 5, 6, 7 demonstrate that with an increase in λ_i, α_i, $i = 1, 2$, and the other fixed, the quality of estimates $\hat{z}_1$, $\hat{z}_2$, $\hat{\gamma}$ improves in sense of decreasing the sample quadratic error. This is explained by the fact that an increase in each of the values λ_i, α_i, $i = 1, 2$, entails a more frequent occurrence of the flow events, which increases the number of observed intervals on which the statistics C_l, $l = \overline{1, 3}$, are based, which naturally improves the quality of the obtained estimates.

It is not difficult to obtain expressions for the mean (first initial moment) and variance (second central moment) of the random variable τ (the interval duration between adjacent events of the correlated doubly stochastic flow) using the known probability density (10): $E[\tau] = \gamma(\lambda_1 + \alpha_1)^{-1} + (1 - \gamma)(\lambda_2 + \alpha_2)^{-1}$, $D[\tau] = 2[\gamma(\lambda_1+\alpha_1)^{-2}+(1-\gamma)(\lambda_2+\alpha_2)^{-2}]-[\gamma(\lambda_1+\alpha_1)^{-1}+(1-\gamma)(\lambda_2+\alpha_2)^{-1}]^2$, where γ is defined in (10).

In the third experiment on the simulation flow model, statistical estimates were obtained for the discussed above probabilistic characteristics of τ: the sample mean $\hat{E}[\tau] = \frac{1}{N}\sum_{k=1}^{N}\tau_k$ and sample variance $\hat{D}[\tau] = \frac{1}{N}\sum_{k=1}^{N}(\tau_k - \hat{E}[\tau])^2$, and the relative error values $\delta = \frac{|\theta-\hat{\theta}|}{\theta}$, $\theta \in \{E[\tau], D[\tau]\}$, $\hat{\theta} \in \{\hat{E}[\tau], \hat{D}[\tau]\}$, were determined. For the fixed event flow realizations defined by the parameter sets of Table 8, the following results shown in Table 9 takes place.

Table 8. Model parameters

$\lambda_1 = 4$	$P_1^{(1)}(\lambda_1|\lambda_1) = 0,2$	$P_1^{(1)}(\lambda_2|\lambda_1) = 0,8$
$\lambda_1/\lambda_2 = 8; 16$	$P_1^{(1)}(\lambda_2|\lambda_2) = 0,4$	$P_1^{(1)}(\lambda_1|\lambda_2) = 0,6$
$\alpha_1 = 4$	$P_1^{(2)}(\lambda_1|\lambda_1) = 0,3$	$P_1^{(2)}(\lambda_2|\lambda_1) = 0,7$
$\alpha_1/\alpha_2 = 8; 16$	$P_1^{(2)}(\lambda_2|\lambda_2) = 0,1$	$P_1^{(2)}(\lambda_1|\lambda_2) = 0,9$

Table 9. Results of the third statistical experiment

	$E[\tau]$	$\hat{E}[\tau]$	$\frac{|E[\tau]-\hat{E}[\tau]|}{E[\tau]}$	$D[\tau]$	$\hat{D}[\tau]$	$\frac{|D[\tau]-\hat{D}[\tau]|}{D[\tau]}$
$\lambda_1/\lambda_2 = 8, \alpha_1/\alpha_2 = 8$	0,5625	0,5584	0,0073	0,6992	0,6897	0,0136
$\lambda_1/\lambda_2 = 8, \alpha_1/\alpha_2 = 16$	0,7500	0,7440	0,0080	1,2917	1,2921	0,0003
$\lambda_1/\lambda_2 = 16, \alpha_1/\alpha_2 = 8$	0,7097	0,7081	0,0023	1,2329	1,2331	0,0002
$\lambda_1/\lambda_2 = 16, \alpha_1/\alpha_2 = 16$	1,0625	1,0593	0,0030	2,8867	2,9324	0,0158

This experiment was performed with different values of the model flow parameters. Analysis of the results, including those reported in Table 9, allows us to speak about the applicability of the statistics C_l, $l = \overline{1,3}$, for solving the estimation problem of the flow parameters by the method of moments. For example, statistics C_1 is close to the analytically obtained $E[\tau]$, $\delta = \frac{|E[\tau]-\hat{E}[\tau]|}{E[\tau]} < 0,009$.

The last example suggests that the constructed flow model, imitating its behavior, is correct and consistent with the input data, as confirmed by the information contained in the statistics C_l, $l = \overline{1,3}$.

The most interesting is the ability of working with real data, implemented in the simulation flow model. To this aim we used the traffic data collected by Leland and Wilson [29–31] over several Ethernet local area network at the Bellcore Morristown Research and Engineering Center and formed as sets, each of which contains one million packet arrivals. We considered the sets **BC-pAug89**, started at 11:25 on August 9, 1989 and lasting for about 3142,82 s (until one million packet were registered), and **BC-pOct89**, started at 11:00 on October 5, 1989 and lasting about 1759,62 s. Note that in this study, packet size (Ethernet data length in bytes) does not matter, only the moment of its arrival is important, which is interpreted as the moment of flow event occurrence, and the fact that previously [29] was used exclusively in the framework of demonstration a self-similarity traffic nature.

Using C_l, $l = \overline{1,3}$, and with formulas (15), (16) for the derived over each of **BC-pAug89** and **BC-pOct89** sequences of the interval durations between the moments of packet arrivals (events), the following estimation results are obtained, reported in Table 10 as $\hat{z}_1^{real}$, $\hat{z}_2^{real}$, $\hat{\gamma}^{real}$, in which, moreover, there are values $\hat{z}_1^{model}$, $\hat{z}_2^{model}$, $\hat{\gamma}^{model}$, determined by the estimation procedure of the corresponding parameters based on the one of the flow realizations, namely the one (with corresponding λ_i, α_i, $P_1^{(1)}(\lambda_j|\lambda_i)$, $P_1^{(2)}(\lambda_j|\lambda_i)$, $i,j = 1,2$) for which the selected quality indicator (the relative error value $\delta[\hat{\theta}] = \frac{|\hat{\theta}^{real} - \hat{\theta}^{model}|}{\hat{\theta}^{real}}$, $\hat{\theta} \in \{\hat{z}_1, \hat{z}_2, \hat{\gamma}\}$) reaches the accuracy $\delta[\hat{\theta}] < \delta = 0,005$. We emphasize that the fluctuations of $\delta[\hat{\theta}]$, due to the choice of a particular data set as a model (each obtained through the simulation process and contains one million values of the inter-event interval durations), are insignificant, which allows to draw conclusions about the modeling quality and the legitimacy of its application to the parameter estimation problems based on the one specific realization.

Table 10. Results of working with the real data

BC-pAug89	$\hat{z}_1^{real} = 390,6379$	$\hat{z}_2^{real} = 45,4911$	$\hat{\gamma}^{real} = 0,9700$
	$\hat{z}_1^{model} = 391,1351$	$\hat{z}_2^{model} = 45,4657$	$\hat{\gamma}^{model} = 0,9696$
	$\delta[\hat{z}_1] = 0,0013$	$\delta[\hat{z}_2] = 0,0006$	$\delta[\hat{\gamma}] = 0,0004$
BC-pOct89	$\hat{z}_1^{real} = 693,5398$	$\hat{z}_2^{real} = 76,9829$	$\hat{\gamma}^{real} = 0,9725$
	$\hat{z}_1^{model} = 692,9417$	$\hat{z}_2^{model} = 77,1840$	$\hat{\gamma}^{model} = 0,9724$
	$\delta[\hat{z}_1] = 0,0009$	$\delta[\hat{z}_2] = 0,0026$	$\delta[\hat{\gamma}] = 0,0001$

In order to establish the adequacy of the data obtained through the model with the real traffic and, as a consequence, the applicability of the simulation apparatus to obtain numerical results of estimation both parameters and flow states, the relative errors of the numerical characteristics estimates were determined (such as mean, median and variance) from the real and model data sets. The results are shown in Table 11.

The results indicate that the flow simulation model used in the current study, as well as in [12,13], not only does not contradict the input data, as was revealed in the third experiment, but also corresponds to the actual arrival processes (in this case, recorded by Leland and Wilson [29–31]), which ensures the consistency of its use to obtain the observed moments of occurrence of events and solve on their basis the previously allocated problems for doubly stochastic flows.

The data of Table 10 at the same time indicate that the method of moments for the estimation of $z_1 = (\lambda_1 + \alpha_1)$ (intensity of occurrence of the flow events at the first state) and $z_2 = (\lambda_2 + \alpha_2)$ (intensity of occurrence of the flow events at

Table 11. Results of working with the real data

	BC-pAug89			BC-pOct89		
	Real	Model	δ	Real	Model	δ
Mean	0,003143	0,003146	0,000955	0,001760	0,001761	0,000568
Median	0,001956	0,001948	0,004090	0,001032	0,001035	0,002907
Variance $\cdot 10^{-5}$	3,184068	3,195083	0,003459	1,023295	1,022307	0,000966

the second state) of a correlated synchronous generalized flow of the second order allows to draw conclusions regarding, for example, the channel capacity as one of the most important characteristics of systems like Ethernet local area networks, quite effectively (in sense of the smallness of the relative error $\delta[\hat{\theta}] < \delta = 0,005$, $\hat{\theta} \in \{\hat{z}_1, \hat{z}_2, \hat{\gamma}\}$, in all cases). In addition, Tables 10, 11 together establish the previously noted adequacy of the mathematical models of arrival processes of data packets in the form of doubly stochastic flows.

6 Conclusion

This paper considered a synchronous generalized event flow of the second order under its complete observability. The probability density of the inter-event interval duration $p(\tau)$ was obtained for the correlated doubly stochastic flow in the form (10), as well as by the method of moments, estimates of the distribution parameters $\hat{z}_1$, $\hat{z}_2$ were found in the form (15) and $\hat{\gamma}$ in the form (16), which allow calculations without using numerical methods.

A number of statistical experiments were carried out on the simulation flow model implemented as a Windows Forms application project using the object-oriented C# programming language in Microsoft Visual Studio environment, the numerical results of which do not contradict the physical interpretation and illustrate the acceptable quality of the estimates $\hat{z}_1$, $\hat{z}_2$, $\hat{\gamma}$ in sense of the smallness of the the sample quadratic error. In addition, this paper illustrates an example with real data, which reflects the feasibility of using doubly stochastic event flows as mathematical models of real information flows in modern packet-switching networks, as well as the correctness and practical applicability of the results obtained by the model, reproducing the considered flows behavior.

References

1. Dudin, A.N., Klimenok, V.I.: Queueing Systems with Correlated Flows. BSU, Minsk (2000). (in Russian)
2. Basharin, G.P., Gaidamaka, Y.V., Samouylov, K.E.: Mathematical theory of tele-traffic and its application to the analysis of multiservice communication of next generation networks. Autom. Control Comput. Sci. **47**(2), 62–69 (2013)

3. Vishnevsky, V.M., Semenova, O.V., Dudin, A.N., Klimenok, V.I.: Approximate method to study M/G/1-type polling system. Qual. Technol. Quant. Manage. (QTQM) **9**(2), 211–228 (2012)
4. Cox, D.R.: The analysis of non-Markovian stochastic processes by the inclusion of supplementary variables. In: Proceedings of the Cambridge Philosophical Society, vol. 51, no. 3, pp. 433–441 (1955)
5. Kingman, J.F.C.: On doubly stochastic Poisson process. Proc. Camb. Philos. Soc. **60**(4), 923–930 (1964)
6. Basharin, G.P., Kokotushkin, V.A., Naumov, V.A.: On the equivalent substitutions method for computing fragments of communication networks. In: Part 1. Proceedings of USSR Academy of Sciences. Technical Cybernetics, vol. 6, pp. 92–99 (1979). (in Russian)
7. Basharin, G.P., Kokotushkin, V.A., Naumov, V.A.: On the equivalent substitutions method for computing fragments of communication networks. In: Part 2. Proceedings of USSR Academy of Sciences. Technical Cybernetics, vol. 1, pp. 55–61 (1980). (in Russian)
8. Neuts, M.F.: A versatile Markov point process. J. Appl. Probab. **16**, 764–779 (1979)
9. Lucantoni, D.M.: New results on the single server queue with a batch Markovian arrival process. Commun. Stat. Stochast. Models **7**, 1–46 (1991)
10. Nezhel'skaya, L.A.: Joint probability density of the intervals duration in modulated MAP event flows and its recurrence conditions. Tomsk State Univ. J. Control Comput. Sci. **1**, 57–67 (2015). (in Russian)
11. Gortsev, A.M., Nezhel'skaya, L.A.: On connection of MC flows and MAP flows of events. Tomsk State Univ. J. Control Comput. Sci. **1**(14), 13–21 (2011). (in Russian)
12. Nezhelskaya, L., Sidorova, E.: Optimal estimation of the states of synchronous generalized flow of events of the second order under its complete observability. Commun. Comput. Inf. Sci. **912**, 157–171 (2018)
13. Nezhelskaya, L.A., Sidorova, E.F.: Optimal estimate of the states of a generalized synchronous flow of second-order events under conditions of incomplete observability. Tomsk State Univ. J. Control Comput. Sci. **45**, 30–41 (2018). (in Russian)
14. Dudin, A.N., Nazarov, A.A.: The MMAP/M/R/0 queueing system with reservation of servers operating in a random environment. Prob. Inf. Transm. **51**(3), 289–298 (2015)
15. Vishnevsky, V.M., Dudin, A.N., Kozyrev, D.V., Larionov, A.A.: Performance evaluation of broadband wireless networks along the long transport routes. In: proceedings of the Eighteenth International Scientific Conference Distributed Computer and Communication Networks: Control, Computation, Communications (DCCN) 2015, pp. 241–256, ICS RAS Moscow (2015)
16. Lucantoni, D.M., Neuts, M.F.: Some steady-state distributions for the MAP/SM/1 queue. Commun. Stat. Stoch. Models **10**, 575–598 (1994)
17. Gortsev, A.M., Nezhel'skaya, L.A., Shevchenko, T.I.: Estimation of the states of an MC-stream of events in the presence of measurements errors. Russ. Phys. J. **36**(12), 1153–1167 (1993)
18. Gortsev, A.M., Nezhel'skaya, L.A., Solov'ev, A.A.: Optimal state estimation in MAP-event flows with unextendable dead time. Autom. Remote Control **73**(8), 1316–1326 (2012)
19. Nezhelskaya, L.: Optimal state estimation in modulated MAP event flows with unextendable dead time. Commun. Comput. Inf. Sci. **487**, 342–350 (2014)

20. Berezin, D.V., Nezhel'skaya, L.A.: Optimal state estimation of generalized MAP under conditions of non-extendable dead time. Tomsk State Univ. J. Control Comput. Sci. **41**, 12–23 (2017). (in Russian)

21. Nezhelskaya, L., Tumashkina, D.: Optimal state estimation of semi-synchronous event flow of the second order under its complete observability. Commun. Comput. Inf. Sci. **912**, 93–105 (2018)

22. Gortsev, A.M., Nezhel'skaya, L.A.: Estimate of parameters of synchronously alternating Poisson stream of events by the moment method. Telecommun. Radio Eng. (Eng. Trans. Elektrosvyaz and Radiotekh.) **50**(1), 56–63 (1996)

23. Gortsev, A.M., Nezhel'skaya, L.A.: Estimation of the parameters of a synchro-alternating Poisson event flow by the method of moments. Radiotekhnika **40**(7–8), 6–10 (1995)

24. Gortsev, A.M., Nezhel'skaya, L.A.: Dead time period and parameter estimation of synchronous alternating flow of events. Tomsk State Univ. J. **S6**, 232–239 (2003). (in Russian)

25. Gortsev, A.M., Kalyagin, A.A., Nezhel'skaya, L.A.: The joint probability density of duration of the intervals in a generalized semisynchronous flow of events with unprolonging dead time. Tomsk State Univ. J. Control Comput. Sci. **2**(27), 19–29 (2014). (in Russian)

26. Okamura, H., Dohi, T., Trivedi, K.S.: Markovian arrival process parameter estimation with group data. IEEE/ACM Trans. Netw. **17**(4), 1326–1339 (2009)

27. Gortsev, A.M., Nezhel'skaya, L.A.: Parameter estimation of synchronous twice-stochastic flow of events using the method of moments. Tomsk State Univ. J. **S1–1**, 24–29 (2002). (in Russian)

28. Sobol', I.M.: Numerical Monte-Carlo Methods. Nauka, Moscow (1973). (in Russian)

29. Leland, W.E., Wilson, D.V.: High time-resolution measurement and analysis of LAN traffic: implications for LAN interconnection. In: Proceedings of the IEEE INFOCOM 1991, pp. 1360–1366. Bal Harbour (1991)

30. Fowler, H.J., Leland, W.E.: Local area network traffic characteristics, with implications for broadband network congestion management. IEEE J. Sel. Areas Commun. **9**(7), 1139–1149 (1991)

31. Leland, W., Taqqu, M., Willinge, R.W., Wilson, D.: On the self-similar nature of Ethernet traffic (extended version). IEEE/ACM Trans. Netw. **2**(1), 1–15 (1994)

On a Two-Server Queue
with Consultation in Random
Environment

Achyutha Krishnamoorthy[1]($\boxtimes$), Resmi Thekkiniyedath[2],
and Balakrishna Lakshmy[3]

[1] Centre for Research in Mathematics, CMS College, Kottayam, India
achyuthacusat@gmail.com
[2] Department of Mathematics, KKTM Government College,
Pullut, Kodungallur, India
[3] 'Balakrishna', Master Road, Vaduthala PO, Cochin 682023, India

Abstract. This paper analyses a two-server queueing model with consultations in random environment given by the main server to the regular server. The main server not only serves customers but also provides consultation to the regular server with a preemptive priority over customers. The customers at the main server undergo interruptions during their service. There are upper bounds for the number of interruptions at the main server and the number of consultations to the regular server. There are K environment factors for consultations. The arrival process and requirement of consultation follow mutually independent Poisson processes. The service times at the main server and the regular server are assumed to follow mutually independent phase type distributions. Duration of threshold clock and the consultation time for each factor are independent and mutually distributed exponential random variables. The stability condition is established. Some performance measures are studied numerically.

Keywords: Main server · Regular server · Consultation ·
Interruption · Environmental factors

1 Introduction

A multi-server queueing system with consultation has many applications in day today life. Such a system is introduced by Chakravarthy [3]. In this paper the author narrates a situation he has experienced in an airport which lead him to the formulation of this model. In multi-server queueing system, the servers may be of different experience level. If some servers are trainees or beginners, they need frequent clarifications for the smooth progress of their work. So an experienced server (namely, main server) helps the fellow servers to clear their doubts together with serving customers. Such queueing systems are common in banks, super market check outs, hospitals, etc. Quality of the service will be improved by consultations.

© Springer Nature Switzerland AG 2019
A. Dudin et al. (Eds.): ITMM 2019, CCIS 1109, pp. 217–229, 2019.
https://doi.org/10.1007/978-3-030-33388-1_18

The queueing system in the above mentioned work consists of c servers. One of these c servers is referred to as the main server and the others as the regular servers. The request for consultation will be attended immediately if the main server is idle. Otherwise, consultation is provided by the main server with preemptive priority over customers to the regular servers. The request for consultation of the regular server is attended by the main server, even if there is a customer being served at the main server. Then that customer at the main server has to wait until the consultation is completed. At this stage the service of the customer at the main server is said to be interrupted. (So the word 'interruption' is associated with the customer at the main server when the main server is providing consultation to the regular server.) The service of the interrupted customer at the main server will be resumed after all consultations are completed. The regular servers who need consultation are queued up at the main sever. The consultations are given in a FCFS basis. The regular servers receive any number of consultations during the service of a customer. The customer at the regular server is not said to be interrupted since consultation is a part of his service. The service times at the main server and the regular servers follow independent exponential distributions.

Krishnamoorthy et al. [8] dealt with a single server queueing model with interruptions to the server controlled by a super clock and a finite number of interruptions. No further interruptions are allowed to that customer after the maximum number of interruptions or after the realisation of the super clock. A threshold clock determines whether the service will be resumed or restarted after each interruption.

Queues with service interruptions are first studied by White and Christie [11]. There are two queues served by a single server. The priority I and II queues are served such that the arrival of a priority I customer preempt the priority II customer. The duration of interruption is an exponentially distributed random variable. The service will be resumed at the end of an interruption. Gaver [5] considers a queueing model with interruption. After the completion of interruption, the service will be repeted or resumed, but there is no particular rule to determine the repetition or resumption. The distributions of the completion time of the job in three cases of repetition, resumption and postponable interruption are computed.

Some of the earlier papers which analyse queueing models with service interruptions, assuming general distributions for the service and interruption durations are by Keilson [7], Ibe and Trivedi [6], Avi-Izhak and Naor [1] and Fiems et al. [4].

In the paper by Chakravarthy [3], the service of the customer at the main server is interrupted at the time of request of the regular server for consultation. There is no an upper bound for the number of interruptions to a customer at the main server. At the same time, the regular server is free to get any number of consultations during the service of a customer. It is not fair to interrupt a customer at the main server too many times or to avail too many consultations by the regular server. So in order to control the number of interruptions and consultations, we impose some upper bounds for them. A maximum of L interruptions are allowed to a customer at the main server. The maximum number of consultations possible to the regular server during the service of a particular customer is M.

A super clock is also introduced to control the number of interruptions. The super clock starts at the beginning of the first interruption to the customer at the main server. At the completion of the interruption, the super clock freezes. When the next interruption to the same customer starts, the super clock again starts from the phase where it has stopped and so on. If the super clock expires during the consultation with one interrupted customer at the main server, then the present consultation is permitted to complete and no more interruption is allowed to befall to that particular customer at the main server. The super clock starts anew when the interruption happens to a new customer at the main server.

If the number of interruptions has reached the upper bound or the super clock has expired and if the regular server needs further consultation, then he/she has to wait until the service at the main server is completed. After the service of the interrupted customer is completed, the main server will immediately attend the consultation before taking a new customer from the queue for service. Since no customer is interrupted at the main server, super clock is not present at this time.

A queueing system in an alternating random environment is discussed by Bhaskar Senguptha [2]. In this paper, the server is subjected to random breakdown. Until it is repaired, the server cannot serve customers. He also assumes that during the break down period, another service facility is provided to some of the arriving customers.

The consultations may be for different matters. For example, the manager in a bank has to provide clarifications to the fellow officers regarding debits, credits, cheques, demand drafts, loans, opening, closing and transfer of accounts, filling cash in ATM's, etc. These are different environments for consultation. In the present paper, we assume that consultations are due to K factors in random environment. Here we consider single factors only, even though combinations of these factors are possible.

2   Model Description

We consider a two-server queueing system with a main server and a regular server. The arrival of customers to the system follows a Poisson process with rate λ. The service time at the main server is a phase type distributed random variable with representation (α, A), where the number of phases is a. The phase type distribution (β, B) with number of phases b represents the service time at the regular server. Here $A^0 = -A\mathbf{e}$ and $B^0 = -B\mathbf{e}$, where $\mathbf{e}$ is a column vector of $1's$ of appropriate order.

The main server offers consultation to the regular server whenever it is needed. Assume that the consultations are provided to the regular server for the K random environmental factors $\phi_1, \phi_2, \ldots\ldots, \phi_K$. The requirement of consultation follows a Poisson process with rate θ. The ith factor of the random environment occurs with probability δ_i, $i = 1, 2, ..., K$. The duration of consultation for the ith factor is exponentially distributed with parameter ξ_i. The upper bounds for interruptions to a customer at the main server and that for the consultations to the regular server are L and M, respectively. The duration of super clock follows a phase type distribution with representation (γ, C), where the number of phases is c. Here $C^0 = -C\mathbf{e}$. The restart or resumption of services at both the servers are determined by the realisation of the threshold clock. The duration of the threshold clock follows an exponential distribution with parameter η.

Notations: We use the following notations in this model.

$$L_0 = L(c+1), L_1 = L_0 + 1, L_2 = L_1 a + (M+1)b + 2MKb,$$
$$L_3 = L_1(M+1)ab + 2MKb + 2L_0MKab + 2MKab,$$
$$\dot{I}_1 = \begin{bmatrix} O & I_{L_0} \end{bmatrix}_{L_0 \times L_1},$$
$$\tilde{\alpha} = e'_{L_1}(1) \otimes \alpha, \tilde{\gamma} = (\gamma, 0), \psi = (1, 0),$$
$$\tilde{C} = \begin{bmatrix} C & C^0 \\ \mathbf{0} & 0 \end{bmatrix}, \tilde{E} = \eta \begin{bmatrix} -1 & 1 \\ 0 & 0 \end{bmatrix},$$
$$\boldsymbol{\xi} = (\xi_1, \xi_2, ..., \xi_K)', \tilde{\xi} = diag(\xi_1, \xi_2, ..., \xi_K).$$

3   The Queueing Model

Consider the queueing model $Z = \{Z(t), t \geq 0\}$,
where $Z(t) = \{N(t), H(t), B_1(t), B_2(t), H_1(t), H_2(t), G(t), J_1(t), J_2(t)\}$.

The variables are defined as follows:

$N(t)$ – the number of customers in the system,

$B_1(t)$ – number of consultations already enjoyed by
the regular server during the service of a particular customer,

$B_2(t)$ – number of interruptions already befell to a customer at the main server,

$H_1(t)$ – phase of the super clock,

$J_1(t)$ – phase of the main server,

$J_2(t)$ – phase of the regular server.

Here $H(t)$ denotes the status of the servers at time t such that

$$H(t) = \begin{cases} \tilde{0}, & \text{if only the regular server is busy} \\ 0, & \text{if the main server together with or without} \\ & \text{the regular server is busy} \\ 1, & \text{if the main server is giving consultation only} \\ 2, & \text{if the main server is giving consultation} \\ & \text{with one interrupted customer at the main server} \\ 3, & \text{if the regular server is waiting for getting consultation} \\ & \text{after the present service at the main server.} \end{cases}$$

The status of the threshold clock is denoted by $H_2(t)$.

$$H_2(t) = \begin{cases} 1, & \text{if the threshold clock is running} \\ 0, & \text{if the threshold clock has expired.} \end{cases}$$

Note that $B_2(t)$ is '0' means the customer at the main server has not interrupted yet and so super clock has not started. In this case the super clock has no role to play. So we do not consider the super clock variable $H_1(t)$ when $B_2(t) = 0$. Also, since super clock is associated with the interruption to a customer at the main server and no customer is present at the main server during the 'consultation only' mode, super clock is not 'present' at this mode.

Here $G(t)$ represents the environmental factor due to which consultation is in progress/ waiting to get consultation.
$G(t) = i$, if the consultation is due to the ith factor.
$\{Z(t), t \geq 0\}$ is a Continuous Time Markov Chain with state space

$$\Omega = \{0\} \cup \bigcup_{i=1}^{\infty} \omega(i).$$

The terms $\omega(i)$'s are defined as

$\omega(1) = \omega(1,0) \cup \omega(1,\tilde{0}) \cup \omega(1,1)$ and
$\omega(i) = \omega(i,0) \cup \omega(i,1) \cup \omega(i,2) \cup \omega(i,3)$, for $i \geq 2$,
where
$\omega(1,0) = \{(1,0,0,t_1)\} \cup \{(1,0,k,l_1,t_1) : 1 \leq k \leq L\}$,
$\omega(1,\tilde{0}) = \{(1,\tilde{0},j,t_2) : 0 \leq j \leq M\}$,
$\omega(1,1) = \{(1,1,j,l_2,l_3,t_2) : 0 \leq j \leq M-1\}$,
and for $i \geq 2$,
$\omega(i,0) = \{(i,0,j,0,t_1,t_2) \cup (i,0,j,k,l_1,t_1,t_2) : 0 \leq j \leq M, 1 \leq k \leq L\}$,
$\omega(i,1) = \{(i,1,j,l_2,l_3,t_2) : 0 \leq j \leq M-1\}$,
$\omega(i,2) = \{(i,2,j,k,l_1,l_2,l_3,t_1,t_2) : 0 \leq j \leq M-1, 0 \leq k \leq L-1\}$,
$\omega(i,3) = \{(i,3,j,l_2,l_3,t_1,t_2) : 0 \leq j \leq M-1\}$,
with $0 \leq l_1 \leq c, l_2 = \{1,0\}, 1 \leq l_3 \leq K, 1 \leq t_1 \leq a$ and $1 \leq t_2 \leq b$.

The infinitesimal generator matrix $\hat{Q}$ is given by

$$\hat{Q} = \begin{bmatrix} -\lambda & S_1 & & & \\ S_2 & S_3 & S_4 & & \\ & S_5 & T_1 & T_0 & \\ & & T_2 & T_1 & T_0 \\ & & & \ddots & \ddots & \ddots \end{bmatrix}, \tag{1}$$

where the block matrices appearing in $\hat{Q}$ are as follows:

$$S_1 = \lambda \begin{bmatrix} \tilde{\boldsymbol{\alpha}} & \underline{\mathbf{0}} \end{bmatrix}, S_2 = \begin{bmatrix} \mathbf{e}_{M_1} \otimes A^0 \\ \mathbf{e}_{K+1} \otimes B^0 \\ \mathbf{0} \end{bmatrix}, S_3 = \begin{bmatrix} I_{M_1} \otimes A & O & O \\ O & S_{31} & S_{32} \\ O & S_{33} & S_{34} \end{bmatrix} - \lambda I,$$

$$S_4 = \lambda \begin{bmatrix} S_{41} & S_{42} & O \end{bmatrix}, S_5 = \begin{bmatrix} S_{51} & S_{52} & S_{53} \end{bmatrix},$$

$$T_0 = \lambda I, T_1 = \begin{bmatrix} T_{11} & O & T_{12} & T_{13} \\ T_{14} & S_{34} & O & O \\ T_{15} & O & T_{16} & O \\ O & O & O & T_{17} \end{bmatrix} - \lambda I, T_2 = \begin{bmatrix} T_{21} & S_{53} & O \end{bmatrix}.$$

Here T_0, T_1 and T_2 are square matrices of order L_3; S_3 is a square matrix of order L_2 and S_1, S_2, S_4, S_5 are matrices of orders $1 \times L_2$, $L_2 \times 1$, $L_2 \times L_3$ and $L_3 \times L_2$, respectively.

Here

$$S_{31} = \begin{bmatrix} I_M \otimes (B - \theta I) & O \\ O & B \end{bmatrix}_{(M+1)b}, S_{32} = \theta \begin{bmatrix} I_M \\ O \end{bmatrix}_{(M+1) \times M} \otimes \boldsymbol{\psi} \otimes \boldsymbol{\delta} \otimes I_b,$$

$$S_{33} = \begin{bmatrix} O & I_M \otimes \Delta_b' \end{bmatrix}_{2MKb \times (M+1)b}, S_{34} = I_M \otimes \tilde{\nabla} \otimes I_b,$$

$$S_{41} = \begin{bmatrix} \mathbf{e}_{M+1}'(1) \otimes I_{L_1} \otimes I_a \beta \\ I_{M+1} \otimes \tilde{\boldsymbol{\alpha}} \otimes I_b \\ O \end{bmatrix}_{L_2 \times (M+1)L_1 ab}, S_{42} = \begin{bmatrix} O \\ I_{2MKb} \end{bmatrix}_{L_2 \times 2MKb},$$

$$S_{51} = \begin{bmatrix} \mathbf{e}_{M+1} \otimes I_{L_1} \otimes I_a \otimes B^0 \\ O \end{bmatrix}_{L_3 \times L_0 a}, S_{52} = \begin{bmatrix} I_{M+1} \otimes \mathbf{e}_{L_1} \otimes A^0 \otimes I_b \\ O \end{bmatrix}_{L_3 \times (M+1)b},$$

$$S_{53} = \begin{bmatrix} O \\ I_{MK} \otimes I_2 \otimes A^0 \otimes I_b \end{bmatrix}_{L_3 \times 2MKb},$$

$$T_{11} = \begin{bmatrix} I_M \otimes I_{L_1} \otimes (A \oplus B - \theta I) & O \\ O & I_{L_1} \otimes (A \oplus B) \end{bmatrix}_{L_1(M+1)ab},$$

$$T_{12} = \theta \begin{bmatrix} I_M \otimes P \\ O \end{bmatrix}_{L_1(M+1) \times L_0 M} \otimes \boldsymbol{\delta} \otimes \boldsymbol{\psi} \otimes I_{ab},$$

$$T_{13} = \theta \begin{bmatrix} I_M \otimes P^* \\ O \end{bmatrix}_{L_1(M+1) \times M} \otimes \boldsymbol{\delta} \otimes \boldsymbol{\psi} \otimes I_{ab},$$

$$T_{14} = \begin{bmatrix} O & I_M \otimes \hat{\Delta} \end{bmatrix}_{2MKb \times L_1(M+1)ab},$$

$$T_{15} = \begin{bmatrix} O & I_M \otimes \hat{I}_1 \otimes \Delta^* \end{bmatrix}_{2L_0MKab \times L_1(M+1)ab},$$

$$T_{16} = I_M \otimes I_L \otimes (\tilde{C} \oplus \tilde{\nabla}) \otimes I_{ab}, T_{17} = I_{MK} \otimes (\tilde{E} \oplus A) \otimes I_b,$$

$$T_{21} = \begin{bmatrix} \tilde{A}^0 + \tilde{B}^0 \\ O \end{bmatrix}_{L_3 \times L_1(M+1)ab}.$$

Here

$$P = \begin{bmatrix} diag(\tilde{\gamma}, I_{L-1} \otimes \hat{I}_c) \\ O \end{bmatrix}_{L_1 \times L_0}, P^* = \begin{bmatrix} 0 \\ \mathbf{e}_{L-1} \otimes \hat{\mathbf{e}}_c \\ \mathbf{e}_{c+1} \end{bmatrix}_{L_1 \times 1},$$

$$\tilde{A}^0 = I_{M+1} \otimes \mathbf{e}_{L_1} \otimes A^0 \otimes \boldsymbol{\alpha} \otimes I_b, \tilde{B}^0 = \mathbf{e}_{M+1} \otimes I_{L_1} \otimes I_a \otimes B^0 \otimes \boldsymbol{\beta},$$

$$\Delta_b' = \begin{bmatrix} \boldsymbol{\xi} \otimes I_b \\ \boldsymbol{\xi} \otimes \mathbf{e}_b \otimes \boldsymbol{\beta} \end{bmatrix}, \tilde{\nabla} = -I_2 \otimes \tilde{\xi} + \tilde{E} \otimes I_K,$$

$$\hat{\Delta} = \begin{bmatrix} \boldsymbol{\xi} \otimes \tilde{\boldsymbol{\alpha}} \otimes I_b \\ \boldsymbol{\xi} \otimes \mathbf{e}_b \otimes \tilde{\boldsymbol{\alpha}} \otimes \boldsymbol{\beta} \end{bmatrix}, \Delta^* = \begin{bmatrix} \boldsymbol{\xi} \otimes I_{ab} \\ \boldsymbol{\xi} \otimes \mathbf{e}_{ab} \otimes \boldsymbol{\alpha} \otimes \boldsymbol{\beta} \end{bmatrix}.$$

4 Steady State Analysis

The steady state analysis of the queueing system under study is performed in this section. Let us first establish the stability condition of the queueing model.

4.1 Stability Condition

Let $\boldsymbol{\pi}$ denote the steady-state probability vector of the generator $T_0 + T_1 + T_2$. That is, $\boldsymbol{\pi}(T_0 + T_1 + T_2) = 0$; $\boldsymbol{\pi}\mathbf{e} = 1$. The LIQBD description of the model indicates that the queueing system is stable (see, Neuts [10]) if and only if

$$\lambda < \boldsymbol{\pi} T_2 \mathbf{e}. \tag{2}$$

That is, the rate of drift to the left has to be higher than that to the right. The vector $\boldsymbol{\pi}$ cannot be obtained explicitly in terms of the parameters of the model.

For future reference, we define the traffic intensity ρ as

$$\rho = \frac{\lambda}{\boldsymbol{\pi} T_2 \mathbf{e}}. \tag{3}$$

Note that the stability condition in Eq. (2) is equivalent to $\rho < 1$. We will discuss the impact of the input parameters of the model on the traffic intensity in Sect. 5.

4.2 Steady State Probability Vector

The present model is studied as a QBD process. Its stationary distribution has a matrix-geometric solution. We assume that the stability condition given by (2) holds. Let the steady-state probability vector of the generator $\hat{Q}$ given in Eq. (1) be denoted by $\mathbf{z}$. That is,

$$\mathbf{z}\hat{Q} = 0; \mathbf{z}\mathbf{e} = 1. \tag{4}$$

Partitioning z as

$$z = (z_0, z_1, z_2, z_3, \dots\dots\dots), \qquad (5)$$

we see that the sub-vectors of z, under the assumption that the stability condition (2) holds, are obtained as (see, Neuts [10])

$$z_j = z_2 R^{j-2}, j \geq 3,$$

where R is the minimal non-negative solution to the matrix quadratic equation

$$R^2 T_2 + R T_1 + T_0 = 0.$$

z_0, z_1 and z_2 are obtained using the boundary equations

$$-\lambda z_0 + z_1 S_2 = 0;$$

$$z_0 S_1 + z_1 S_3 + z_2 S_5 = 0;$$

$$z_1 S_4 + z_2 (T_1 + R T_2) = 0.$$

The normalizing condition given by (4) results in

$$z_0 + z_1 \mathbf{e} + z_2 (I - R)^{-1} \mathbf{e} = 1.$$

Once the rate matrix R is obtained, we compute the vector z by exploiting the special structure of the coefficient matrices.

4.3 Performance Measures

After calculating the steady state probability vector, we now calculate some key performance measures of the system and their formulae for computation. This helps to bring out the qualitative aspects of the present model.

Towards this end, we further partition the vectors z_i as $z_1 = (z_{10}, z_{1\bar{0}}, z_{11})$ and $z_i = (z_{i0}, z_{i1}, z_{i2}, z_{i3})$, for $i \geq 2$.

Note that z_0 is a scalar, $z_{10}, z_{1\bar{0}}, z_{11}, z_{i0}, z_{i1}, z_{i2}, z_{i3}$, for $i \geq 2$ are vectors of dimensions $M_1 a$, $(K+1)b$, $2KLb$, $(K+1)M_1 ab$, $2KLb$, $2M_0 KLab$ and $2KLab$, respectively.

(a) Mean number of customers in the system

$$\mu_1 = \sum_{i=1}^{\infty} i z_i \mathbf{e}.$$

(b) Mean number of customers in the queue

$$\mu_2 = \sum_{i=2}^{\infty} (i-1) z_{i1} \mathbf{e} + \sum_{i=3}^{\infty} (i-2)(z_{i0} \mathbf{e} + z_{i2} \mathbf{e} + z_{i3} \mathbf{e}).$$

(c) Effective rate of consultation

$$E_1 = \theta \sum_{j=0}^{M-1} \boldsymbol{z}_{1\tilde{0}j}\mathbf{e} + \theta \sum_{i=2}^{\infty} \sum_{j=0}^{M-1} \boldsymbol{z}_{i0j}\mathbf{e}.$$

(d) Effective rate of interruption

$$E_2 = \theta \sum_{i=2}^{\infty} \sum_{j=0}^{M-1} \boldsymbol{z}_{i0j0}\mathbf{e} + \theta \sum_{i=2}^{\infty} \sum_{j=0}^{M-1} \sum_{k=1}^{L-1} \sum_{l_1=1}^{c} \boldsymbol{z}_{i0jkl_1}\mathbf{e}.$$

(e) Fraction of time the main server is idle

$$F_1 = \boldsymbol{z}_0\mathbf{e} + \boldsymbol{z}_{1\tilde{0}}\mathbf{e}.$$

(f) Fraction of time the regular server is idle

$$F_2 = \boldsymbol{z}_0\mathbf{e} + \boldsymbol{z}_{10}\mathbf{e}.$$

(g) Fraction of time the main server is busy serving a customer

$$F_3 = \boldsymbol{z}_{10}\mathbf{e} + \sum_{i=2}^{\infty} \boldsymbol{z}_{i0}\mathbf{e} + \sum_{i=2}^{\infty} \boldsymbol{z}_{i3}\mathbf{e}.$$

(h) Fraction of time the regular server is busy serving a customer

$$F_4 = \boldsymbol{z}_{1\tilde{0}}\mathbf{e} + \sum_{i=2}^{\infty} \boldsymbol{z}_{i0}\mathbf{e}.$$

(i) Fraction of time main server remains interrupted

$$F_5 = \sum_{i=2}^{\infty} \boldsymbol{z}_{i2}\mathbf{e}.$$

(j) Fraction of time regular server is getting consultation

$$F_6 = \sum_{i=1}^{\infty} \boldsymbol{z}_{i1}\mathbf{e} + \sum_{i=2}^{\infty} \boldsymbol{z}_{i2}\mathbf{e}.$$

(k) Fraction of time regular server is waiting to get consultation

$$F_7 = \sum_{i=2}^{\infty} \boldsymbol{z}_{i3}\mathbf{e}.$$

(l) Rate at which interruption completion takes place before threshold is realised

$$R_1 = \sum_{i=2}^{\infty} \sum_{j=0}^{M-1} \sum_{k=0}^{L-1} \sum_{l_1=0}^{c} \sum_{l_3=1}^{K} \xi_{l_3} \boldsymbol{z}_{i2jkl_1 1l_3}\mathbf{e}.$$

(m) Rate at which interruption completion takes place after threshold is realised

$$R_2 = \sum_{i=2}^{\infty} \sum_{j=0}^{M-1} \sum_{k=0}^{L-1} \sum_{l_1=0}^{c} \sum_{l_3=1}^{K} \xi_{l_3} \mathbf{z}_{i2jkl_10l_3} \mathbf{e}.$$

(n) Rate at which consultation completion takes place before threshold is realised

$$R_3 = \sum_{i=1}^{\infty} \sum_{j=0}^{M-1} \sum_{l_3=1}^{K} \xi_{l_3} \mathbf{z}_{i1j1l_3} \mathbf{e} + \sum_{i=2}^{\infty} \sum_{j=0}^{M-1} \sum_{k=0}^{L-1} \sum_{l_1=0}^{c} \sum_{l_3=1}^{K} \xi_{l_3} \mathbf{z}_{i2jkl_11l_3} \mathbf{e}.$$

(o) Rate at which consultation completion takes place after the threshold is realised

$$R_4 = \sum_{i=1}^{\infty} \sum_{j=0}^{M-1} \sum_{l_3=1}^{K} \xi_{l_3} \mathbf{z}_{i1j0l_3} \mathbf{e} + \sum_{i=2}^{\infty} \sum_{j=0}^{M-1} \sum_{k=0}^{L-1} \sum_{l_1=0}^{c} \sum_{l_3=0}^{K} \xi_{l_3} \mathbf{z}_{i2jkl_10l_3} \mathbf{e}.$$

(p) Rate at which service completion at the main server takes place without any interruption

$$R_5 = \sum_{t_1=1}^{a} A_{t_1}^0 \mathbf{z}_{100t_1} \mathbf{e} + \sum_{i=2}^{\infty} \sum_{j=0}^{M} \sum_{t_1=1}^{a} A_{t_1}^0 \mathbf{z}_{i0j0t_1} \mathbf{e}.$$

(q) Rate at which service completion (with at least one interruption) at the main server takes place before super clock is realised

$$R_6 = \sum_{i=2}^{\infty} \sum_{j=0}^{M} \sum_{k=1}^{L} \sum_{l_1=1}^{c} \sum_{t_1=1}^{a} A_{t_1}^0 \mathbf{z}_{i0jkl_1t_1} \mathbf{e} + \sum_{j=0}^{M} \sum_{k=1}^{L} \sum_{l_1=1}^{c} \sum_{t_1=1}^{a} A_{t_1}^0 \mathbf{z}_{10jkl_1t_1} \mathbf{e}.$$

(r) Rate at which service completion (with at least one interruption) at the main server takes place after super clock is realised

$$R_7 = \sum_{i=2}^{\infty} \sum_{j=0}^{M} \sum_{k=1}^{L} \sum_{t_1=1}^{a} A_{t_1}^0 \mathbf{z}_{i0jk0t_1} \mathbf{e} + \sum_{j=0}^{M} \sum_{k=1}^{L} \sum_{t_1=1}^{a} A_{t_1}^0 \mathbf{z}_{10jk0t_1} \mathbf{e}.$$

(s) Rate at which service completion at the regular server takes place without any consultation

$$R_8 = \sum_{t_2=1}^{b} B_{t_2}^0 \mathbf{z}_{1\tilde{0}0t_2} \mathbf{e} + \sum_{i=2}^{\infty} \sum_{t_1=1}^{a} \sum_{t_2=1}^{b} B_{t_2}^0 \mathbf{z}_{i000t_1t_2} \mathbf{e}$$

$$+ \sum_{i=2}^{\infty} \sum_{k=1}^{L} \sum_{l_1=0}^{c} \sum_{t_1=1}^{a} \sum_{t_2=1}^{b} B_{t_2}^0 \mathbf{z}_{i00kl_1t_1t_2} \mathbf{e}.$$

(t) Rate at which service completion (with at least one consultation) at the regular server takes place

$$R_9 = \sum_{j=1}^{M} \sum_{t_2=1}^{b} B_{t_2}^0 z_{1\tilde{0}jt_2} e + \sum_{i=2}^{\infty} \sum_{j=1}^{M} \sum_{t_1=1}^{a} \sum_{t_2=1}^{b} B_{t_2}^0 z_{i0j0t_1t_2} e$$

$$+ \sum_{i=2}^{\infty} \sum_{j=1}^{M} \sum_{k=1}^{L} \sum_{l_1=0}^{c} \sum_{t_1=1}^{a} \sum_{t_2=1}^{b} B_{t_2}^0 z_{i0jkl_1t_1t_2} e.$$

5 Numerical Results

In this section, we present some examples numerically to describe the system characteristics of the queueing model under study. The effect of the parameters λ and θ on the key performance measures are analysed here.

Let us choose the following data so that the system is stable.

$$A = \begin{bmatrix} -12 & 6 \\ 5 & -10 \end{bmatrix}; B = \begin{bmatrix} -9 & 3 \\ 2 & -8 \end{bmatrix}; C = \begin{bmatrix} -12 & 8 \\ 8 & -12 \end{bmatrix};$$

$$\alpha = \begin{bmatrix} 0.3 & 0.7 \end{bmatrix}; \beta = \begin{bmatrix} 0.4 & 0.6 \end{bmatrix}; \gamma = \begin{bmatrix} 0.6 & 0.4 \end{bmatrix};$$

$$\delta = \begin{bmatrix} 0.3 & 0.4 & 0.3 \end{bmatrix}; \xi = \begin{bmatrix} 1 & 1.5 & 2 \end{bmatrix}^T; L = 3; M = 3.$$

Table 1. Effect of θ on various performance measures

$\lambda = 2$						
θ	3	3.5	4	4.5	5	5.5
ρ	0.5309	0.5803	0.6266	0.6699	0.7103	0.7481
μ_1	1.6367	1.9958	2.4136	2.8976	3.4560	4.0969
μ_2	0.9242	1.2177	1.5694	1.9869	2.4778	3.0482
E_1	0.2906	0.3432	0.3960	0.4482	0.4993	0.5489
E_2	0.1537	0.1784	0.2026	0.2261	0.2487	0.2702
F_1	0.5321	0.4974	0.4626	0.4281	0.3944	0.3618
F_2	0.6797	0.6358	0.5918	0.5482	0.5054	0.4639
F_3	0.2574	0.2537	0.2501	0.2463	0.2426	0.2388
F_4	0.1017	0.1046	0.1073	0.1099	0.1121	0.1141
F_5	0.1348	0.1600	0.1855	0.2108	0.2358	0.2602
F_6	0.2105	0.2489	0.2874	0.3255	0.3629	0.3991
F_7	0.0081	0.0107	0.0134	0.0164	0.0195	0.0226

Table 2. Effect of λ on various performance measures

$\theta = 2$						
λ	2	2.5	3	3.5	4	4.5
ρ	0.4224	0.5280	0.6336	0.7392	0.8448	0.9504
μ_1	1.0672	1.8760	3.2097	5.3483	8.6509	13.510
μ_2	0.4818	1.0897	2.2077	4.0546	6.6089	9.3050
E_1	0.1871	0.2755	0.3701	0.4617	0.5380	0.5840
E_2	0.1035	0.1611	0.2281	0.2983	0.3646	0.4212
F_1	0.6003	0.4952	0.3929	0.2992	0.2196	0.1566
F_2	0.7652	0.6536	0.5337	0.4162	0.3115	0.2258
F_3	0.2644	0.3057	0.3394	0.3654	0.3820	0.3843
F_4	0.0957	0.1410	0.1895	0.2365	0.2756	0.2991
F_5	0.0861	0.1340	0.1890	0.2453	0.2940	0.3242
F_6	0.1352	0.1992	0.2676	0.3337	0.3874	0.4172
F_7	0.0039	0.0062	0.0090	0.0119	0.0144	0.0160

Referring to Table 1, the traffic intensity ρ increases as the rate of consultation θ increases. Also the effective rate of consulations and interruptions, E_1 and E_2 will increase with the increase of θ. Then the duration of time they spend for consultation increases. This results in an increase in F_5 and F_6. As θ increases, there are more frequent consultations, and so the upper bounds of number of interruptions to the customer at the main server will reach sooner or the super clock may realise faster. Then the main server has to complete the service of the customer at him before further consultations and thus regular server has to wait more time to get further consultation. So F_7 increases. In this case restart of the service at the regular server is more frequent and F_4 increases. Since there is increase in F_5, F_6 and F_7, the customers compel to stay in the system and in the queue for longer time. This results in a hike in μ_1 and μ_2. Thus the main server and regular server get lesser time to be idle to make a decrease in F_1 and F_2. As θ increases, main server spends more time in consultation. So the main server gets much less time to serve its customers. Thus F_3 decreases.

Referring to Table 2, there is an increase in traffic intensity ρ as the arrival rate λ increases. More and more customers enter into the system and therefore accumulation of customers enhances. So μ_1 and μ_2 increase. As the number of customers increases, the servers get lesser time to be idle. Thus F_1 and F_2 decrease. In this case they have to serve customers for longer time and so F_3 and F_4 increase. Since there are more service, consultations are more frequent. Thus there is an increase in effective rates of consultations and interruptions. Therefore E_1 and E_2 increse. So the servers have to spend more time in consultations which results in a hike in F_5 and F_6. The possibility for the number of interruptions to a customer at the main server to reach the upper bound or that for the super clock to be expired is high. So the duration of time at which the regular server is waiting to avail consultation, F_7 increses.

6   Conclusion

In this paper we analyse a two-server queueing model with consultation in random environment by main server to the regular server. Quality of service is enhanced by consultations. The interruptions to a customer at the main server are controlled by upper bounds of interruptions and consultations and a super clock. We establish stability condition and provide numerical illustrations. As an extension of the model discussed, we can consider consultation in Markovian environment. In this case the environmental factors will be related to each other by a transition probability matrix.

Acknowledgement. The authors thank the reviewer(s) for comments that improved the presentation of the paper.

References

1. Avi-Itzhak, B., Naor, P.: Some queueing problems with the service station subject to breakdowns. Oper. Res. **11**(3), 303–320 (1963)
2. Sengupta, B.: A queue with service interruptions in an alternating random environment. Oper. Res. **38**(2), 308–318 (1990)
3. Chakravarthy, S.R.: A multi-server queueing model with server consultations. Eur. J. Oper. Res. **233**(3), 625–639 (2014)
4. Fiems, D., Maertens, T., Bruneel, H.: Queueing systems with different types of interruptions. Eur. J. Oper. Res. **188**(3), 838–845 (2008)
5. Gaver, D.P.: A waiting line with interrupted service including priorities. J. Royal Stat. Soc. **24**(1), 73–90 (1962)
6. Ibe, O.C., Trivedi, K.S.: Two queues with alternating service and server breakdown. Queueing Syst. **7**(3), 253–268 (1990)
7. Keilson, J.: Queues subject to service interruptions. Ann. Math. Stat. **33**(4), 1314–1322 (1962)
8. Krishnamoorthy, A., Pramod, P.K., Chakravarthy, S.R.: A note on characterizing service interruptions with phase type distribution. Stoch. Anal. Appl. **31**(4), 671–683 (2013)
9. Krishnamoorthy, A., Pramod, P.K., Deepak, T.G.: On a queue with interruptions and repeat/resumption of service. Non-linear Anal. Theory, Methods Appl. **71**, e1673–e1683 (2009)
10. Neuts, M.F.: Matrix-Geometric Solutions in Stochastic Models: An Algorithmic Approach. The Johns Hopkins University Press, Baltimore (1981)
11. White, H., Christie, L.S.: Queueing with preemptive priorities or with breakdown. Oper. Res. **6**(1), 79–95 (1958)

Asymptotic Analysis of Retrial Queueing System M/GI/1 with Collisions and Impatient Calls

Elena Danilyuk[✉], Svetlana Moiseeva, and Anatoly Nazarov

National Research Tomsk State University, Lenina Avenue, 36, 634050 Tomsk, Russia
daniluc_elena@sibmail.com, smoiseeva@mail.ru, nazarov.tsu@gmail.com

Abstract. In the paper, the retrial queueing system of M/GI/1 type with input Poison flow of events, collisions and impatient calls is considered. The delay time of calls in the orbit and the impatience time of calls in the orbit have exponential distribution. Service time on server is with any distribution function. Asymptotic analysis method is proposed for the solving problem of finding distribution of the number of calls in the orbit under a long delay of calls in orbit and long time patience of calls in the orbit condition. The theorem about the Gauss form of the asymptotic probability distribution of the number of calls in the orbit is formulated and proved. Numerical illustrations, results are also given.

Keywords: Retrial queueing system · Collisions · Impatient calls · Asymptotic analysis

1 Introduction

Nowadays one of the modern problems connected with traffic growth is the problem of analysis and optimal designing of communication systems. Any business process today is related to exchange of information. Therefore, developing of appropriate mathematical models of modern telecommunication systems and modifying of existing ones are important. Queueing systems with repeated calls, or retrial queueing systems, adequately describe such telecommunication systems, networks, mobile networks, call-centres and etc. This is evidenced by numerous papers and books devoted their study [1–11]. The main feature of RQ-systems is that in these queueing systems unserved calls are not lost when there are not available service devices (servers are busy or broken). So, the calls that don't get a service repeat to occupy server after a random time.

The present paper is advancing the results achieved in [12] and their consolidating in a way. The problem statement is common. We find the stationary distribution of the number of calls in orbit for the system under consideration. We continue to study retrial queueing system with collisions and impatient calls as well. Collisions in the model usually arise in the task of studying communication networks and suggest the emergence of situations when another message is

© Springer Nature Switzerland AG 2019
A. Dudin et al. (Eds.): ITMM 2019, CCIS 1109, pp. 230–242, 2019.
https://doi.org/10.1007/978-3-030-33388-1_19

transmitted during the transmission of a message. Such messages collide, they are considered distorted and go into the orbit, from where they ask the device for servicing again after a random delay [12–15]. Impatience of calls in the orbit is understood as case when a call in the orbit can leave the orbit after a random time without server recalling. This approach was used in our previous paper [16, 17] and by others [14, 18–25]. But there is one else way to specify impatience, for example, in papers [18, 19, 25] authors use non-persistence.

In this study we use not exponential distribution of service time but any one with distribution function $B(x)$. The same discipline of service is described by many scientists [13, 26–29], etc.

In literature, the primary methods of studying RQ-systems are matrix methods [7, 11, 30, 31], numerical methods [5, 32, 33], and simulation modeling, since one can obtain exact analytic formulas only for the simplest models [6, 7]. To solve the problem we use asymptotic analysis method that is widely applied for RQ-systems research. The method makes it possible to produce analytical result for different types of queueing systems and networks under given asymptotic condition. More information about the asymptotic analysis method is provided in [6, 16, 17, 22, 26, 27, 29, 31, 34], etc.

The general information about mathematical model of the retrial queueing system discussed in the paper and the problem statement are presented in the Sect. 2. In the Sect. 3 the detailed derivation of the model and the system of Kolmogorov equations for the stationary state probabilities are cited. The Sect. 4 consists of the decision of the problem under study by the asymptotic analysis method. As a result of the section the Theorem about stationary probability distribution of the calls number in the orbit for Retrial queueing system of M/GI/1 type with collisions and impatient calls in the orbit under a long delay of calls in orbit and long time patience of calls in the orbit condition is formulated and proved. Some numerical results, graphs, that proved the theoretical results, are performed in the Sect. 5. Section 6 concludes the paper.

2   Mathematical Model

We consider an RQ-system with one servicing device with Poisson arrival process with intensity λ. A call that has found the device free takes it for service for a random time, which has any distribution with function of distribution $B(x)$. If the device is busy, calls that arrive and are on the device enter into a "collision" and both go into orbit, or a source of repeated calls. On the orbit, each call, independently of others, waits for a random time whose duration has an exponential distribution with parameter σ, and then again accesses the device with a second attempt to obtain servicing. If the device is free, then the call from orbit occupies it for a random servicing time, and if the device is busy we again have a "collision," and both calls immediately go into orbit and wait there once more for a random time interval. Moreover, a call in orbit leaves the system after a random time, which has an exponential distribution with parameter α, demonstrating the "impatience" property.

The structure of the model of the RQ-system M/GI/1 with collisions and impatient calls is presented in Fig. 1.

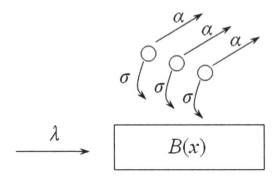

Fig. 1. Retrial queue M/GI/1 with collisions and impatient calls in the orbit

The problem is to find the stationary distribution of the number of calls in orbit for the described system.

3   Process of the System States: Stationary Distribution

Let us consider Markovian process $\{k(t), i(t), z(t)\}$ determined states of the Retrial queue M/GI/1 with collisions and impatient calls in the orbit where the random process $i(t)$ is the number of calls in the orbit at the moment t, $i(t) = 0, 1, 2, 3, \ldots,$ $z(t)$ is the interval duration from the moment t to the end of service of call on device, the random process $k(t)$ defines device state at the moment t and takes one of the following values

$$k(t) = \begin{cases} 0, & \text{if server is free at the moment } t; \\ 1, & \text{if server is busy at the moment } t. \end{cases}$$

Denote as $P_0(i, t) = P\{k(t) = 0, i(t) = i\}$ the probability that, at the moment t, there are i calls in the orbit, $i = 0, 1, 2, \ldots,$ and the service device is free. The probability that the server is busy at time t, there are i calls in the orbit, $i = 0, 1, 2, \ldots,$ and the time that remains before the end of service is shorter than z is denoted as $P_1(i, z, t) = P\{k(t) = 1, i(t) = i, z(t) < z\}$. When we say about probability that the device is busy at time t, there are i calls in the orbit, $i = 0, 1, 2, \ldots,$ and the time that remains before the end of service is unknown, we use notation $P_1(i, t) = P\{k(t) = 1, i(t) = i, z(t) < \infty\}$.

To obtain the probability distribution $P_0(i, t), P_1(i, t), P_1(i, z, t)$ for the states of the considered RQ-system, we construct a system of Kolmogorov differential equations

$$
\begin{cases}
\dfrac{\partial P_0(i,t)}{\partial t} = \dfrac{\partial P_1(i,0,t)}{\partial z} - (\lambda + i\sigma + i\alpha)\, P_0(i,t) \\
+(i+1)\alpha P_0(i+1,t) + \lambda P_1(i-2,t) + (i-1)\sigma P_1(i-1,t), \\
\dfrac{\partial P_1(i,z,t)}{\partial t} = \dfrac{\partial P_1(i,z,t)}{\partial z} - \dfrac{\partial P_1(i,0,t)}{\partial z} - (\lambda + i\sigma + i\alpha)\, P_1(i,z,t) \\
+(i+1)\alpha P_1(i+1,z,t) + \lambda B(z) P_0(i,t) + (i+1)\sigma B(z) P_0(i+1,t),
\end{cases}
\tag{1}
$$

where $\Pi_k(i) = \lim\limits_{t\to\infty} P_k(i,t)$, $k = 0,1$, $\Pi_1(i,z) = \lim\limits_{t\to\infty} P_1(i,z,t)$. Then the system of Kolmogorov equations for the stationary state probabilities $\Pi_k(i)$, $\Pi_1(i,z)$, $k = 0,1$, of the process $\{k(t), i(t), z(t)\}$ is written as follows

$$
\begin{cases}
\dfrac{\partial \Pi_1(i,0)}{\partial z} - (\lambda + i\sigma + i\alpha)\,\Pi_0(i) + (i+1)\alpha \Pi_0(i+1) \\
+\lambda \Pi_1(i-2) + (i-1)\sigma \Pi_1(i-1) = 0, \\
\dfrac{\partial \Pi_1(i,z)}{\partial z} - \dfrac{\partial \Pi_1(i,0)}{\partial z} - (\lambda + i\sigma + i\alpha)\,\Pi_1(i,z) \\
+(i+1)\alpha \Pi_1(i+1,z) + \lambda B(z)\Pi_0(i) + (i+1)\sigma B(z)\Pi_0(i+1) = 0.
\end{cases}
\tag{2}
$$

We get in (2) the indefinite dimensional system of difference equations with variable coefficients. In common case it is not possible to produce the exact solution of this system. To find solution of (2), we use the method of asymptotic analysis under a long delay of calls in orbit and long time patience of calls in the orbit condition.

4   Asymptotic Analysis Method

The method of asymptotic analysis in queueing theory is the method of research of the equations determining some characteristics of an queueing system under some limit (asymptotic) condition, which is specific for any model and solving problem.

We introduce the partial characteristic functions

$$
H_k(u) = \sum_{i=0}^{\infty} e^{jui} \Pi_k(i), \quad H_1(u,z) = \sum_{i=0}^{\infty} e^{jui} \Pi_1(i,z),
\tag{3}
$$

$$
H_k(0) = \sum_{i=0}^{\infty} \Pi_k(i) = R_k, \quad H_1(0,z) = \sum_{i=0}^{\infty} \Pi_1(i,z) = R_1(z),
\tag{4}
$$

where $j = \sqrt{-1}$, $k = 0,1$, and R_k are stationary state probabilities of the process $k(t)$, and $\lim\limits_{z\to\infty} R_1(z) = R_1$.

Using (3), (4), $\dfrac{\partial H_k(u)}{\partial u} = j \sum_{i=0}^{\infty} i e^{jui} \Pi_k(i)$, $\dfrac{\partial H_1(u,z)}{\partial u} = j \sum_{i=0}^{\infty} i e^{jui} \Pi_1(i,z)$,

$\dfrac{\partial H_1(u,z)}{\partial z} = \sum_{i=0}^{\infty} e^{jui} \dfrac{\partial \Pi_1(i,z)}{\partial z}$, $k = 0, 1$, we can write the system (2) as

$$
\begin{cases}
-\lambda H_0(u) + j\left(\sigma + \alpha\left(1 - e^{-ju}\right)\right) H_0'(u) + \lambda e^{2ju} H_1(u) \\
\quad - j\sigma e^{ju} H_1'(u) + \dfrac{\partial H_1(u,0)}{\partial z} = 0, \\
\lambda B(z) H_0(u) - j\sigma B(z) e^{-ju} H_0'(u) - \lambda H_1(u,z) \\
\quad + j\left(\sigma + \alpha\left(1 - e^{-ju}\right)\right) \dfrac{\partial H_1(u,z)}{\partial u} + \dfrac{\partial H_1(u,z)}{\partial z} - \dfrac{\partial H_1(u,0)}{\partial z} = 0.
\end{cases}
\tag{5}
$$

In adding the first equation by the second equation of (5) we get the system below

$$
\begin{cases}
-\lambda H_0(u) + j\left(\sigma + \alpha\left(1 - e^{-ju}\right)\right) H_0'(u) + \lambda e^{2ju} H_1(u) \\
\quad - j\sigma e^{ju} H_1'(u) + \dfrac{\partial H_1(u,0)}{\partial z} = 0, \\
\lambda B(z) H_0(u) - j\sigma B(z) e^{-ju} H_0'(u) - \lambda H_1(u,z) \\
\quad + j\left(\sigma + \alpha\left(1 - e^{-ju}\right)\right) \dfrac{\partial H_1(u,z)}{\partial u} + \dfrac{\partial H_1(u,z)}{\partial z} - \dfrac{\partial H_1(u,0)}{\partial z} = 0, \\
j\left(\sigma + \alpha\right) e^{-ju} H_0'(u) + \lambda\left(1 + e^{ju}\right) H_1(u) - j\left(\sigma - \alpha e^{-ju}\right) H_1'(u) = 0.
\end{cases}
\tag{6}
$$

The system in (6) is the basic system for analysis of Retrial queueing system of M/GI/1 type with collisions and impatient calls in the orbit under a long delay of calls in orbit ($\sigma \to 0$) and long time patience of calls in the orbit ($\alpha \to 0$) condition. We summarize the results of our study in Theorem 1.

Theorem 1. *The stationary probability distribution of the calls number in the orbit for Retrial queueing system of M/GI/1 type with collisions and impatient calls in the orbit under a long delay of calls in orbit and long time patience of calls in the orbit condition (with the Poisson arrival process of intensity λ, any servicing distribution with function of distribution $B(x)$, exponential distribution law of the random delay parameter σ, exponential distribution of a call's impatience with parameter $\alpha = q\sigma$, and constant $q > 0$) can be approximated by the Gaussian distribution with mean and variance equal to κ_1/σ and κ_2/σ respectively, where κ_1, and κ_2 are determined by equations*

$$
2\lambda\left[1 - B^*(\lambda + \kappa_1)\right] - \kappa_1\left[B^*(\lambda + \kappa_1) - q\left(2 - B^*(\lambda + \kappa_1)\right)\right] = 0,
$$

$$
\kappa_2 = \frac{R_0 \kappa_1 + q\kappa_1 + \lambda R_1 - 2\left(\lambda + \kappa_1\right) a}{R_0 - R_1 + q + 2\left(\lambda + \kappa_1\right) b},
$$

$$
a = \frac{R_0 \kappa_1 \left(1 - B^*\left(\lambda + \kappa_1\right)\right) + q\kappa_1 \left(R_1 - R_1^*\left(\lambda + \kappa_1\right)\right)}{\left(\lambda + \kappa_1\right)\left(2 - B^*\left(\lambda + \kappa_1\right)\right)},
$$

$$
b = \frac{R_1 - R_1^*\left(\lambda + \kappa_1\right) - R_0\left(1 - B^*\left(\lambda + \kappa_1\right)\right)}{\left(\lambda + \kappa_1\right)\left(2 - B^*\left(\lambda + \kappa_1\right)\right)},
$$

where $R_0 = \dfrac{1}{2 - B^*(\lambda + \kappa_1)}$, $R_1 = \dfrac{1 - B^*(\lambda + \kappa_1)}{2 - B^*(\lambda + \kappa_1)}$, $B^*(s) = \int\limits_{0}^{\infty} e^{-sz} dB(z)$,

$$R_1^*(\lambda + \kappa_1) = -\dfrac{dB^*(s)}{ds}\bigg|_{s=\lambda+\kappa_1} \cdot (\lambda + \kappa_1)\, R_0.$$

Proof. The Theorem 1 proving will carried out in two stages.

Stage 1. Finding First-Order Asymptotic. In the basic system of (6), we make the substitutions $\sigma = \varepsilon$, $\alpha = q\varepsilon$, $u = \varepsilon w$, $H_0(u) = F_0(w, \varepsilon)$, $H_1(u) = F_1(w, \varepsilon)$, $H_1(u, z) = F_1(w, \varepsilon, z)$, where ε is infinitesimal value ($\varepsilon \to 0$).

Since $H'_k(u) = \dfrac{1}{\varepsilon}\dfrac{\partial F_k(w, \varepsilon)}{\partial w}$, $k = 0, 1$, $H'_1(u, z) = \dfrac{1}{\varepsilon}\dfrac{\partial F_1(w, \varepsilon, z)}{\partial w}$, $\dfrac{\partial H_1(u, z)}{\partial z} = \dfrac{\partial F_1(w, \varepsilon, z)}{\partial z}$, the equations system (6) can be written as

$$
\begin{cases}
- \lambda F_0(w, \varepsilon) + j\,(1 + qjw\varepsilon)\,\dfrac{\partial F_0(w, \varepsilon)}{\partial w} + \lambda\,(1 + 2jw\varepsilon)\,F_1(w, \varepsilon) \\
\quad - j\,(1 + jw\varepsilon)\,\dfrac{\partial F_1(w, \varepsilon)}{\partial w} + \dfrac{\partial F_1(w, \varepsilon, 0)}{\partial z} = o(\varepsilon^2), \\[2mm]
\lambda B(z) F_0(w, \varepsilon) - jB(z)\,(1 - jw\varepsilon)\,\dfrac{\partial F_0(w, \varepsilon)}{\partial w} - \lambda F_1(w, \varepsilon, z) \\
\quad + j\,(1 + qjw\varepsilon)\,\dfrac{\partial F_1(w, \varepsilon, z)}{\partial w} + \dfrac{\partial F_1(w, \varepsilon, z)}{\partial z} - \dfrac{\partial F_1(w, \varepsilon, 0)}{\partial z} = o(\varepsilon^2), \\[2mm]
j\,(1 + q)\,(1 - jw\varepsilon)\,\dfrac{\partial F_0(w, \varepsilon)}{\partial w} + \lambda\,(2 + jw\varepsilon)\,F_1(w, \varepsilon) \\
\quad - j\,(1 - q + qjw\varepsilon)\,\dfrac{\partial F_1(w, \varepsilon)}{\partial w} = o(\varepsilon^2).
\end{cases}
\tag{7}
$$

The transformation of equations of (7) under $\varepsilon \to 0$ with $F_k(w) = \lim\limits_{\varepsilon \to 0} F_k(w, \varepsilon)$, $k = 0, 1$, $F_1(w, z) = \lim\limits_{\varepsilon \to 0} F_1(w, \varepsilon, z)$, leads to equations system as follows

$$
\begin{cases}
- \lambda F_0(w) + j\dfrac{dF_0(w)}{dw} + \lambda F_1(w) - j\dfrac{dF_1(w)}{dw} + \dfrac{\partial F_1(w, 0)}{\partial z} = 0, \\[2mm]
\lambda B(z) F_0(w) - jB(z)\dfrac{dF_0(w)}{dw} - \lambda F_1(w, z) \\
\quad + j\dfrac{\partial F_1(w, z)}{\partial w} + \dfrac{\partial F_1(w, z)}{\partial z} - \dfrac{\partial F_1(w, 0)}{\partial z} = 0, \\[2mm]
j\,(1 + q)\dfrac{dF_0(w)}{dw} + 2\lambda F_1(w) - j\,(1 - q)\dfrac{dF_1(w)}{dw} = 0.
\end{cases}
\tag{8}
$$

We suggest to find the Eq. (8) solution $F_k(w)$, $k = 0, 1$, $F_1(w, z)$ in the form

$$F_k(w) = R_k \Phi(w), \quad k = 0, 1, \quad F_1(w, z) = R_1(z)\Phi(w), \tag{9}$$

where $R_k = H_k(0)$, $k = 0, 1$, $\lim\limits_{z \to \infty} R_1(z) = R_1$.

Substituting (9) in (8) we have

$$
\begin{cases}
\left(\lambda R_1 - \lambda R_1 + \dfrac{dR_1(0)}{dz} \right) \Phi(w) + j \left(R_0 - R_1 \right) \Phi'(w) = 0, \\[2mm]
\left(\lambda B(z) R_0 - \lambda R_1(z) + \dfrac{dR_1(z)}{dz} - \dfrac{dR_1(0)}{dz} \right) \Phi(w) \\[2mm]
+ j \left(R_1(z) - B(z) R_0 \right) \Phi'(w) = 0, \\[2mm]
2\lambda R_1 \Phi(w) + j \left(R_0 - R_1 + q \right) \Phi'(w) = 0.
\end{cases}
\tag{10}
$$

According to Eq. (10) we can found their solution

$$
\Phi(w) = \exp\left\{ j \kappa_1 w \right\},
\tag{11}
$$

where κ_1 is defined below.

Using (9), (11) in (10) we get

$$
\begin{cases}
(\lambda + \kappa_1)(R_1 - R_0) + \dfrac{dR_1(0)}{dz} = 0, \\[2mm]
\dfrac{dR_1(z)}{dz} - \dfrac{dR_1(0)}{dz} - (\lambda + \kappa_1)(R_1(z) - B(z) R_0) = 0, \\[2mm]
2\lambda R_1 + \kappa_1 (R_1 - R_0) - q\kappa_1 = 0.
\end{cases}
\tag{12}
$$

Laplace and Stieltjes transformation the second equation of (12) lets to obtain

$$
s R_1^*(s) - \frac{dR_1(0)}{dz} - (\lambda + \kappa_1)(R_1^*(s) - R_0 B^*(s)) = 0,
\tag{13}
$$

where $B^*(s) = \int\limits_0^\infty e^{-sz} dB(z)$, $R_1^*(s) = \int\limits_0^\infty e^{-sz} dR_1(z)$.

If in (13) $s = 0$, then $\dfrac{dR_1(0)}{dz} = - (\lambda + \kappa_1)(R_1 - R_0)$. If in (13) $s = \lambda + \kappa_1$,

then using $\dfrac{dR_1(0)}{dz}$ and $R_0 + R_1 = 1$ we finally get

$$
R_0 = \frac{1}{2 - B^*(\lambda + \kappa_1)}, \quad R_1 = \frac{1 - B^*(\lambda + \kappa_1)}{2 - B^*(\lambda + \kappa_1)}.
\tag{14}
$$

From the third equation of (12) and (14) we found equation that defines κ_1

$$
2\lambda \left[1 - B^*(\lambda + \kappa_1) \right] - \kappa_1 \left[B^*(\lambda + \kappa_1) - q \left(2 - B^*(\lambda + \kappa_1) \right) \right] = 0.
\tag{15}
$$

Pre-limit characteristic function $h(u)$ is approximately equal to

$$
H(u) = H_0(u) + H_1(u) \approx h_1(u).
$$

So, the first-order asymptotic characteristic function $h_1(u)$ of the probability distribution of the number of calls in the orbit under the assumption of a long delay of calls in orbit and their high "patience" can be presented as

$$
h_1(u) = \exp\left\{ \frac{\kappa_1}{\sigma} ju \right\}.
\tag{16}
$$

Stage 2. Finding the Second-Order Asymptotics . In the basic system of Eqs. (6) with (16) we let

$$H_k(u) = \exp\left\{\frac{\kappa_1}{\sigma}ju\right\}H_k^{(2)}(u), \quad H_1(u,z) = \exp\left\{\frac{\kappa_1}{\sigma}ju\right\}H_1^{(2)}(u,z), \quad (17)$$

$k = \{0;1\}$.

Let $\sigma = \varepsilon^2, \alpha = q\varepsilon^2, u = \varepsilon w, H_1^{(2)}(u,z) = F_1^{(2)}(w,\varepsilon,z), H_k^{(2)}(u) = F_k^{(2)}(w,\varepsilon), k = \{0;1\}$, where ε is an infinitesimal, then (6) with some transformations can be rewritten as

$$\begin{cases} -[\lambda + (1+qjw\varepsilon)\kappa_1]F_0^{(2)}(w,\varepsilon) + [\lambda + \kappa_1 + jw\varepsilon(2\lambda + \kappa_1)]F_1^{(2)}(w,\varepsilon) \\ + j\varepsilon\dfrac{\partial F_0^{(2)}(w,\varepsilon)}{\partial w} - j\varepsilon\dfrac{\partial F_1^{(2)}(w,\varepsilon)}{\partial w} + \dfrac{\partial F_1^{(2)}(w,\varepsilon,0)}{\partial z} = o(\varepsilon^2), \\ B(z)[\lambda + (1-jw\varepsilon)\kappa_1]F_0^{(2)}(w,\varepsilon) - [\lambda + (1+qjw\varepsilon)\kappa_1]F_1^{(2)}(w,\varepsilon,z) \\ - jB(z)\varepsilon\dfrac{\partial F_0^{(2)}(w,\varepsilon)}{\partial w} + j\varepsilon\dfrac{\partial F_1^{(2)}(w,\varepsilon,z)}{\partial w} \\ + \dfrac{\partial F_1^{(2)}(w,\varepsilon,z)}{\partial z} - \dfrac{\partial F_1^{(2)}(w,\varepsilon,0)}{\partial z} = o(\varepsilon^2), \\ [2\lambda + \kappa_1(1-q) + jw\varepsilon(\lambda + q\kappa_1)]F_1^{(2)}(w,\varepsilon) - j(1-q)\varepsilon\dfrac{\partial F_1^{(2)}(w,\varepsilon)}{\partial w} \\ + (1+q)(jw\varepsilon-1)\kappa_1 F_0^{(2)}(w,\varepsilon) + j(1+q)\varepsilon\dfrac{\partial F_0^{(2)}(w,\varepsilon)}{\partial w} = o(\varepsilon^2). \end{cases} \quad (18)$$

When $\varepsilon \to 0$ in (18) and $\lim\limits_{\varepsilon\to 0} F_k^{(2)}(w,\varepsilon) = F_k^{(2)}(w)$, $\lim\limits_{\varepsilon\to 0} F_1^{(2)}(w,\varepsilon,z) = F_1^{(2)}(w,z)$, $k = \{0;1\}$, we get

$$\begin{cases} -(\lambda + \kappa_1)F_0^{(2)}(w) + (\lambda + \kappa_1)F_1^{(2)}(w) + \dfrac{\partial F_1^{(2)}(w,0)}{\partial z} = 0, \\ B(z)[\lambda + \kappa_1]F_0^{(2)}(w) - [\lambda + \kappa_1]F_1^{(2)}(w,z) \\ + \dfrac{\partial F_1^{(2)}(w,z)}{\partial z} - \dfrac{\partial F_1^{(2)}(w,0)}{\partial z} = 0, \\ -(1+q)\kappa_1 F_0^{(2)}(w) + [2\lambda + \kappa_1 - q\kappa_1]F_1^{(2)}(w) = 0. \end{cases} \quad (19)$$

The solution of equations system (18) has the following form

$$\begin{cases} F_k^{(2)}(w,\varepsilon) = (R_k + jw\varepsilon f_k)\Phi_2(w) + o(\varepsilon^2), \quad k = \{0;1\}, \\ F_1^{(2)}(w,\varepsilon,z) = (R_1(z) + jw\varepsilon f_1(z))\Phi_2(w) + o(\varepsilon^2), \\ R_0 + R_1 = 1, \end{cases} \quad (20)$$

where $R_0, R_1, R_1(z)$ are defined above, $f_0, f_1, f_1(z)$ are constants, and function $\Phi_{(2)}(w)$ is to be determined.

Substituting (20) into (18) and taking into account (19), we write the system under $\varepsilon \to 0$ as

$$\begin{cases} A_1 w \Phi_2(w) + B_1 \Phi_2'(w) = 0, \\ A_2 w \Phi_2(w) + B_2 \Phi_2'(w) = 0, \end{cases} \tag{21}$$

where constants A_1, A_2, B_1, B_2 are defined by (22)

$$\begin{cases} A_1 = (\lambda + \kappa_1)(f_1 - f_0) + (2\lambda + \kappa_1) R_1 - q\kappa_1 R_0 + f_1'(0), \\ B_1 = R_0 - R_1, \\ A_2 = (\lambda + \kappa_1)[B(z)f_0 - f_1(z)] - \kappa_1 [B(z)R_0 + qR_1(z)] + f_1'(z) - f_0'(z), \\ B_2 = R_1(z) - B(z)R_0. \end{cases} \tag{22}$$

The solution of system (21) has the form

$$\Phi_2(w) = \exp\left\{\kappa_2 \frac{(jw)^2}{2}\right\}, \tag{23}$$

where $\kappa_2 = A_1/B_1 = A_2/B_2$.

Using the same transformation as for the first-order asymptotics and additional condition $f_0 + f_1 = 0$, we finally obtain expression for κ_2

$$\kappa_2 = \frac{R_0 \kappa_1 + q\kappa_1 + \lambda R_1 - 2(\lambda + \kappa_1)a}{R_0 - R_1 + q + 2(\lambda + \kappa_1)b}, \tag{24}$$

$$a = \frac{R_0 \kappa_1 (1 - B^*(\lambda + \kappa_1)) + q\kappa_1 (R_1 - R_1^*(\lambda + \kappa_1))}{(\lambda + \kappa_1)(2 - B^*(\lambda + \kappa_1))},$$

$$b = \frac{R_1 - R_1^*(\lambda + \kappa_1) - R_0 (1 - B^*(\lambda + \kappa_1))}{(\lambda + \kappa_1)(2 - B^*(\lambda + \kappa_1))},$$

where R_0, R_1 and κ_1 are determined in (14), (15). Expression for $R_1^*(\lambda + \kappa_1)$ we obtain from (13) with differentiating by s and then letting $s = \lambda + \kappa_1$. So,

$$R_1^*(\lambda + \kappa_1) = -\left.\frac{dB^*(s)}{ds}\right|_{s=\lambda+\kappa_1} \cdot (\lambda + \kappa_1) R_0.$$

Making the reverse substitutions, we get under $z \to \infty$ with (19)

$$H_k^{(2)}(u) = F_k^{(2)}(w, \varepsilon) = (R_k + jw\varepsilon f_k) \exp\left\{\kappa_2 \frac{(jw)^2}{2}\right\} + o(\varepsilon^2)$$

$$\approx R_k \exp\left\{\frac{\kappa_2}{\sigma} \frac{(ju)^2}{2}\right\}, \tag{25}$$

then using (25) expressions (17) can be written as

$$H_k(u) = \exp\left\{\frac{\kappa_1}{\sigma} ju\right\} H_k^{(2)}(u) \approx R_k \exp\left\{\frac{\kappa_1}{\sigma} ju + \frac{\kappa_2}{\sigma} \frac{(ju)^2}{2}\right\}. \tag{26}$$

Taking into account (26), the characteristic function $H(u) = H_0(u) + H_1(u)$, provided that the calls in orbit have long delays and the "patience" is high, is a Gaussian

$$h_2(u) = \exp\left\{\frac{\kappa_1}{\sigma}ju + \frac{\kappa_2}{\sigma}\frac{(ju)^2}{2}\right\}. \tag{27}$$

The Theorem 1 is proved.

5   Numerical Results

In this section we give some notes to Theorem 1 and several numerical examples.

As we consider retrial queueing systems where the service time has any distribution function $B(x)$ we can not produce exact solution of Eq. (15) and as well as it is impossible to prove that root κ_1 of the (15) is single. Let $B(x) = \beta^\gamma / (\beta + x)^\gamma$, where β, γ are shape and scale parameters of gamma distribution $B(x)$. It can be numerically shown that in most cases (for most commonly used parameters of system) there is one root κ_1 of the (15), but we were able to find such system's parameters set when (15) has three solutions: $\kappa_1^1 = 0.201$, $\kappa_1^2 = 4.544$, $\kappa_1^3 = 16.628$ when $\lambda = 0.217$, $q = 0.01$, $\beta = \gamma = 2.5$. Figure 2 demonstrates behaviour of function of κ_1.

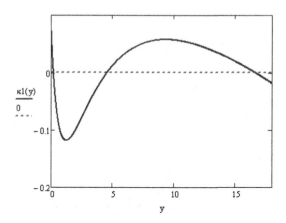

Fig. 2. Function of the κ_1 when $\lambda = 0.217$, $q = 0.01$, $\beta = \gamma = 2.5$

In this case we obtain distribution of number of call in the orbit consisting of the three parts, each of which has Gaussian distribution form. And this question requires further consideration.

Preliminary calculations suggest that theoretical results are consistent with simulation ones. To compare the probability distribution of the number of calls in the orbit of considered queueing system $P(i)$ calculated via simulation and its approximation $P_{asympt}(i)$ constructed by using the asymptotic method for different values of the system parameters we use Kolmogorov distance Δ between respective distribution functions: $\Delta = \max\limits_{i \geq 0} \left| \sum\limits_{l=0}^{i} [P(l) - P_{asympt}(l)] \right|$.

The comparison of the distributions is shown in Figs. 3 and 4.

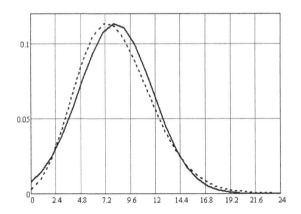

Fig. 3. Comparison of the asymptotic (solid line) and the simulated (dashed line) distributions for $\sigma = 0.01$, $\lambda = 0.4$, $\gamma = \beta = 1.5$, $\Delta = 0.032$.

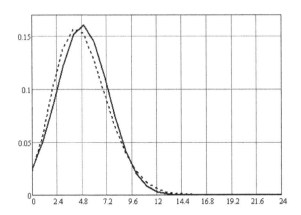

Fig. 4. Comparison of the asymptotic (solid line) and the simulated (dashed line) distributions for $\sigma = 0.1$, $\lambda = 1.3$, $\gamma = \beta = 1.5$, $\Delta = 0.043$.

6   Conclusion

In the present paper, retrial queueing system of M/GI/1 type with collisions and impatient calls in the orbit is considered. It is proved that the probability distribution of the calls number in the orbit can be approximated by the Gaussian distribution under a long delay of calls in orbit and a long time patience of calls in the orbit condition. More detailed numerical results that allow to draw a conclusion about an applicability area of the asymptotic result is the purpose of the future studies.

Acknowledgments. This work is financially supported by the Russian Foundation for Basic Research according to the research project No. 19-41-703002.

References

1. Wilkinson, R.I.: Theories for toll traffic engineering in the USA. Bell Syst. Tech. J. **35**(2), 421–507 (1956)
2. Cohen, J.W.: Basic problems of telephone traffic and the influence of repeated calls. Philips Telecommun. Rev. **18**(2), 49–100 (1957)
3. Gosztony, G.: Repeated call attempts and their effect on traffic engineering. Budavox Telecommun. Rev. **2**, 16–26 (1976)
4. Elldin, A., Lind, G.: Elementary Telephone Traffic Theory. Ericsson Public Telecommunications, Stockholm (1971)
5. Artalejo, J.R., Gomez-Corral, A.: Retrial Queueing Systems. A Computational Approach. Springer, Stockholm (2008). https://doi.org/10.1007/978-3-540-78725-9
6. Falin, G.I., Templeton, J.G.C.: Retrial Queues. Chapman & Hall, London (1997)
7. Artalejo, J.R., Falin, G.I.: Standard and retrial queueing systems: a comparative analysis. Rev. Mat. Complut. **15**, 101–129 (2002)
8. Roszik, J., Sztrik, J., Kim, C.: Retrial queues in the performance modelling of cellular mobile networks using MOSEL. Int. J. Simul. **6**, 38–47 (2005)
9. Aguir, S., Karaesmen, F., Askin, O.Z., Chauvet, F.: The impact of retrials on call center performance. OR Spektrum **26**, 353–376 (2004)
10. Phung-Duc, T., Kawanishi, K.: An efficient method for performance analysis of blended call centres with redial. Asia-Pac. J. Oper. Res. **31**(2), 1–39 (2014)
11. Dudin, A.N., Klimenok, V.I.: Queueing system BMAP/G/1 with repeated calls. Math. Comput. Model. **30**(3–4), 115–128 (1999)
12. Danilyuk, E.Y., Fedorova, E.A., Moiseeva, S.P.: Asymptotic analysis of an retrial queueing system M/M/1 with collisions and impatient calls. Autom. Remote Control **79**(12), 2136–2146 (2018)
13. Nazarov, A., Sztrik, J., Kvach, A.: Comparative analysis of methods of residual and elapsed service time in the study of the closed retrial queuing system M/GI/1//N with collision of the customers and unreliable server. In: Dudin, A., Nazarov, A., Kirpichnikov, A. (eds.) ITMM 2017. CCIS, vol. 800, pp. 97–110. Springer, Cham (2017). https://doi.org/10.1007/978-3-319-68069-9_8
14. Kim, J.: Retrial queueing system with collision and impatience. Commun. Korean Math. Soc. **4**, 647–653 (2010)
15. Bérczes, T., Sztrik, J., Tóth, Á., Nazarov, A.: Performance modeling of finite-source retrial queueing systems with collisions and non-reliable server using MOSEL. In: Vishnevskiy, V.M., Samouylov, K.E., Kozyrev, D.V. (eds.) DCCN 2017. CCIS, vol. 700, pp. 248–258. Springer, Cham (2017). https://doi.org/10.1007/978-3-319-66836-9_21
16. Vygovskaya, O., Danilyuk, E., Moiseeva, S.: Retrial queueing system of MMPP/M/2 type with impatient calls in the orbit. In: Dudin, A., Nazarov, A., Moiseev, A. (eds.) ITMM/WRQ-2018. CCIS, vol. 912, pp. 387–399. Springer, Cham (2018). https://doi.org/10.1007/978-3-319-97595-5_30

17. Danilyuk, E., Vygoskaya, O., Moiseeva, S.: Retrial queue M/M/N with impatient customer in the orbit. In: Vishnevskiy, V.M., Kozyrev, D.V. (eds.) DCCN 2018. CCIS, vol. 919, pp. 493–504. Springer, Cham (2018). https://doi.org/10.1007/978-3-319-99447-5_42
18. Yang, T., Posner, M., Templeton, J.: The M/G/1 retrial queue with non-persistent customers. Queueing Syst. **7**(2), 209–218 (1990)
19. Krishnamoorthy, A., Deepak, T.G., Joshua, V.C.: An M/G/1 retrial queue with non-persistent customers and orbital search. Stoch. Anal. Appl. **23**, 975–997 (2005)
20. Martin, M., Artalejo, J.: Analysis of an M/G/1 queue with two types of impatient units. Adv. Appl. Probab. **27**, 647–653 (1995)
21. Kumar, M., Arumuganathan, R.: Performance analysis of single server retrial queue with general retrial time, impatient subscribers, two phases of service and Bernoulli schedule. Tamkang J. Sci. Eng. **13**(2), 135–143 (2010)
22. Fedorova, E., Voytikov, K.: Retrial queue M/G/1 with impatient calls under heavy load condition. In: Dudin, A., Nazarov, A., Kirpichnikov, A. (eds.) ITMM 2017. CCIS, vol. 800, pp. 347–357. Springer, Cham (2017). https://doi.org/10.1007/978-3-319-68069-9_28
23. Klimenok, V.I., Orlovsky, D.S., Dudin, A.N.: BMAP/PH/N system with impatient repeated calls. Asia-Pac. J. Oper. Res. **24**(3), 293–312 (2007)
24. Kim, C.S., Klimenok, V., Dudin, A.: Retrial queueing system with correlated input, finite buffer, and impatient customers. In: Dudin, A., De Turck, K. (eds.) ASMTA 2013. LNCS, vol. 7984, pp. 262–276. Springer, Heidelberg (2013). https://doi.org/10.1007/978-3-642-39408-9_19
25. Dudin, A., Klimenok, V.: Retrial queue of BMAP/PH/N type with customers balking, impatience and non-persistence. In: 2013 Conference on Future Internet Communications (CFIC), pp. 1–6. IEEE (2013)
26. Lisovskaya, E., Moiseeva, S., Pagano, M.: Multiclass GI/GI/∞ queueing systems with random resource requirements. In: Dudin, A., Nazarov, A., Moiseev, A. (eds.) ITMM/WRQ-2018. CCIS, vol. 912, pp. 129–142. Springer, Cham (2018). https://doi.org/10.1007/978-3-319-97595-5_11
27. Pankratova, E., Moiseeva, S.: Queueing system GI|GI|∞ with n types of customers. In: Dudin, A., Nazarov, A., Yakupov, R. (eds.) ITMM 2015. CCIS, vol. 564, pp. 216–225. Springer, Cham (2015). https://doi.org/10.1007/978-3-319-25861-4_19
28. Decreusefond, L., Moyal, P.: Stochastic Modeling and Analysis of Telecom Networks. ISTE Ltd., London (2012)
29. Nazarov, A., Baymeeva, G.: The M/GI/∞ system subject to semi-Markovian random environment. In: Dudin, A., Nazarov, A., Yakupov, R. (eds.) ITMM 2015. CCIS, vol. 564, pp. 128–140. Springer, Cham (2015). https://doi.org/10.1007/978-3-319-25861-4_11
30. Stepanov, S.N.: Algorithms approximate design systems with repeated calls. Autom. Remote Control **44**(1), 63–71 (1983)
31. Nazarov, A.A., Lyubina, T.V.: The non-Markov dynamic RQ system with the incoming MMP flow of requests. Autom. Remote Control **74**(7), 1132–1143 (2013)
32. Artalejo, J.R., Pozo, M.: Numerical calculation of the stationary distribution of the main multiserver retrial queue. Annu. Oper. Res. **116**, 41–56 (2002)
33. Neuts, M.F., Rao, B.M.: Numerical investigation of a multiserver retrial model. Queueing Syst. **7**(2), 169–189 (1990)
34. Borovkov, A.A.: Asymptotic Methods in Queueing Theory. Wiley, New York (1984)

Production Inventory System with Positive Service Time Under Local Purchase

Achyutha Krishnamoorthy[1(✉)], Resmi Varghese[2],
and Balakrishna Lakshmy[3]

[1] Centre for Research in Mathematics, CMS College, Kottayam, India
achyuthacusat@gmail.com
[2] Department of Mathematics, St. Xavier's College for Women, Aluva, India
[3] "Balakrishna", Master Road, Vaduthala P.O., Kochi 682023, India

Abstract. In this paper, we consider an (s, S) production inventory model in which addition taking place to stock one at a time through a production process. The model involves positive service time and there is only one server. The time taken to produce an item and processing time of items are assumed to follow exponential distribution. Arrivals are according to a Poisson process. Local purchase is incorporated in the model to ensure customer satisfaction and goodwill. The model has high significance in real life as it can be applied to several small scale cottage industries, which result in upliftment of less privileged, hence having much social relevance. The problem is modelled as a continuous time Markov chain and we obtained stochastic decomposition of system states. Several performance measures are obtained and convexity of cost function is established numerically.

Keywords: Production inventory · Local purchase · Stochastic decomposition

1 Introduction

Sigman and Simchi-Levi [14] presented notable work on $M/G/1$ queue with inventory with positive service time in 1992. In 1993, Berman et al. [1] considered inventory management problem which involved deterministic and constant demand and service rates. A survey by Krishnamoorthy et al. [3] presents more details on inventory with positive service time. Krishnamoorthy and Raju [4, 5] introduced the concept of local purchase in (s, S) inventory systems with negligible service time. Krishnamoorthy et al. [6] considered N-policy in (s, Q) inventory systems with positive service time and lead time, where they arrive at a product form solution for the system state distribution, which owed to a local purchase of items. A continuous review (s, S) production inventory system was studied by Doshi et al. [2], which involved a compound Poisson arrival of demands. An (s, S) production inventory system which involve failure and repair of machine was considered by Sharafali [13]. Here, the repair time is assumed to follow general distribution.

The first work on $M/M/1$ queueing inventory system with positive lead time that involve product form solution was carried out by Schwarz et al. [12]. Product form

© Springer Nature Switzerland AG 2019
A. Dudin et al. (Eds.): ITMM 2019, CCIS 1109, pp. 243–256, 2019.
https://doi.org/10.1007/978-3-030-33388-1_20

solution in an $M/G/1$ queueing inventory system with arbitrarily distributed lead time was obtained by Saffari et al. [11].

Production inventory with positive service time with MAP arrivals and Markovian production process (MPP) was first studied by Krishnamoorthy and Viswanath [7]. Stochastic decomposition in production inventory with positive service time was obtained by Krishnamoorthy and Viswanath [8], owing to the assumption that no customer joins the system when the inventory level falls to zero. Neuts [10] introduced matrix analytic methods in M/G/1 and GI/M/1 type stochastic queueing models, where an algorithmic analysis of the models was carried out. Latouche and Ramaswami [9] also presents an in-depth discussion on matrix-analytic methods.

In this paper, we consider an (s, S) production inventory model which involves local puchase. Arrival of demands is according to a Poisson process with parameter λ. The production process is switched on when the inventory level reaches s. The items are produced one by one, and the time to produce an item follows exponential distribution with parameter θ. The produced item requires a processing time, which follows exponential distribution with parameter μ. We keep the production process in "on" mode, till the inventory level reaches S, and we switch it off, as and when the inventory level reaches S. It is also assumed that as and when the inventory level reaches zero, a local purchase of 1 unit is made at a higher cost, to avoid customer loss. The supply of item is instantaneous in local purchase. We carry out local purchase to avoid loss of customer, thereby ensuring customer satisfaction and goodwill. The model can be applied to small scale cottage industries which are run at homes and which involve cost-effective methods. Examples include making of garment, umbrella, pickle or other food items, soap, soap powder, jewellery etc. Analysis of such a model has high social relevance as it helps to reduce the total expected cost of production process, which results in improving the profit in cottage industry.

We can model the situation as a continuous time Markov chain which is given below.

2   Model Formulation and Analysis

Let $X(t)$ = Number of customers in the system at time t,
$I(t)$ = Inventory level at time t,
$K(t)$ = Status of the production process:

$$K(t) = \begin{cases} 1, & \text{if production process is on at time } t \\ 0, & \text{if production process is off at time } t \end{cases}$$

$\tilde{Y}(t) = \{(X(t), I(t), K(t)), t \geq 0\}$ is a continuous-time Markov chain with state space

$$\{(i,j) \,|\, i \geq 0, \; 1 \leq j \leq s\} \cup \{(i, j, k)/i \geq 0; s+1 \leq j \leq S-1; k = 0, 1\} \cup \{(i, S) \,|\, i \geq 0\}$$

2.1   Infinitesimal Generator $\hat{A}$

Infinitesimal generator $\hat{A}$ of this model is obtained as

$$\hat{A} = \begin{bmatrix} B_1 & A_0 & & \\ A_2 & A_1 & A_0 & \\ & A_2 & A_1 & A_0 \\ & & \ddots & \ddots & \ddots \end{bmatrix}$$

where $A_0 = \lambda I_{2S-s-1}$ and

$$A_1 = \begin{bmatrix} \tilde{J}_1 & \theta & & & & & & & \\ & \tilde{J}_1 & \theta & & & & & & \\ & & \ddots & \ddots & & & & & \\ & & & \tilde{J}_1 & \theta & & & & \\ & & & & \tilde{J}_1 & \tilde{P}_1 & & & \\ & & & & & \tilde{J}_2 & \tilde{P}_2 & & \\ & & & & & & \ddots & \ddots & \\ & & & & & & & \tilde{J}_2 & \tilde{P}_2 \\ & & & & & & & & \tilde{J}_2 & \tilde{P}_3 \\ & & & & & & & & & \tilde{J}_3 \end{bmatrix}$$

where $\tilde{J}_1 = [-(\lambda + \theta + \mu)]$, $\tilde{J}_2 = \begin{bmatrix} -(\lambda+\mu) & \\ & -(\lambda+\theta+\mu) \end{bmatrix}$ and $\tilde{J}_3 = [-(\lambda + \mu)]$

$$A_2 = \begin{bmatrix} \mu & & & & & & & & \\ & \mu & & & & & & & \\ & & \ddots & & & & & & \\ & & & \mu & & & & & \\ & & & & \mu & & & & \\ & & & & & M_1 & & & \\ & & & & & & M_2 & & \\ & & & & & & & \ddots & \\ & & & & & & & & M_2 & \\ & & & & & & & & & M_3 & 0 \end{bmatrix}$$

where $M_1 = \begin{bmatrix} \mu \\ \mu \end{bmatrix}$, $M_2 = \mu I_2$ and $M_3 = [\mu \quad 0]$.

$$B_1 = \begin{bmatrix} J_1 & \theta & & & & & & & & \\ & J_1 & \theta & & & & & & & \\ & & \ddots & \ddots & & & & & & \\ & & & J_1 & \theta & & & & & \\ & & & & J_1 & \tilde{P}_1 & & & & \\ & & & & & J_2 & \tilde{P}_2 & & & \\ & & & & & & \ddots & \ddots & & \\ & & & & & & & J_2 & \tilde{P}_2 & \\ & & & & & & & & J_2 & \tilde{P}_3 \\ & & & & & & & & & J_3 \end{bmatrix}$$

where $J_1 = [-(\lambda + \theta)]$, $J_2 = \begin{bmatrix} -\lambda & \\ & -(\lambda + \theta) \end{bmatrix}$, $J_3 = [-\lambda]$,

$\tilde{P}_1 = [0 \quad \theta]$, $\tilde{P}_2 = \begin{bmatrix} 0 & 0 \\ 0 & \theta \end{bmatrix}$ and $\tilde{P}_3 = \begin{bmatrix} 0 \\ \theta \end{bmatrix}$.

A_0, A_1, A_2 and B_1 are square matrices of order $2S - s - 1$.

2.2 Steady-State Analysis

Let $A = A_0 + A_1 + A_2$ be the generator matrix. Then A is a square matrix of order $2S - s - 1$ and is obtained as

$$A = \begin{bmatrix} -\theta & \theta & & & & & & & & & \\ \mu & \hat{J}_1 & \theta & & & & & & & & \\ & \mu & \hat{J}_1 & \theta & & & & & & & \\ & & \ddots & \ddots & \ddots & & & & & & \\ & & & \mu & \hat{J}_1 & \theta & & & & & \\ & & & & \mu & \hat{J}_1 & \tilde{P}_1 & & & & \\ & & & & & M_1 & \hat{J}_2 & \tilde{P}_2 & & & \\ & & & & & & M_2 & \hat{J}_2 & \tilde{P}_2 & & \\ & & & & & & & \ddots & \ddots & \ddots & \\ & & & & & & & & M_2 & \hat{J}_2 & \tilde{P}_2 \\ & & & & & & & & & M_2 & \hat{J}_2 & \tilde{P}_3 \\ & & & & & & & & & & M_3 & \hat{J}_3 \end{bmatrix}$$

where $\hat{J}_1 = [-(\mu + \theta)]$, $\hat{J}_2 = \begin{bmatrix} -\mu & \\ & -(\mu + \theta) \end{bmatrix}$, $\hat{J}_3 = [-\mu]$.

The stationary distribution of A is obtained in the following theorem.

Theorem 1. The steady-state probability vector $\overline{X}$ of A is $\overline{X} = (x_1, x_2, \ldots\ldots, x_{s-1},$ $x_s, \tilde{x}_{s+1}, \tilde{x}_{s+2}, \ldots\ldots \tilde{x}_{S-1}, \tilde{x}_S)$ where

$$\tilde{x}_{l_1} = (x_{l_1,0}, \ x_{l_1,1}), \quad l_1 = s+1, \ldots \ldots, S-1$$

$$x_l = \left(\frac{\mu}{\theta}\right)^{s-l}\left(\frac{\mu}{\theta} + \left(\frac{\mu}{\theta}\right)^2 + \left(\frac{\mu}{\theta}\right)^3 + \ldots \left(\frac{\mu}{\theta}\right)^Q\right) x_S \ and \ l = 1, 2, \ldots \ldots, s.$$

$$x_{l,0} = x_S, \ where \quad l = s+1, \ldots \ldots, S-1.$$

$$x_{l,1} = \left(\frac{\mu}{\theta} + \left(\frac{\mu}{\theta}\right)^2 + \left(\frac{\mu}{\theta}\right)^3 + \ldots \left(\frac{\mu}{\theta}\right)^{S-1}\right) x_S, \quad l = s+1, \ldots \ldots, S-1.$$

When $\mu \neq \theta$, x_S can be obtained by solving $\bar{X}\bar{e} = 1$ as

$$x_S = \frac{\left(1 - \frac{\mu}{\theta}\right)^2}{\frac{\mu}{\theta}\left[\left(\frac{\mu}{\theta}\right)^S - \left(\frac{\mu}{\theta}\right)^s - Q\right] + Q - \left(1 - \frac{\mu}{\theta}\right)^2}.$$

Proof. We can prove the result using $\bar{X}A = 0$ and $\bar{X}\bar{e} = 1$.

Remark. When $\mu = \theta$ we get the following:

$$x_l = Q\,x_S, \quad where \ l = 1, 2, \ldots \ldots, s$$

$$x_{l,0} = x_S, \quad where \ l = s+1, s+2, \ldots \ldots, S-1$$

$$x_{l,1} = (S-1)x_S, \quad where \ l = s+1, s+2, \ldots \ldots \ldots, S-1 \ and$$

$$x_S = \left[sQ + (Q-1)\left(1 + \frac{Q}{2}\right) + 1\right]^{-1}.$$

2.3 Stability Condition

Theorem 2. The process under study is stable iff $\lambda < \mu$.

Proof. Since the process under consideration is level-independent quasi-birth-death process, it is stable iff $\bar{X}A_0\bar{e} < \bar{X}A_2\bar{e}$ (Neuts [10]), where $\bar{X}$ is the steady-state distribution of the generator matrix A. Substituting A_0 and A_2 in the above equation, we get the required result after some algebra.

Remark. After getting the stability condition, we are now in a position to analyse the probability distribution of the system states in long-run. This will enable us to show that the joint distribution of the system state can be expressed as the product of the marginal distribution of its components.

3 The Steady-State Probability Distribution of $\hat{A}$

3.1 Stationary Distribution When Service Time Is Negligible

Let $\hat{Q}$ be the generator matrix associated with the Markov chain of the inventory process, when service time is negligible, and $\Delta = (\pi_1, \pi_2, \ldots \pi_S)$ be the stationary probability vector corresponding to $\hat{Q}$. We get $\hat{Q}$ as

$$\hat{Q} = \begin{bmatrix} -\theta & \theta \\ \lambda & J_1 & \theta \\ & \lambda & J_1 & \theta \\ & & \ddots & \ddots & \ddots \\ & & & \lambda & J_1 & \theta \\ & & & & \lambda & J_1 & \tilde{P}_1 \\ & & & & & \hat{M}_1 & J_2 & \tilde{P}_2 \\ & & & & & & \hat{M}_2 & J_2 & \tilde{P}_2 \\ & & & & & & & \ddots & \ddots & \ddots \\ & & & & & & & & \hat{M}_2 & J_2 & \tilde{P}_2 \\ & & & & & & & & & \hat{M}_2 & J_2 & \tilde{P}_3 \\ & & & & & & & & & & \hat{M}_3 & J_3 \end{bmatrix}$$

where $J_1 = [-(\lambda + \theta)]$, $J_2 = \begin{bmatrix} -\lambda & \\ & -(\lambda+\theta) \end{bmatrix}$, $J_3 = [-\lambda]$, $\hat{M}_1 = \begin{bmatrix} \lambda \\ \lambda \end{bmatrix}$, $\hat{M}_2 = \lambda I_2$

and $\hat{M}_3 = [\lambda \quad 0]$.

Theorem 3. The steady-state probability vector Δ of $\hat{Q}$ is

$$\Delta = (\pi_1, \pi_2, \ldots, \pi_{s-1}, \pi_s, \tilde{\pi}_{s+1}, \tilde{\pi}_{s+2}, \ldots \tilde{\pi}_{S-1}, \tilde{\pi}_S)$$

where

$$\tilde{\pi}_{l_1} = (\pi_{l_1,0}, \pi_{l_1,1}), \quad l_1 = s+1, \ldots, S-1$$
$$\pi_l = \left(\tfrac{\lambda}{\theta}\right)^{s-l} \left(\tfrac{\lambda}{\theta} + \left(\tfrac{\lambda}{\theta}\right)^2 + \left(\tfrac{\lambda}{\theta}\right)^3 + \ldots \left(\tfrac{\lambda}{\theta}\right)^Q\right) \pi_S \text{ where } l = 1, 2, \ldots, s.$$

$$x_{l,0} = \pi_S, \text{ where } l = s+1, \ldots, S-1.$$

$$\pi_{l,1} = \left(\tfrac{\lambda}{\theta} + \left(\tfrac{\lambda}{\theta}\right)^2 + \left(\tfrac{\lambda}{\theta}\right)^3 + \ldots \left(\tfrac{\lambda}{\theta}\right)^{S-1}\right) \pi_S,$$
$$l = s+1, \ldots, S-1.$$

When $\lambda \neq \theta$, π_S can be obtained by solving $\Delta \bar{e} = 1$ as

$$\pi_S = \frac{\left(1 - \frac{\lambda}{\theta}\right)^2}{\frac{\lambda}{\theta}\left[\left(\frac{\lambda}{\theta}\right)^S - \left(\frac{\lambda}{\theta}\right)^s - Q\right] + Q - \left(1 - \frac{\lambda}{\theta}\right)^2}.$$

Proof. We can prove the result using $\Delta \hat{Q} = \bar{0}$ and $\Delta \bar{e} = 1$.

Remark. When $\lambda = \theta$ we get the following:

$$\pi_l = Q \pi_S, \quad \text{where } l = 1, 2, \ldots, s$$

$$\pi_{l,0} = \pi_S, \quad \text{where } l = s + 1, s + 2, \ldots, S - 1$$

$$\pi_{l,1} = (S - 1) \pi_S, \quad \text{where } l = s + 1, s + 2, \ldots, S - 1 \text{ and}$$

$$\pi_S = \left[sQ + (Q - 1)\left(1 + \frac{Q}{2}\right) + 1 \right]^{-1}.$$

3.2 Stochastic Decomposition of System States

Theorem 4. Let $\bar{Z}$ be the steady-state probability vector of $\hat{A}$ and

$$\bar{Z} = \left(z^{(0)}, z^{(1)}, z^{(2)}, \ldots \right)$$

where

$$z^{(i)} = \left(z^{(i,1)}, z^{(i,2)}, z^{(i,3)}, \ldots z^{(i,s)}, z^{(i,s+1,0)}, z^{(i,s+1,1)}, \ldots, z^{(i,S-1,0)}, z^{(i,S-1,1)}, z^{(i,S)} \right)$$

where $i = 0, 1, 2, \ldots$ and $z^{(i,j)} = \lim_{t \to \infty} P(X(t) = i, I(t) = j)$, $z^{(i,j,k)} = \lim_{t \to \infty} P(X(t) = i, I(t) = j, K(t) = k)$. Then

$$z^{(i)} = K\rho^i \Delta, \quad i \geq 0 \tag{1}$$

where $\Delta = (\pi_1, \pi_2, \ldots, \pi_{s-1}, \pi_s, \tilde{\pi}_{s+1}, \tilde{\pi}_{s+2}, \ldots \tilde{\pi}_{S-1}, \pi_S)$ is the steady-state probability vector when service time is negligible, K is a constant to be determined and $\rho = \frac{\lambda}{\mu}$.

Proof. $\bar{Z} A = \bar{0}$ gives

$$z^{(0)}B_1 + z^{(1)}A_2 = \bar{0} \tag{2}$$

$$z^{(i+2)}A_2 + z^{(i+1)}A_1 + z^{(i)}A_0 = \bar{0}, i = 0, 1, \ldots \tag{3}$$

When (1) is substituted in (2) and (3), we get $\Delta \hat{Q} = \bar{0}$, which is true. Hence stochastic decomposition of system states is verified.

3.3 Determination of K

On substitution of $\bar{Z} = \left(z^{(0)}, z^{(1)}, z^{(2)}, \ldots\right)$ and using stochastic decomposition given by (1), $\bar{Z}\,\bar{e} = 1$ gives $K = 1 - \rho$ where $\rho = \frac{\lambda}{\mu}$.

3.4 Explicit Solution

The steady-state probability vector can be obtained explicitly as follows:

Theorem 5. Let $\bar{Z}$ be the steady-state probability vector of $\hat{A}$ and $\bar{Z} = \left(z^{(0)}, z^{(1)}, z^{(2)}, \ldots\right)$ where

$$z^{(i)} = \left(z^{(i,1)}, z^{(i,2)}, z^{(i,3)}, \ldots z^{(i,s)}, z^{(i,s+1,0)}, z^{(i,s+1,1)}, \ldots, z^{(i,S-1,0)}, z^{(i,S-1,1)}, z^{(i,S)}\right)$$

where $i = 0, 1, 2, \ldots$. Then $z^{(i)} = (1 - \rho)\rho^i \Delta$, $\quad i \geq 0$
where $\Delta = (\pi_1, \pi_2, \ldots, \pi_{s-1}, \pi_s, \tilde{\pi}_{s+1}, \tilde{\pi}_{s+2}, \ldots \tilde{\pi}_{S-1}, \tilde{\pi}_S)$ is the steady-state probability vector when service time is negligible and is as given by Theorem 3, and $\rho = \frac{\lambda}{\mu}$.

Remark. The result given in above Theorem indicates that the system possess stochastic decomposition. We also get that the system state distribution is the product of the distribution of its marginal, which means, one component is that of the classical $M/M/1$ queue having long run distribution for i customers in the system as $(1 - \rho)\rho^i$, $i \geq 0$ and the other component is the probability of j items in the inventory.

4 Performance Measures of the System

When $\lambda \neq \theta$, we get the following measures:

(a) Expected number of customers in the system,

$$L = \sum_{i=0}^{\infty} i(1 - \rho)\rho^i$$

$$= \frac{\lambda}{\mu - \lambda}$$

(b) Expected inventory held in the system,

$$E(I) = \sum_{i=0}^{\infty}\sum_{j=1}^{s} jz^{(i,j)} + \sum_{i=0}^{\infty}\sum_{j=s+1}^{S-1} j(z^{(i,j,0)} + z^{(i,j,1)}) + \sum_{i=0}^{\infty} Sz^{(i,S)} \text{ and is obtained as}$$

$$E(I) = \left\{\tfrac{\frac{\lambda}{\theta}}{1-\frac{\lambda}{\theta}}\left(s - (s+1)\tfrac{\lambda}{\theta} + \left(\tfrac{\lambda}{\theta}\right)^{s+1} - \left(\tfrac{\lambda}{\theta}\right)^{S+1} - s\left(\tfrac{\lambda}{\theta}\right)^{Q} + (S+1)\left(\tfrac{\lambda}{\theta}\right)^{Q+1}\right)\right.$$
$$+ (1 - \tfrac{\lambda}{\theta})\left[(Q-1)(s + \tfrac{Q}{2}) + \tfrac{\frac{\lambda}{\theta}}{1-\frac{Q}{\lambda}}\left[\left(\tfrac{\lambda}{\theta}\right)^{Q-1}\left((1-\tfrac{\lambda}{\theta})^{-1} - (s+1)\right)\right.\right.$$
$$\left.\left.+ (s+Q) - (1-\tfrac{\lambda}{\theta})^{-1}\right]\right]$$
$$\left. + S(1 - \tfrac{\lambda}{\theta})^2\right\}\left[\tfrac{\lambda}{\theta}\left[\left(\tfrac{\lambda}{\theta}\right)^{S} - \left(\tfrac{\lambda}{\theta}\right)^{s} - Q\right] + Q - (1-\tfrac{\lambda}{\theta})^2\right]^{-1}.$$

(c) Expected rate at which production process is switched 'on',

$$R_{ON} = \mu\sum_{i=1}^{\infty} z^{(i,s+1,0))}$$

$$= \frac{\lambda\left(1 - \tfrac{\lambda}{\theta}\right)^2}{\tfrac{\lambda}{\theta}\left[\left(\tfrac{\lambda}{\theta}\right)^{S} - \left(\tfrac{\lambda}{\theta}\right)^{s} - Q\right] + Q - (1 - \tfrac{\lambda}{\theta})^2}.$$

(d) Expected production rate,

$$R_P = \theta\left(\sum_{i=0}^{\infty}\sum_{j=1}^{s} z^{(i,j)} + \sum_{i=0}^{\infty}\sum_{j=s+1}^{S-1} z^{(i,j,1)}\right)$$

$$= \frac{\theta\left[\left(\tfrac{\lambda}{\theta}\right)^{S+1} - \left(\tfrac{\lambda}{\theta}\right)^{s+1} + Q\left(\tfrac{\lambda}{\theta}\right) - \left(\tfrac{\lambda}{\theta}\right)^{2}\right]}{\tfrac{\lambda}{\theta}\left[\left(\tfrac{\lambda}{\theta}\right)^{S} - \left(\tfrac{\lambda}{\theta}\right)^{s} - Q\right] + Q - (1 - \tfrac{\lambda}{\theta})^2}.$$

(e) Expected local purchase rate,

$$R_{LP} = \mu\sum_{i=1}^{\infty} z^{(i,1)}$$

$$= \frac{\lambda\left[\left(\tfrac{\lambda}{\theta}\right)^{s} - \left(\tfrac{\lambda}{\theta}\right)^{S} + \left(\tfrac{\lambda}{\theta}\right)^{s+1} + \left(\tfrac{\lambda}{\theta}\right)^{S+1}\right]}{\tfrac{\lambda}{\theta}\left[\left(\tfrac{\lambda}{\theta}\right)^{S} - \left(\tfrac{\lambda}{\theta}\right)^{s} - Q\right] + Q - (1 - \tfrac{\lambda}{\theta})^2}.$$

(f) Mean waiting time of customers in the system,

$$W_S = \frac{L}{\lambda} \text{ and can be obtained as } W_S = \frac{1}{\mu - \lambda} \text{ on substituting } L.$$

Remark. When $\lambda = \theta$, we get the performance measures as follows:

$$E(I) = \left[\frac{Qs(s + 1) + (S + 1)(Q - 1)(S + s)}{2} - \sum_{j=s+1}^{S-1} j^2 + S \right] \pi_S$$

$$R_P = \frac{Q}{2}[s(s + 1) + Q - 1]\theta\, \pi_S$$

$$R_{ON} = \lambda\, \pi_S$$

$$R_{LP} = \lambda Q\, \pi_S$$

5 Cost Analysis

To obtain a cost function, first we consider the various costs involved in the model as follows:

H_C: Inventory holding cost per unit item per unit time.
S_C: Fixed cost to start the production process.
P_C: Cost of production per unit time per unit inventory.
LP_C: Local purchase cost per unit inventory per unit time.
W_C: Cost of waiting time per customer per unit time.

When $\lambda \neq \theta$, we get,

$$TC = \left[H_C \left\{ \frac{\lambda}{\theta} \left(1 - \frac{\lambda}{\theta} \right)^{-1} \left[s - (s + 1)\frac{\lambda}{\theta} + \left(\frac{\lambda}{\theta} \right)^{s+1} - \left(\frac{\lambda}{\theta} \right)^{S+1} \right. \right. \right.$$

$$\left. \left. \left. - s\left(\frac{\lambda}{\theta} \right)^{Q} + (s + 1)\left(\frac{\lambda}{\theta} \right)^{Q+1} \right] \right. \right]$$

$$+ \left(1 - \frac{\lambda}{\theta}\right)\left[(Q - 1)\left(s + \frac{Q}{2}\right) + \frac{\lambda}{\theta}\left(1 - \frac{\theta}{\lambda}\right)^{-1}\left[\left(\frac{\lambda}{\theta}\right)^{Q-1}\left(\left(1 - \frac{\lambda}{\theta}\right)^{-1} - (s + 1)\right) + S - \left(1 - \frac{\lambda}{\theta}\right)^{-1}\right]\right]\right\}$$

$$+ (H_C S + S_C \lambda)\left(1 - \frac{\lambda}{\theta}\right)^2 + (P_C \theta + L P_C \lambda)\left[\left(\frac{\lambda}{\theta}\right)^{S+1} - \left(\frac{\lambda}{\theta}\right)^{s+1}\right]$$

$$+ P_C \theta \left(Q\frac{\lambda}{\theta} - \left(\frac{\lambda}{\theta}\right)^2\right) + L P_C \lambda \left(\left(\frac{\lambda}{\theta}\right)^s - \left(\frac{\lambda}{\theta}\right)^S\right)\right]$$

$$\times \left[\frac{\lambda}{\theta}\left[\left(\frac{\lambda}{\theta}\right)^S - \left(\frac{\lambda}{\theta}\right)^s - Q\right] + Q - \left(1 - \frac{\lambda}{\theta}\right)^2\right]^{-1}$$

When $\lambda = \theta$, we get the cost function as

$$TC = \left\{H_C\left[Q\frac{s(s + 1)}{2} + (S + 1)(Q - 1)\left(\frac{S + s}{2}\right) - \sum_{j=s+1}^{S-1} j^2 + S\right]\right.$$

$$\left. + P_C\frac{Q\theta}{2}[s(s + 1) + Q - 1] + S_C\lambda + L P_C \lambda Q\right\}\pi_S + W_C\left(\frac{1}{\mu - \lambda}\right).$$

5.1 Numerical Analysis

Case 1. Analysis of TC as a function of s (when $\lambda \neq \theta$).
Input Data:

$$H_C = 50, \ P_C = 200, \ S_C = 2000, \ L P_C = 220, \ W_C = 2250,$$
$$S = 25, \ \lambda = 2, \ \mu = 3, \ \theta = 2.5$$

It is numerically verified that TC function is convex with respect to s from Table 1, since we get that TC values decrease, reach a minimum at $s = 5$ and then increase, as s varies from 2 to 10.

Case 2. Analysis of TC as a function of S (when $\lambda \neq \theta$).
Input Data:

$$H_C = 50, \ P_C = 200, \ S_C = 2000, \ L P_C = 220, \ W_C = 2250,$$
$$s = 2, \ \lambda = 2, \ \mu = 3, \ \theta = 2.5$$

It is numerically verified that TC function is convex with respect to S from Table 2, since we get that TC values decrease, reach a minimum at $S = 15$ and then increase, as S increases from 10 to 18.

Table 1. Effect of s on TC (when $\lambda \neq \theta$)

s	TC
2	5182.2
3	5160.6
4	5149.6
5	**5147.0**
6	5151.3
7	5161.2
8	5175.9
9	5194.6
10	5216.9

Table 2. Effect of S on TC (when $\lambda \neq \theta$)

S	TC
10	5147.5
11	5125.0
12	5109.1
13	5098.5
14	5092.3
15	**5089.9**
16	5090.6
17	5094.0
18	5099.6

Table 3. Effect of (s, S) on TC (when $\lambda \neq \theta$)

(s, S)	TC
(2, 25)	5182.2
(3, 26)	**5178.1**
(4, 27)	5187.2
(5, 28)	5206.5
(6, 29)	5233.5
(7, 30)	5266.6

Case 3. Analysis of TC when (s, S) varies simultaneously (when $\lambda \neq \theta$).
  Input Data:

$$H_C = 50, \ P_C = 200, \ S_C = 2000, \ LP_C = 220, \ W_C = 2250,$$
$$\lambda = 2, \ \mu = 3, \ \theta = 2.5$$

Table 4. Effect of S on TC (when $\lambda \neq \theta$)

S	TC
28	432.9668
29	394.0952
30	355.5556
31	317.3181
32	279.3567
33	241.6484

It is numerically verified that TC function is convex from Table 3, since we get thatTC values decrease, reach a minimum at the values $(3, 26)$ of (s, S) and then increase, as (s, S) increases simultaneously.

Case 4. Analysis of TC as a function of S (when $\lambda = \theta$).
Input Data:

$$H_C = 50, \ P_C = 200, \ S_C = 2000, \ LP_C = 220, \ W_C = 2250,$$
$$\lambda = \theta = 1.5, \ \mu = 6, \ s = 5.$$

It is numerically verified that TC function is convex with respect to S from Table 4, since we get that TC values monotonically decrease as S increases.

6 Conclusion

In this paper, we considered an (s, S) production inventory model with positive service time that follows exponential distribution. Arrivals are according to a Poisson process. Local purchase is incorporated in the model to ensure customer satisfaction and goodwill. The problem is modelled as a continuous time Markov chain and we obtained stochastic decomposition of system states. Explicit cost functions are analysed encouraged by the stochastic decomposition property. Several performance measures are obtained and convexity of cost function is established numerically. The result can be extended to more general situations, such as MAP arrivals and phase type service time.

References

1. Berman, O., Kaplan, E.H., Shimshack, D.G.: Deterministic approximations for inventory management at service facilities. IIE Trans. **25**(5), 98–104 (1993)
2. Doshi, B.T., Schouten, F.A., Talman, A.J.J.: A production- inventory control model with a mixture of back orders and lost sales. Manage. Sci. **24**, 1078–1086 (1978)
3. Krishnamoorthy, A., Lakshmy, B., Manikandan, R.: A survey on inventory models with positive service time. Opsearch **48**(2), 153–169 (2011)
4. Krishnamoorthy, A., Raju, N.: N-policy for (s, S) perishable inventory system with positive lead time. Korean J. Comput. Appl. Math. **5**, 253–261 (1998)

5. Krishnamoorthy, A., Raju, N.: (s, S) inventory with lead time-the N-policy. Inf. Manag. Sci. **9**(4), 45–52 (1998)
6. Krishnamoorthy, A., Resmi, V., Lakshmi, B.: An (s, Q) inventory system with positive lead time and service time under N-policy. Calcutta Stat. Assoc. Bull. **66**(3–4), 241–260 (2014)
7. Krishnamoorthy, A., Viswanath, C.N.: Production inventory with service time and vacation to the server. IMA J. Manag. Math. **22**(1), 33–45 (2011)
8. Krishnamoorthy, A., Viswanath, C.N.: Stochastic decomposition in production inventory with service time. Eur. J. Oper. Res. **228**(2), 358–366 (2013)
9. Latouche, G., Ramaswami, V.: Introduction to Matrix Analytic Methods in Stochastic Modeling. ASA-SIAM, Philadelphia (1999)
10. Neuts, M.F.: Matrix-Geometric Solutions in Stochastic Models-An Algorithmic Approach. Johns-Hopkins University Press, Baltimore (1981). Also published by Dover (2002)
11. Saffari, M., Asmussen, S., Haji, R.: The M/M/1 queue with inventory, lost sale, and general lead times. Queueing Syst.: Theory Appl. **75**(1), 65–77 (2013)
12. Schwarz, M., Sauer, C., Daduna, H., Kulik, R., Szekli, R.: M/M/1queueing system with inventory. Queueing Syst.: Theory Appl. **54**(1), 55–78 (2006)
13. Sharafali, M.: On a continuous review production-inventory problem. Oper. Res. Lett. **3**(4), 199–204 (1984)
14. Sigman, K., Simchi-Levi, D.: Light traffic heuristic for an M/G/1 queue with limited in ventory. Ann. Oper. Res. **40**(1), 371–380 (1992)

On the Total Amount of the Occupied Resources in the Multi-resource QS with Renewal Arrival Process

Anastasia Galileyskaya[1]([✉])[ID], Ekaterina Lisovskaya[1][ID], and Michele Pagano[2][ID]

[1] Tomsk State University, 36 Lenina Avenue, Tomsk 634050, Russia
n.galileyskaya@bk.ru, ekaterina_lisovs@mail.ru
[2] Department of Information Engineering, University of Pisa,
Via Caruso 16, 56122 Pisa, Italy
michele.pagano@iet.unipi.it

Abstract. In this paper we analyse a multi-resource queueing system with renewal arrival process and arbitrary service time distribution. In more detail, we apply the dynamic screening method to obtain the asymptotic expression for the stationary probability distribution describing the process of the total volume of the occupied resource in the system. Finally we verify the goodness of the obtained Gaussian approximation by means of discrete event simulation.

Keywords: Queuing system · Asymptotic analysis method · Arbitrary service time

1 Introduction

The methods of queuing theory are widely used to describe different economic problems, to process large data in technical systems, as well as cloud computing. Modern computer networks are characterized by the integration of heterogeneous streams, including phone calls, text messages, video sources, etc., which require the use of a more complex flow model [1–3]. To study such models, it is necessary to take into account different kinds of resources needed for the transmission and processing of the transmitted information. Thus, the development of computer and mobile communication networks has led to the need of developping new *"resource"* models that would allow us to estimate the amount of the occupied resource. Queuing systems with resources, in which customers require a server and a certain amount of resources for the duration of their service, allow modeling any peculiarities of resource allocation in modern wireless networks. However, the application of the queuing systems leads to complex computations [6,7].

Most often the analisys is limited to systems with incoming stationary Poisson flow and exponential service time. But the fact is that the Poisson flow does not always accurately describe real flows and the service time is not always exponential [8–10]. Therefore, it is very relevant in practice to consider a system with an incoming non-Poisson (for example, renewal arrival process) flow and an arbitrary service time.

© Springer Nature Switzerland AG 2019
A. Dudin et al. (Eds.): ITMM 2019, CCIS 1109, pp. 257–269, 2019.
https://doi.org/10.1007/978-3-030-33388-1_21

2   Mathematical Model

Consider a queuing system with infinite number of servers and arbitrary service time. Renewal arrival process is determined by the distribution function $A(z)$ of the interarrival times.

Each arriving customer instantly occupies the first free server, with service time distribution $B(\tau)$, and different resources $(i = 1,\ldots,n)$ with distribution $G_i(y)$, depending on the type i of the resource. When the service is completed, the customer leaves the system. Resource amounts and service times are mutually independent and do not depend on the epochs of customer arrivals. Figure 1 shows the structure of the system.

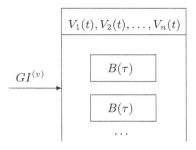

Fig. 1. Resource queuing system with infinite number of servers and renewal arrival process

Denote by $V_i(t)$ the total amount of i-th type resources $(i = 1,\ldots,n)$ occupied at time t. Our goal is to derive the probabilistic characterization of the n-dimensional process $V(t) = [V_1(t),\ldots,V_n(t)]$. This process is, in general, not Markovian and, therefore, we use the dynamic screening method for its investigation. Consider two time axes that are numbered as 0 and 1 (see Fig. 2). Let axis 0 show the epochs of customers' arrivals, while axis 1 corresponds to the screened process.

We introduce the function (dynamic probability) $S(t)$ that satisfies the condition $0 \le S(t) \le 1$. The incoming flow event can be screened on the axis 1 with probability $S(t)$ and not screened with probability $1 - S(t)$. Let the system be empty at moment t_0, and let us fix some arbitrary moment T in the future. $S(t)$ represents the probability that a customer arriving at the time t will be serviced in the system by moment T. It is easy to show that $S(t) = 1 - B(T - t)$ for $t_0 \le t \le T$.

Denote by $W_i(t)$ the total amount of i-th type resource screened on axis i. It easy to prove that

$$P\left\{\mathbf{V}(t) < \mathbf{x}\right\} = P\left\{\mathbf{W}(t) < \mathbf{x}\right\}, \tag{1}$$

for all $\mathbf{x} = \{x_1,\ldots,x_n\}$, where the inequalities $\mathbf{V}(T) < \mathbf{x}$ and $\mathbf{W}(T) < \mathbf{x}$ mean that $V_1(T) < x_1,\ldots,V_n(T) < x_n$ and $W_1(T) < x_1,\ldots,W_n(T) < x_n$, respectively.

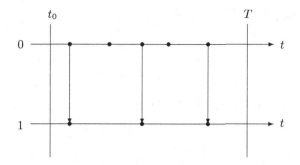

Fig. 2. Screening of the customers arrivals

We use (1) to investigate the process $\{\mathbf{V}(t)\}$ via the analysis of the process $\{\mathbf{W}(t)\}$.

3   Kolmogorov Integro-Differential Equations

Let us consider the $(n+1)$–dimensional Markovian process $\{z(t), \mathbf{W}(t)\}$, where $z(t)$ is the residual time from t to the next arrival (in the renewal input process). Denoting the probability distribution of this process by

$$P\{z(t) < z, \mathbf{W}(t) < \mathbf{w}\} = P(z, \mathbf{w}, t)$$

and taking into account the formula of total probability, we can write the following system of Kolmogorov integro-differential equations

$$\frac{\partial P(z, \mathbf{w}, t)}{\partial t} = \frac{\partial P(z, \mathbf{w}, t)}{\partial z} + \frac{\partial P(0, \mathbf{w}, t)}{\partial z}(A(z) - 1) +$$

$$A(z)S(t)\left[\int_0^{w_1} \cdots \int_0^{w_n} \frac{\partial P(0, \mathbf{w} - \mathbf{y}, t)}{\partial z} dG_n(y_n) \ldots dG_1(y_1) - \frac{\partial P(0, \mathbf{w}, t)}{\partial z}\right],$$

where $\mathbf{y} = \{y_1, \ldots y_n\}$, with the initial conditions

$$P(z, \mathbf{w}, t_0) = \begin{cases} R(z), \mathbf{w} = \mathbf{0} \\ 0, \text{ otherwise,} \end{cases}$$

where $R(z)$ denotes the stationary probability distribution of the random variable, which is determined by equality

$$R(z) = \lambda \int_0^z (1 - A(x))\, dx,$$

where

$$\lambda = \frac{1}{\int_0^\infty (1 - A(x))\, dx}.$$

We introduce the partial characteristic function

$$h(z, \mathbf{v}, t) = \int\limits_0^\infty e^{j v_1 w_1} \ldots \int\limits_0^\infty e^{j v_n w_n} P(z, d\mathbf{w}, t),$$

where $j = \sqrt{-1}$ is the imaginary unit. Then, we can write the following differential equation

$$\frac{\partial h(z, \mathbf{v}, t)}{\partial t} = \frac{\partial h(z, \mathbf{v}, t)}{\partial z} + \frac{\partial h(0, \mathbf{v}, t)}{\partial z} \Big[A(z) - 1 + A(z) S(t) (G^*(\mathbf{v}) - 1) \Big], \quad (2)$$

where

$$G^*(\mathbf{v}) = \int\limits_0^\infty e^{j v_1 y_1} dG_1(y_1) \ldots \int\limits_0^\infty e^{j v_n y_n} dG_n(y_n),$$

with the initial condition

$$h(z, \mathbf{v}, t_0) = R(z). \quad (3)$$

4   Asymptotic Analysis Method

In general, the exact solution of Eq. (2) is not available, but it may be found under asymptotic conditions. In this paper, we consider the case of infinitely growing arrival rate. Let us write the distribution function of the interarrival times as $A(Nz)$, where N is some parameter used for the asymptotic analysis ($N \to \infty$ in theoretical analysis [4,5]).

Then, the Eq. (2) takes the form

$$\frac{1}{N} \frac{\partial h(z, \mathbf{v}, t)}{\partial t} = \frac{\partial h(z, \mathbf{v}, t)}{\partial z} + \frac{\partial h(0, \mathbf{v}, t)}{\partial z} \Big[A(z) - 1 + A(z) S(t) (G^*(\mathbf{v}) - 1) \Big], \quad (4)$$

with the initial condition (3).

Theorem 1. *The first-order asymptotic characteristic function of the probability distribution of the process $\{z(t), \mathbf{W}(t)\}$ has the form*

$$h(z, \mathbf{v}, t) = R(z) \exp \left\{ N\lambda \sum_{i=1}^n j v_i a_1^{(i)} \int_{t_0}^t S(\tau) d\tau \right\},$$

where $a_1^{(i)}$ is the mean amount of i-th type occupied resource.

Proof. By performing the substitutions

$$\frac{1}{N} = \varepsilon, \mathbf{v} = \varepsilon \mathbf{y}, h(z, \mathbf{v}, t) = f_1(z, \mathbf{y}, t, \varepsilon), \quad (5)$$

in expression (4), one obtains

$$\varepsilon \frac{\partial f_1(z,\mathbf{y},t)}{\partial t} = \frac{\partial f_1(z,\mathbf{y},t)}{\partial z} +$$

$$\frac{\partial f_1(0,\mathbf{y},t)}{\partial z}\Big[A(z) - 1 + A(z)S(t)\left(G^*(\varepsilon\mathbf{y}) - 1\right)\Big], \tag{6}$$

with the initial condition

$$f_1(z,\mathbf{y},t_0,\varepsilon) = R(z). \tag{7}$$

Let us find the asymptotic solution of Problem (6)–(7) in two steps.

Step 1. Let $\varepsilon = 0$ in (6), then we obtain the following equation

$$\frac{\partial f_1(z,\mathbf{y},t)}{\partial z} + \frac{\partial f_1(0,\mathbf{y},t)}{\partial z}(A(z) - 1) = 0.$$

We can conclude that $f_1(z,\mathbf{y},t)$ can be expressed as

$$f_1(z,\mathbf{y},t) = R(z)\varPhi_1(\mathbf{y},t), \tag{8}$$

where $\varPhi_1(\mathbf{y},t)$ is some scalar function that satisfies the condition

$$\varPhi_1(\mathbf{y},t_0) = 1.$$

Step 2. Let $z \to \infty$ in (6). We obtain

$$\varepsilon \frac{\partial f_1(\infty,\mathbf{y},t)}{\partial t} = \frac{\partial f_1(0,\mathbf{y},t)}{\partial z}S(t)\left(G^*(\varepsilon\mathbf{y}) - 1\right),$$

We substitute here the expression (8), use the expansion

$$e^{j\varepsilon x} = 1 + j\varepsilon x + o(\varepsilon^2),$$

divide by ε and perform the limit as $\varepsilon \to \infty$. Taking into account that $R'(0) = \lambda$, we obtain the following differential equation

$$\frac{\partial \varPhi_1(\mathbf{y},t)}{\partial t} = \varPhi_1(\mathbf{y},t)\lambda S(t)\sum_{i=1}^{n} jy_i a_1^{(i)}, \tag{9}$$

where $a_1^{(i)} = \int\limits_0^\infty y\, dG_i(y)$.

Taking into account the initial condition, the solution of (9) is

$$\varPhi_1(\mathbf{y},t) = \exp\left\{\lambda \sum_{i=1}^{n} jy_i a_1^{(i)} \int\limits_{t_0}^{t} S(\tau)d\tau\right\}.$$

By substituting $\varPhi_1(\mathbf{y},t)$ from (8) and performing replacements opposite to (5) we obtain

$$h(z,\mathbf{v},t) = f_1(z,\mathbf{y},t,\varepsilon) \approx f_1(z,\mathbf{y},t) = R(z)\exp\left\{\lambda \sum_{i=1}^{n} jy_i a_1^{(i)} \int\limits_{t_0}^{t} S(\tau)d\tau\right\} =$$

$$R(z)\exp\left\{N\lambda \sum_{i=1}^{n} jv_i a_1^{(i)} \int\limits_{t_0}^{t} S(\tau)d\tau\right\}.$$

The proof is complete.

Theorem 2. *The second-order asymptotic characteristic function of the probability distribution of the process $\{z(t), \mathbf{W}(t)\}$ has the form*

$$h(z, \mathbf{v}, t) = R(z) \exp\left\{ N\lambda \sum_{i=1}^{n} jv_i a_1^{(i)} \int_{t_0}^{t} S(\tau)d\tau + N\sum_{i=1}^{n} \frac{(jv_i)^2}{2} \left(\lambda a_2^{(i)} \int_{t_0}^{t} S(\tau)d\tau + \right.\right.$$

$$\left.\left. \kappa \left(a_1^{(i)}\right)^2 \int_{t_0}^{t} S(\tau)d\tau\right) + \kappa N \sum_{i=1}^{n} \sum_{\substack{l=1 \\ l\neq i}}^{n} \frac{jv_i jv_l}{2} a_1^{(i)} a_1^{(l)} \int_{t_0}^{t} S^2(\tau)d\tau \right\},$$

where $\kappa = \lambda^3(\sigma^2 - a^2)$, a and σ^2 being the mean and the variance of the random variable with distribution function $A(z)$, respectively.

Proof. In Eq. (4) perform the replacement

$$h(z, \mathbf{v}, t) = h_2(z, \mathbf{v}, t) \exp\left\{ N\lambda \sum_{i=1}^{n} jv_i a_1^{(i)} \int_{t_0}^{t} S(\tau)d\tau \right\}. \tag{10}$$

Then, we can write

$$\frac{1}{N}\frac{\partial h_2(z, \mathbf{v}, t)}{\partial t} + h_2(z, \mathbf{v}, t)\lambda S(t) \sum_{i=1}^{n} jv_i a_1^{(i)} = \frac{\partial h_2(z, \mathbf{v}, t)}{\partial z} +$$

$$\frac{\partial h_2(0, \mathbf{v}, t)}{\partial z}\left[A(z) - 1 + A(z)S(t)\left(G^*(\mathbf{v}) - 1\right) \right], \tag{11}$$

with the initial condition
$$h_2(z, \mathbf{v}, t_0) = R(z). \tag{12}$$

Let us perform the following substitutions

$$\varepsilon^2 = \frac{1}{N}, \mathbf{v} = \varepsilon \mathbf{y}, h_2(z, \mathbf{v}, t) = f_2(z, \mathbf{y}, t, \varepsilon). \tag{13}$$

Substituting these expressions into (11) and (12), we obtain the following problem

$$\varepsilon^2 \frac{\partial f_2(z, \mathbf{y}, t, \varepsilon)}{\partial t} + f_2(z, \mathbf{y}, t, \varepsilon)\lambda S(t) \sum_{i=1}^{n} j\varepsilon y_i a_1^{(i)} =$$

$$\frac{\partial f_2(z, \mathbf{y}, t, \varepsilon)}{\partial z} + \frac{\partial f_2(0, \mathbf{y}, t, \varepsilon)}{\partial z}\left[A(z) - 1 + A(z)S(t)\left(G^*(\mathbf{v}) - 1\right) \right], \tag{14}$$

with the initial condition
$$f_2(z, \mathbf{y}, t, \varepsilon) = R(z). \tag{15}$$

Let us find the asymptotic solution of this problem

$$f_2(z, \mathbf{y}, t) = \lim_{\varepsilon \to 0} f_2(z, \mathbf{y}, t, \varepsilon).$$

Step 1. Letting $\varepsilon = 0$ in (14), we obtain the following equation

$$\frac{\partial f_2(z, \mathbf{y}, t, \varepsilon)}{\partial z} + \frac{\partial f_2(0, \mathbf{y}, t, \varepsilon)}{\partial z}(A(z) - 1) = 0.$$

Then, we can write

$$f_2(z, \mathbf{y}, t) = R(z)\Phi_2(\mathbf{y}, t), \tag{16}$$

where $\Phi_2(\mathbf{y}, t)$ is some scalar function, which satisfies the condition

$$\Phi_2(\mathbf{y}, t_0) = 1.$$

Step 2. The solution $f_2(z, \mathbf{y}, t)$ can be represented in the expansion form

$$f_2(z, \mathbf{y}, t, \varepsilon) = \Phi_2(\mathbf{y}, t)\left\{ R(z) + g(z)S(t)\sum_{i=1}^{n} j\varepsilon y_i a_1^{(i)} \right\} + o(\varepsilon^2), \tag{17}$$

where $g(z)$ is a suitable function. By substituting this expression in (14), using the following decomposition

$$e^{j\varepsilon x} = 1 + j\varepsilon x + o(\varepsilon^2).$$

We obtained

$$\lambda \sum_{i=1}^{n} j\varepsilon y_i a_1^{(i)} S(t) R(z)\Phi_2(\mathbf{y}, t) = \Phi_2(\mathbf{y}, t)\Big[R'(z) + \sum_{i=1}^{n} j\varepsilon y_i a_1^{(i)} S(t) g'(z) +$$

$$\lambda(A(z) - 1) + \lambda S(t) A(z) \sum_{i=1}^{n} j\varepsilon y_i a_1^{(i)} + g'(0)S(t)(A(z) - 1) \sum_{i=1}^{n} j\varepsilon y_i a_1^{(i)} \Big].$$

Taking into account that

$$R'(z) = \lambda(1 - A(z)),$$

and dividing both parts by $\sum_{i=1}^{n} j\varepsilon y_i a_1^{(i)} S(t)$, we obtain the differential equation for the unknown function $g(z)$

$$\lambda R(z) = g'(z) + g'(0)(A(z) - 1) + \lambda A(z),$$

whose solution is

$$g(z) = \lambda \int_0^z (A(u) - R(u))du + g'(0)\int_0^z (A(u) - 1)du.$$

It is easy to show that

$$g'(0) = \lambda g(\infty) + \frac{\kappa}{2}, \tag{18}$$

where $\kappa = \lambda^3(\sigma^2 - a^2)$, a and σ^2 being the mean and the variance of the random variable with distribution function $A(z)$.

Step 3. Letting $z \to \infty$ in (14) and taking advantage of the definition of the function $f_2(z, \mathbf{y}, t, \varepsilon)$, we can write

$$\lim_{z \to \infty} \frac{\partial f_2(\infty, \mathbf{y}, t, \varepsilon)}{\partial z} = 0.$$

Using this result and

$$e^{j\varepsilon x} = 1 + j\varepsilon x + \frac{(j\varepsilon x)^2}{2} + o(\varepsilon^3),$$

we obtain

$$\varepsilon^2 \frac{\partial f_2(\infty, \mathbf{y}, t, \varepsilon)}{\partial t} + f_2(\infty, \mathbf{y}, t, \varepsilon)\lambda S(t) \sum_{i=1}^{n} j\varepsilon y_i a_1^{(i)} =$$

$$\frac{\partial f_2(z, \mathbf{y}, t, \varepsilon)}{\partial z} S(t) \Big[\sum_{i=1}^{n} j\varepsilon y_i a_1^{(i)} + \sum_{i=1}^{n} \frac{(j\varepsilon y_i)^2}{2} a_2^{(i)} + \sum_{i=1}^{n} \sum_{\substack{l=1 \\ l \neq i}}^{n} \frac{j\varepsilon y_i j\varepsilon y_l}{2} a_1^{(i)} a_1^{(l)} \Big].$$

By substituting here the expansion (17) and considering the limit as $z \to \infty$, we can write

$$\varepsilon^2 \frac{\partial \Phi_2(\mathbf{y}, t)}{\partial t} + \Phi_2(\mathbf{y}, t)\lambda S(t) \sum_{i=1}^{n} j\varepsilon y_i a_1^{(i)} + \Phi_2(\mathbf{y}, t)g(\infty) \sum_{i=1}^{n} j\varepsilon y_i a_1^{(i)} \sum_{l=1}^{n} j\varepsilon y_l a_1^{(l)} S^2(t)\lambda =$$

$$\Phi_2(\mathbf{y}, t) \Big(\lambda S(t) \Big[\sum_{i=1}^{n} j\varepsilon y_i a_1^{(i)} + \sum_{i=1}^{n} \frac{(j\varepsilon y_i)^2}{2} a_2^{(i)} + \sum_{i=1}^{n} \sum_{\substack{l=1 \\ l \neq i}}^{n} \frac{j\varepsilon y_i j\varepsilon y_l}{2} a_1^{(i)} a_1^{(l)} \Big] \Big) + o(\varepsilon^3).$$

We divide by ε^2, taking into account (18) and passing to the limit as $\varepsilon \to \infty$, we obtain the differential equation for the unknown function $\Phi_2(\mathbf{y}, t)$

$$\frac{\partial \Phi_2(\mathbf{y}, t)}{\partial t} = \Phi_2(\mathbf{y}, t) \Big[\lambda S(t) \sum_{i=1}^{n} \frac{(j\varepsilon y_i)^2}{2} a_2^{(i)} + \kappa S^2(t) \sum_{i=1}^{n} \frac{(j\varepsilon y_i)^2}{2} \big(a_1^{(i)}\big)^2$$

$$\kappa S^2(t) \sum_{i=1}^{n} \sum_{\substack{l=1 \\ l \neq i}}^{n} \frac{j\varepsilon y_i j\varepsilon y_l}{2} a_1^{(i)} a_1^{(l)} \Big].$$

The solution of the latter equation with the available initial condition gives the expression

$$\Phi_2(\mathbf{y}, t) = \exp\left\{\sum_{i=1}^{n} \frac{(jy_i)^2}{2}\left(\lambda a_2^{(i)} \int_{t_0}^{t} S(\tau)d\tau + \kappa(a_1^{(i)})^2 \int_{t_0}^{t} S^2(\tau)d\tau\right) + \right.$$

$$\left. \kappa \sum_{i=1}^{n}\sum_{\substack{l=1 \\ l \neq i}}^{n} \frac{jy_i jy_l}{2} a_1^{(i)} a_1^{(l)} \int_{t_0}^{t} S^2(\tau)d\tau\right\}.$$

Substituting this expression in (16), we can write

$$f_2(z, \mathbf{y}, t) = R(z)\exp\left\{\sum_{i=1}^{n} \frac{(jy_i)^2}{2}\left(\lambda a_2^{(i)} \int_{t_0}^{t} S(\tau)d\tau + \kappa(a_1^{(i)})^2 \int_{t_0}^{t} S^2(\tau)d\tau\right) + \right.$$

$$\left. \kappa \sum_{i=1}^{n}\sum_{\substack{l=1 \\ l \neq i}}^{n} \frac{jy_i jy_l}{2} a_1^{(i)} a_1^{(l)} \int_{t_0}^{t} S^2(\tau)d\tau\right\}. \tag{19}$$

Substituting this expression into (19) and performing the substitutions that are inverse to (10) and (13), we obtain the following expression for the asymptotic characteristic function of the process $\{z(t), \mathbf{W}(t)\}$

$$h(z, \mathbf{v}, t) \approx R(z)\exp\left\{N\lambda \sum_{i=1}^{n} jv_i a_1^{(i)} + \right.$$

$$N\sum_{i=1}^{n} \frac{(jv_i)^2}{2}\left(\lambda a_2^{(i)} \int_{t_0}^{t} S(\tau)d\tau + \kappa(a_1^{(i)})^2 \int_{t_0}^{t} S^2(\tau)d\tau\right) + $$

$$\left. N\sum_{i=1}^{n}\sum_{\substack{l=1 \\ l \neq i}}^{n} \frac{jv_i jv_l}{2} a_1^{(i)} a_1^{(l)} \int_{t_0}^{t} S^2(\tau)d\tau\right\}.$$

Corollary. When $z \to \infty$, $t = T$ and $t_0 \to -\infty$, we obtain the characteristic function of the process $\{\mathbf{V}(t)\}$ in the steady state regime. From the structure of function (19) it is clear that the n-dimensional process $\{\mathbf{V}(t)\}$ is asymptotically Gaussian with mean

$$\mathbf{a} = N\lambda \left[a_1^{(1)}\ a_1^{(2)}\ \cdots\ a_1^{(n)}\right] b,$$

where

$$b = \int_{0}^{\infty} (1 - B(\tau))d\tau,$$

and covariance matrix

$$\mathbf{K} = N\left(\lambda \mathbf{K}^{(1)}b + \kappa \mathbf{K}^{(2)}\beta\right),$$

where

$$\mathbf{K}^{(1)} = \begin{bmatrix} a_2^{(1)} & 0 & \dots & 0 \\ 0 & a_2^{(2)} & \dots & 0 \\ \dots & \dots & \dots & \dots \\ 0 & 0 & \dots & a_2^{(n)} \end{bmatrix}, \mathbf{K}^{(2)} = \begin{bmatrix} \left(a_1^{(1)}\right)^2 & a_1^{(1)}a_1^{(2)} & \dots & a_1^{(1)}a_1^{(n)} \\ a_1^{(2)}a_1^{(1)} & \left(a_1^{(2)}\right)^2 & \dots & a_1^{(2)}a_1^{(n)} \\ \dots & \dots & \dots & \dots \\ a_1^{(n)}a_1^{(1)} & a_1^{(n)}a_1^{(2)} & \dots & \left(a_1^{(n)}\right)^2 \end{bmatrix},$$

$$\beta = \int_0^\infty \left(1 - B(\tau)\right)^2 d\tau.$$

5 Numerical Example

The Gaussian distribution is obtained under the asymptotic condition $N \to \infty$. Therefore, the result may be used just as an approximation and it is applicable when N is great enough. So, we need to determine a lower bound of the parameter N, which causes the approximation (19) be applicable. To do this we make series of simulation experiments, considering two types of resources and compare asymptotic distributions with empiric ones by using the Kolmogorov distance

$$\Delta = \max_x \mid F(x) - A(x) \mid \tag{20}$$

as an accuracy measure. Here $F(x)$ is the cumulative distribution function of total capacity of customers, constructed on the basis of simulation results, and $A(x)$ is the corresponding Gaussian approximation (its parameters for the two classes are given in Table 1).

To investigate the goodness of the approximation, we consider the following numerical example. In particular, we assume that the input renewal process is characterized by the following distribution function

$$A(z) = \begin{cases} 0, & z < 0.5; \\ z - 0.5, & z \in [0.5, 1.5]; \\ 1, & z > 1.5. \end{cases}$$

Hence, the fundamental rate of arrivals is $\lambda = 1$ customers per time unit. Moreover, each arriving customer occupies 2 types of resources and the corresponding customer capacities have uniform distribution in the range $[0; 1]$ and $[0; 2]$, respectively. Service time has gamma distribution with parameters $\alpha = \beta = 0.5$ and so the fundamental rate of arrivals is N times the service rate.

Table 1. Parameters of Gaussian approximations

Type of resource	Mean	Variance
First	$0.5N$	$0.25N$
Second	$1N$	$1N$

In the following we consider both the marginal and the bidimensional distributions of the amount of occupied resources at the system. In more detail, Tables 2 and 3 report the values of the Kolmogorov distance for the two types of resource, highlighting that the goodness of the approximation depends not only on N, but also on the different statistical features of the considered types of customer.

Table 2. Kolmogorov distance for the first type of resource

N	1	3	5	7	10	20	50	100
Δ	0.293	0.073	0.034	0.022	0.016	0.011	0.007	0.005

Table 3. Kolmogorov distance for the second type of resource

N	1	3	5	7	10	20	50	100
Δ	0.299	0.079	0.039	0.026	0.019	0.012	0.008	0.006

This conclusion is confirmed by Figs. 3 and 4, which compare the asymptotic approximations with the empirical results for the total resource amount of each type for two different values of N. A similar conclusion can be drawn also for the joint distribution, as highlighted by Table 4 and Fig. 5.

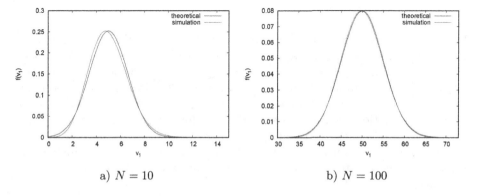

a) $N = 10$ b) $N = 100$

Fig. 3. Distributions of the total resource amount for the first type

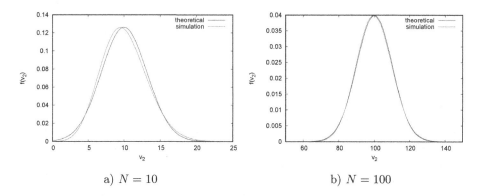

a) $N = 10$ b) $N = 100$

Fig. 4. Distributions of the total resource amount for the second type

Table 4. Kolmogorov distance for bidimensional distributions

N	1	3	5	7	10	20	50	100
Δ	0.304	0.085	0.04	0.026	0.019	0.013	0.008	0.006

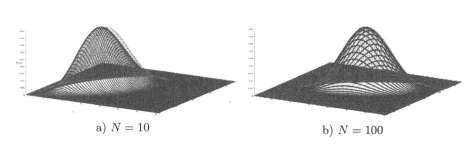

a) $N = 10$ b) $N = 100$

Fig. 5. Distributions of the total resource amount first and second type

6   Conclusions

In this paper we presented the analysis of Multi-resource $GI^{(\nu)}/GI/\infty$ queueing system with renewal arrival process and arbirtary service time. We applied dinamic screening method to obtain asymptotic expression for the stationary probability distribution of the process describing the total volume of the occupied resource in the system. In more detail we derived first and second-order asymptotic approximations under the assumption of infinitely growing arrival rate, and we showed that the n-dimensional probability distribution of the total resource amount is asymptotically n-dimensional Gaussian. Numerical experiments and simulations allow us to determine the applicability area of the asymptotic result for different classes of users.

References

1. Baskett, F., Chandy, K.M., Muntz, R.R., Palacios, F.G.: Open, closed, and mixed networks of queues with different classes of customers. J. ACM **2**(22), 248–260 (1975)
2. Brown, L., Gans, N., Mandelbaum, A., Sakov, A.: Statistical analysis of a telephone call center. A queueing-science perspective. J. Am. Stat. Assoc. **100**, 36–50 (2005)
3. Lisovskaya, E., Moiseeva, S., Pagano, M.: Multiclass $GI/GI/\infty$ queueing systems with random resource requirements. In: Dudin, A., Nazarov, A., Moiseev, A. (eds.) ITMM/WRQ -2018. CCIS, vol. 912, pp. 129–142. Springer, Cham (2018). https://doi.org/10.1007/978-3-319-97595-5_11
4. Moiseev, A., Nazarov, A.: Asymptotic analysis of the infinite-server queueing system with high-rate semi-Markov arrivals. In: 2014 6th International Congress on Ultra Modern Telecommunications and Control Systems and Workshops (ICUMT), pp 507–513 (2014)
5. Moiseev, A., Nazarov, A.: Queueing network with hight-rate arrivals. Eur. J. Oper. Res. **254**(1), 161–168 (2016)
6. Naumov, V., Samouylov, K., Sopin, E., Andreev, S.: Two approaches to analysis of queuing systems with limited resources. In: Ultra-Modern Telecommunications and Control Systems and Workshops Proceedings, pp. 485–489 (2014)
7. Pankratova, E., Moiseeva, S.: Queueing system $GI/GI/\infty$ with n types of customers. Commun. Comput. Inf. Sci. **564**, 216–225 (2015)
8. Tikhonenko, O., Kempa, W.M.: Queue-size distribution in $M/G/1$-type system with bounded capacity and packet dropping. In: Dudin, A., Klimenok, V., Tsarenkov, G., Dudin, S. (eds.) BWWQT 2013. CCIS, vol. 356, pp. 177–186. Springer, Heidelberg (2013). https://doi.org/10.1007/978-3-642-35980-4_20
9. Tikhonenko, O., Kempa, W.M.: The generalization of AQM algorithms for queueing systems with bounded capacity. In: Wyrzykowski, R., Dongarra, J., Karczewski, K., Waśniewski, J. (eds.) PPAM 2011. LNCS, vol. 7204, pp. 242–251. Springer, Heidelberg (2012). https://doi.org/10.1007/978-3-642-31500-8_25
10. Wang, R., Jouini, O., Benjaafar, S.: Service systems with finite and heterogeneous customer arrivals. Manuf. Serv. Oper. Manag. **3**(16), 365–380 (2014)

Simulation of Finite-Source Retrial Queues with Two-Way Communications to the Orbit

János Sztrik$^{(\boxtimes)}$, Ádám Tóth, Ákos Pintér, and Zoltán Bács

University of Debrecen, Debrecen 4032, Hungary
{sztrik.janos,toth.adam}@inf.unideb.hu
apinter@science.unideb.hu, bacs.zoltan@econ.unideb.hu

Abstract. In this paper we investigate a single-server two-way communication system by the help of retrial queuing systems with finite source. From the finite source incoming primary calls enter into the system according to an exponential distribution. If the server is idle then the service of incoming customer starts immediately. Alternatively, if an incoming customer discovers the server in busy state it is directed towards the orbit, where after some exponentially distributed time retries to reach the server again. As soon as the server becomes idle it can generate an outgoing call to the customers in the orbit after an exponentially distributed time. In case of two-way communication after the service of an outgoing call it returns to the source. In this work we concentrate on emphasizing a phenomena of outgoing call on the mean waiting time of incoming customers. The novelty of this paper is to carry out a sensitivity analysis comparing various distributions of service time of primary customers on the performance measures like utilization of the server or mean waiting time. By the use of simulation several graphical results and comparison of the applied systems are illustrated.

Keywords: Retrial queues · Two-way communication · Sensitivity analysis · Finite-source queuing systems · Simulation

1 Introduction

Finite-source retrial queues are effective and commonly used systems for modeling real life problems arising in main telecommunication systems like cellular mobile communication networks, computer networks, local-area networks with random access protocols, call centers and CSMA-based wire-less networks. With the decrement of the rate of generation of new calls the number of customers in the system increases in case of many practical situation. This can be performed

The research was financed by the Higher Education Institutional Excellence Programme of the Ministry of Human Capacities in Hungary, within the framework of the 20428-3/2018/FEKUTSTRAT thematic programme of the University of Debrecen.

© Springer Nature Switzerland AG 2019
A. Dudin et al. (Eds.): ITMM 2019, CCIS 1109, pp. 270–284, 2019.
https://doi.org/10.1007/978-3-030-33388-1_22

with the help of finite-source or quasi-random input models. Their importance can be viewed by the reader in the following works, for example [4, 10, 11, 14, 15].

Systems with retrial feature are identified by a specific feature of arriving customers when the server is occupied. These customers stay in the system and spend their time in a virtual waiting room called orbit. Customers in the orbit attempt to be served after a random time. Because the number of calls are finite, the assumption of working with finite-source queuing systems follows real circumstances. In this paper we examine two-way communication retrial queuing system which is quite popular topic in the recent years. This can be explained by the fact that using two-way communication scheme is very helpful in many application fields to model real life problems. Especially in case of call centers where service unit can perform certain other work in idle state like selling, advertising and promoting products including serving incoming calls. In such systems utilization of the service unit is always pivotal, see for example in [1, 2, 6, 9, 13, 16, 20, 22].

Once the server becomes idle it calls for customers inside and outside of the system which is called an outgoing call. This is a typical feature of two-way communication system. In our investigated model the idle service unit can generate a call only from the orbit which arrives after a random time. It will only be served if no customers from the finite source or from the orbit come. Otherwise this outgoing call will be canceled. Papers dealing with two-way communication systems by the help of retrial queues, where the source is infinite, are found in [3, 5, 7, 8, 17–19, 21].

Our aim is to study the operation of the system where the service unit is reliable and can perform outgoing call from the orbit. The novelty of this paper is to compare this system with the common finite source retrial system using various distribution of service on performance measures like mean waiting time of an incoming call or utilization of the server. We are mainly interested in how the different distributions modify the characteristics of the system. To achieve this goal a simulation program has been developed using the base of SimPack [12] which contains a number of C/C++ libraries and executable programs. One of the main reasons for its usage is that the user has the freedom what performance measure are calculated and how the model is built up. SimPack toolkit also provides a set of utilities that demonstrate how to build a working simulation from a model description.

2   System Model

We consider a retrial queuing system of type $M/G/1//N$ with a reliable server which is capable to produce outgoing calls to the customers residing in the orbit. N customers are located in the source, where all of them can generate incoming, primary calls towards the server. The distribution of the inter-request times is exponential with rate λ/N. In default of waiting queue an incoming customer either from the source or orbit finds the server in an idle state then its service begins instantly. The service times of incoming customers are assumed to be gamma, hypo-exponentially, hyper-exponentially, Pareto and lognormal distributed with different parameters but with the same mean value. Customers return to the source after their service is terminated. If the server is busy, meaning that a request is under service, an incoming customer remains in the system and enters into the orbit. Customers located in the orbit are able to attempt to access the server again after an exponentially distributed time with parameter σ/N. In the other hand, when the server becomes idle it can make outgoing call towards the customers in the orbit. It is performed after an exponentially distributed time with parameter ν. The service time of these outgoing customers follows gamma distribution with parameters α_2 and β_2. In a consecutive paper we aim to investigate the same system by the help of asymptotic methods when N tends to infinity and that is the reason we use λ/N and σ/N parameters. All the random variables involved in the model construction are assumed to be totally independent of each other.

3   Applied Distributions and Its Parameters

In this Section the reader gets an insight of the parameters of the applied distributions and the process how to select them in order to execute a valid comparison. To do so our program is integrated with random number generators according to gamma, hyper-exponential, hypo-exponential, lognormal and Pareto distribution. These random number generators need input parameters which are different in every distribution, thus parameter selection is crucial. For valid comparison we use the same mean and variance in case of every distribution hence we take over every distribution and how the fitting process is accomplished.

3.1   Gamma Distribution

Gamma distribution is a general type of statistical distribution and a random variable X has a gamma distribution if its density function is the following:

$$f(x) = \begin{cases} 0 & \text{if } x < 0 \\ \frac{\beta(\beta x)^{\alpha-1}e^{-\beta x}}{\Gamma(\alpha)} & \text{if } x \geq 0 \end{cases}$$

where $\beta > 0$ and $\alpha > 0$.

$$\Gamma(\alpha) = \int_0^\infty t^{\alpha-1}\,e^{-t}\,dt$$

This is the so-called complete gamma function, which has two parameters: α is called the shape parameter and β is called the scale parameter. These two parameters are also the input parameters of the random number generator.

The coefficient $C_X^2 = \frac{Var(X)}{(EX)^2}$ is defined as the squared coefficient of variation of random variable X.

The mean value, variation and the squared coefficient of variation can be calculated:

$$\overline{X} = \frac{\alpha}{\beta}, \qquad Var(X) = \frac{\alpha}{\beta^2}, \qquad C_X^2 = \frac{1}{\alpha}$$

For a predetermined mean value and variance to obtain parameters α and β the next calculation has to be done:

$$\alpha = \frac{1}{C_X^2}, \qquad \beta = \frac{\alpha}{\overline{X}}$$

3.2   Pareto Distribution

A random variable X has a Pareto distribution if its density function is the following:

$$f(x) = \begin{cases} 0 & \text{if } x < k \\ \alpha k^\alpha x^{-\alpha-1} & \text{if } x \geq k \end{cases}$$

Hence the distribution function is:

$$F(x) = \begin{cases} 0 & \text{if } x < k \\ 1 - \left(\frac{k}{x}\right)^\alpha & \text{if } x \geq k \end{cases}$$

where $\alpha, k > 0$.

It has two parameters: α is called the shape parameter and k is called the location parameter. These two parameters are the input parameters of the random number generator.

The mean value, variation and the squared coefficient of variation can be calculated as follows:

$$\overline{X} = \begin{cases} \frac{k\alpha}{\alpha-1} & \text{if } \alpha > 1 \\ \infty & \text{if } \alpha \leq 1 \end{cases}$$

$$Var(X) = \frac{k^2 \alpha}{\alpha - 2} - \left(\frac{k\alpha}{\alpha - 1}\right)^2, \qquad C_X^2 = \frac{(\alpha - 1)^2}{\alpha(\alpha - 2)} - 1, \qquad \alpha > 2.$$

For a predetermined mean value and variance to obtain parameters α and k the following interrelation is used:

$$\alpha = 1 + \frac{\sqrt{1 + C_X^2}}{\sqrt{C_X^2}}, \qquad k = \frac{\alpha - 1}{\alpha} \times \overline{X}$$

3.3 Lognormal Distribution

Let $Y \in N(m, \sigma)$ a random variable with normal distribution, lognormal is a continuous distribution in which the logarithm of a variable having a normal distribution, namely $X = e^Y$ has lognormal distribution with parameters (m, σ). Its distribution and density function are the following:

$$F_x(x) = \Phi\left(\frac{\ln(x) - m}{\sigma}\right), \qquad x > 0.$$

$$f_x(x) = \frac{1}{\sigma x}\varphi\left(\frac{\ln(x) - m}{\sigma}\right), \qquad x > 0.$$

The mean value, variance and the squared coefficient of variation can be calculated:

$$\overline{X} = e^{m + \frac{\sigma^2}{2}}, \qquad Var(X) = e^{2m + \sigma^2}(e^{\sigma^2} - 1), \qquad C_X^2 = e^{\sigma^2} - 1.$$

To obtain the two parameters of the lognormal distribution the following interrelation is applied:

$$\sigma = \sqrt{\ln(1 + C_X^2)}, \qquad m = \ln(\overline{X}) - \frac{\sigma^2}{2}$$

3.4 Hypo-exponential Distribution

Continuous statistical distribution, let $X_i \in Exp(\mu_i)(i = 1, ..., n)$ be independent exponentially distributed random variables. Then $Y_n = X_1 + ... + X_n$ has n-phase hypo-exponential distribution. Its density function is given by

$$f_{Y_n}(x) = \begin{cases} 0 & \text{if } x < 0 \\ (-1)^{n-1}\left[\prod_{i=1}^{n}\mu_i\right]\sum_{j=1}^{n}\frac{e^{-\mu_j x}}{\prod_{k=1, k \neq j}^{n}(\mu_j - \mu_k)} & \text{if } x \geq 0. \end{cases}$$

The mean value, variance and the squared coefficient of variation can be calculated:

$$\overline{Y_n} = \sum_{i=1}^{n}\frac{1}{\mu_i}, \qquad Var(Y_n) = \sum_{i=1}^{n}\frac{1}{\mu_i^2}, \qquad C_{Y_n}^2 = \frac{\sum_{i=1}^{n}\left(\frac{1}{\mu_i}\right)^2}{\left(\sum_{i=1}^{n}\frac{1}{\mu_i}\right)^2}.$$

In our simulation program we used the 2-phase hypo-exponential distribution where the parameters are the parameters of the two independent exponential distribution (μ_1, μ_2). For a predetermined mean value and variance to obtain parameters μ_1 and μ_2 the next equation system has to be solved:

$$\overline{X} = \frac{1}{\mu_1} + \frac{1}{\mu_2}, \qquad Var(X) = \frac{1}{\mu_1^2} + \frac{1}{\mu_2^2}$$

3.5   Hyper-exponential Distribution

Suppose $X_1, X_2, \cdots, X_n$ are independent exponential random variables, where the rate parameter of X_i is λ_i. The random variable X can be one of the n independent exponential random variables $X_1, X_2, \cdots, X_n$ such that X is X_i with probability p_i with $p_1 + \cdots + p_n = 1$. Such a random variable X is said to follow a hyper-exponential distribution. Its density function is given by

$$f_X(x) = \begin{cases} 0 & \text{if } x < 0 \\ \sum_{i=1}^n p_i \lambda_i e^{-\lambda_i x} & \text{if } x \geq 0. \end{cases}$$

Its distribution function is

$$F_X(x) = \begin{cases} 0 & \text{if } x < 0 \\ 1 - \sum_{i=1}^n p_i e^{-\lambda_i x} & \text{if } x \geq 0. \end{cases}$$

In the case when for a random variable $X, C_X^2 > 1$ then the the following two-moment fit is suggested

$$f_Y(t) = p\lambda_1 e^{-\lambda_1 t} + (1 - p)\lambda_2 e^{-\lambda_2 t}.$$

Y is a 2-phase hyper-exponentially distributed random variable. The most commonly used procedure is the balanced mean method, that is

$$\frac{p}{\lambda_1} = \frac{1-p}{\lambda_2}.$$

To obtain the three parameters of the hyper-exponential distribution the following calculation is used:

$$p = \frac{1}{2}\left(\sqrt{\frac{C_X^2 - 1}{C_X^2 + 1}}\right), \qquad \lambda_1 = \frac{2p}{\overline{X}}, \qquad \lambda_2 = \frac{2(1 - p)}{\overline{X}}.$$

4   Simulation Results

4.1   Squared Coefficient of Variation is Greater than One

The values of the input parameters are shown in Table 1. In this section these results are in connection with the effect of different service time distributions of incoming customers where the mean and variance are equal, respectively. We use hyper-exponential distribution if the squared coefficient of variation is greater than one, Table 2 shows the exact values of parameters of service time of incoming customers. Besides hyper-exponential, gamma, lognormal and Pareto distributions are also used for comparisons.

Figure 1 shows the mean waiting time in function of arrival intensity of incoming customers. For these values of parameters regardless of the applied distribution a maximum value of the mean waiting time can be seen. This maximum

Table 1. Numerical values of model parameters

N	σ/N	ν	α_2	β_2
100	0.01	0.02	1	1.1

Table 2. Parameters of service time of incoming customers

Distribution	Gamma	Hyper-exponential	Pareto	Lognormal
Parameters	$\alpha = 0.04$	$p = 0.48$	$\alpha = 2.02$	$m = -1.629$
	$\beta = 0.04$	$\lambda_1 = 0.961$	$k = 0.505$	$\sigma = 1.805$
		$\lambda_2 = 1.04$		
Mean	1			
Variance	25			
Squared coefficient of variation	25			

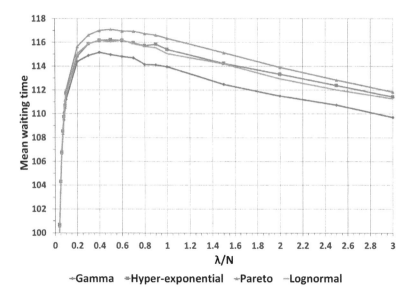

Fig. 1. Mean waiting time vs. arrival intensity using various distributions

feature occurs for finite-source retrial queues, see for example [4,9,10,16]. Differences can be observed among the values of mean waiting time especially in the case of using gamma and Pareto distribution, despite the fact that the mean and variance are the same. On this figure the effect of different distributions is clearly observable.

Figures 2 and 3 illustrates how the utilization of the server grows with the increment of the arrival intensity of incoming customers. The highest values can be found at gamma distribution but the differences of the applied distributions

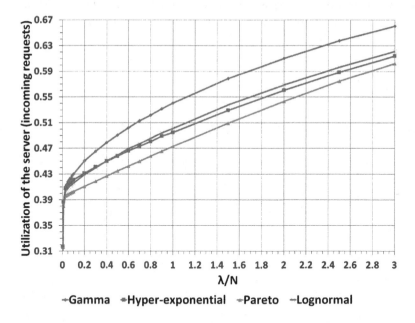

Fig. 2. Utilization of server vs. arrival intensity using various distributions

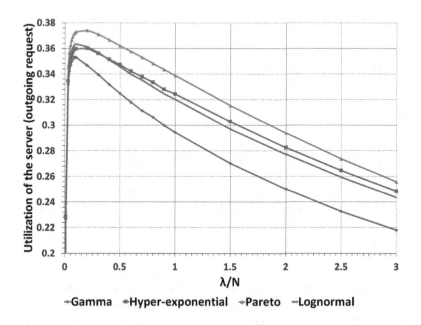

Fig. 3. Utilization of server vs. arrival intensity using various distributions

are as commensurable as in case of Fig. 1. As the arrival intensity increases the probability of performing outgoing call become less so outgoing requests spend less time at the service unit.

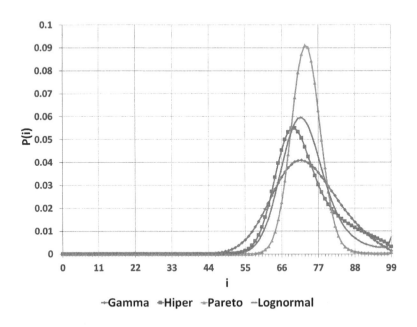

Fig. 4. Comparison of steady-state distributions

On Fig. 4 the comparison of steady-state distribution can be seen when the distribution of service time of the incoming customers is different. It represents the probability of how many customers residing in the orbit. Exploring the curves in more detail they correspond to normal distribution. The same parameter setting is used what Table 1 demonstrates where λ/N is 0.03.

To emphasize the importance of outgoing calls we compare our investigated model to the model without outgoing calls. This model is named as the classical retrial queuing system. On Fig. 5 comparison of the mean waiting time can be seen and due to the phenomena of outgoing call customers spend less time in the system, which is obvious looking at the curves. However, in our investigated model the utilization of the service unit (Fig. 6) is much higher compared to the classical retrial queuing system therefore it spends less time in idle state. In this way the efficiency of the server grows such that the mean waiting time decreases substantively. The distribution of service time of the incoming customer is gamma at Figs. 5 and 6, but the ratio of difference is also true for the other distributions, too. On this figure under total utilization of server we mean the service of both the incoming and outgoing requests at the curve of with outgoing call.

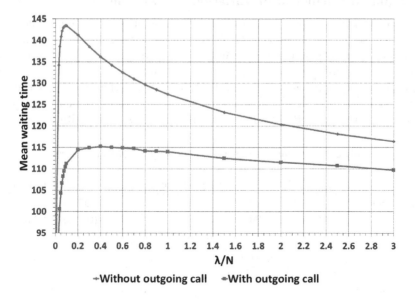

Fig. 5. Comparison of our investigated model and the classical retrial queuing model on the mean waiting time

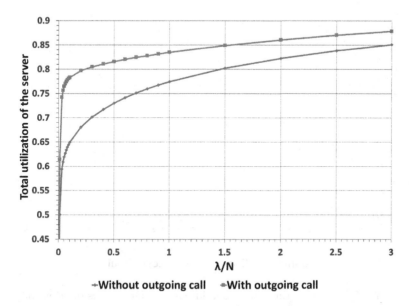

Fig. 6. Comparison of our investigated model and the classical retrial queuing model on the utilization of server

4.2 Squared Coefficient of Variation is Less than One

The same input parameters are used as in the previous section, see Table 1. The results are also in connection with the effect of different service time distributions of incoming customers where the mean and variance are equal. Instead of hyper-exponential distribution hypo-exponential distribution is used if the squared coefficient of variation is less than one. Table 3 illustrates the values of parameters of service time of incoming customers. In addition to hypo-exponential, we apply gamma, lognormal and Pareto distributions to perform sensitivity analysis.

Table 3. Parameters of service time of incoming customers

Distribution	Gamma	Hypo-exponential	Pareto	Lognormal
Parameters	$\alpha = 1.5504$	$\mu_1 = 1.3$	$\alpha = 2.597$	$m = -0.249$
	$\beta = 1.5504$	$\mu_2 = 4.333$	$k = 0.615$	$\sigma = 0.705$
Mean	1			
Variance	0.6449704142			
Squared coefficient of variation	0.6449704142			

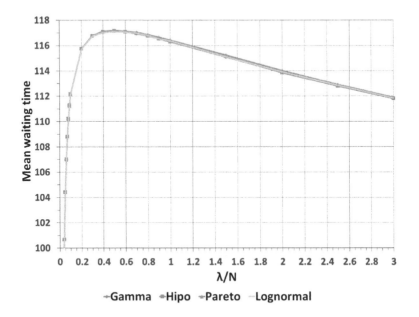

Fig. 7. Mean waiting time vs. arrival intensity using various distributions

Figure 7 demonstrates the mean waiting time in the function of arrival intensity of incoming calls. Taking closer look at the curves it can be stated that the values of mean waiting time are almost identical regardless of the applied distribution. With this parameter setting the interesting maximum value of the mean waiting time appears as in the previous section.

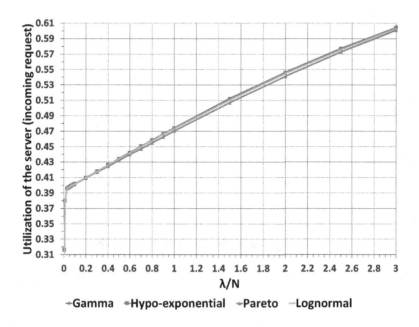

Fig. 8. Utilization of server vs. arrival intensity using various distributions

Figures 8 and 9 illustrates how the utilization of the server increases with the increment of the arrival intensity of incoming customers. As in case of mean waiting time here using different distributions result the same utilization. It seems that when the squared coefficient of variation is less than one using different distributions have no effect on the performance measures and the obtained results are nearly identical.

Similarly to the previous section we compare the results between the classical retrial queuing system and our investigated model. On Figs. 8 and 9 the same tendency can be observed like when the the squared coefficient of variation is greater than one, namely values of mean waiting time is lesser when the server can make outgoing calls. But this also affects the utilization of the service unit because with the help of outgoing calls server spend less time without satisfying the needs of the customers. As in the previous section the service time of the incoming customer follows gamma distribution at Fig. 10 and 11, but the ratio of difference is also true for the other distributions, too.

From Figs. 5, 6, 10 and 11 it can be said that the utilization of service unit escalates when outgoing calls are performed, but it also results lesser mean waiting time of incoming customers. With a proper parameter setting in the case of outgoing calls the utilization of the server is much higher in a way that customers

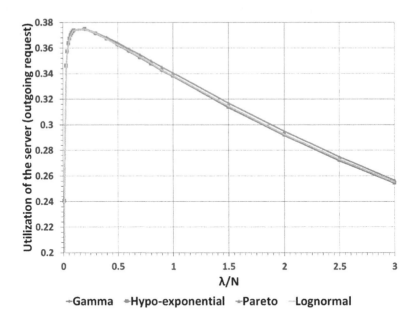

Fig. 9. Utilization of server vs. arrival intensity using various distributions

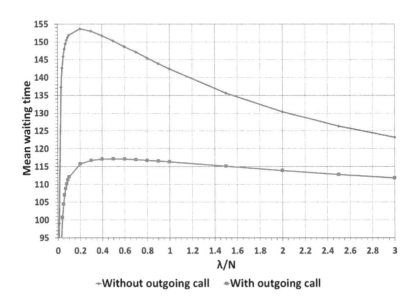

Fig. 10. Comparison of our investigated model and the classical retrial queuing model on the mean waiting time

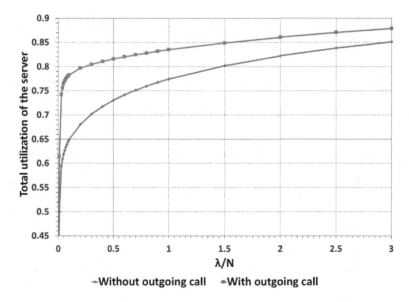

Fig. 11. Comparison of our investigated model and the classical retrial queuing model on the utilization of server

spend less time in the orbit. In the case of with outgoing calls total utilization of the server includes both incoming and outgoing requests occupying the service unit.

5   Conclusion

A finite-source retrial queueing system is introduced where the server can produce outgoing calls towards the customers of the orbit. Several figures present the effect of the applied distributions on the mean waiting time and on the utilization of the server. Using stochastic simulation method results clearly indicate that when the squared coefficient of variation is greater than one then the contrast of the values of the performance measures is quite high having the same mean and variance.

References

1. Aguir, S., Karaesmen, F., Akşin, O.Z., Chauvet, F.: The impact of retrials on call center performance. OR Spectr. **26**(3), 353–376 (2004)
2. Aksin, Z., Armony, M., Mehrotra, V.: The modern call center: a multi-disciplinary perspective on operations management research. Prod. Oper. Manag. **16**(6), 665–688 (2007)
3. Artalejo, J.R., Phung-Duc, T.: Markovian retrial queues with two way communication. J. Ind. Manag. Optim. **8**(4), 781–806 (2012)
4. Artalejo, J., Corral, A.G.: Retrial Queueing Systems: A Computational Approach. Springer, Heidelberg (2008). https://doi.org/10.1007/978-3-540-78725-9

5. Artalejo, J., Phung-Duc, T.: Single server retrial queues with two way communication. Appl. Math. Modell. **37**(4), 1811–1822 (2013)
6. Brown, L., et al.: Statistical analysis of a telephone call center: a queueing-science perspective. J. Am. Stat. Assoc. **100**(469), 36–50 (2005)
7. Dimitriou, I.: A retrial queue to model a two-relay cooperative wireless system with simultaneous packet reception. In: Wittevrongel, S., Phung-Duc, T. (eds.) ASMTA 2016. LNCS, vol. 9845, pp. 123–139. Springer, Cham (2016). https://doi.org/10.1007/978-3-319-43904-4_9
8. Dragieva, V., Phung-Duc, T.: Two-way communication M/M/1 retrial queue with server-orbit interaction. In: Proceedings of the 11th International Conference on Queueing Theory and Network Applications, p. 11. ACM (2016)
9. Dragieva, V., Phung-Duc, T.: Two-way communication M/M/1//N retrial queue. In: Thomas, N., Forshaw, M. (eds.) ASMTA 2017. LNCS, vol. 10378, pp. 81–94. Springer, Cham (2017). https://doi.org/10.1007/978-3-319-61428-1_6
10. Falin, G., Artalejo, J.: A finite source retrial queue. Eur. J. Oper. Res. **108**, 409–424 (1998)
11. Fiems, D., Phung-Duc, T.: Light-traffic analysis of random access systems without collisions. Ann. Oper. Res. **277**(2), 1–17 (2017). https://doi.org/10.1007/s10479-017-2636-7
12. Fishwick, P.A.: Simpack: getting started with simulation programming in C and C++. In: WSC 1992 Proceedings of the 24th Conference on Winter simulation, pp. 154–162. ACM (1992)
13. Gans, N., Koole, G., Mandelbaum, A.: Telephone call centers: tutorial, review, and research prospects. Manuf. Serv. Oper. Manag. **5**(2), 79–141 (2003)
14. Gómez-Corral, A., Phung-Duc, T.: Retrial queues and related models. Ann. Oper. Res. **247**(1), 1–2 (2016). https://doi.org/10.1007/s10479-016-2305-2
15. Kim, J., Kim, B.: A survey of retrial queueing systems. Ann. Oper. Res. **247**(1), 3–36 (2016). https://doi.org/10.1007/s10479-015-2038-7
16. Kuki, A., Sztrik, J., Tóth, Á., Bérczes, T.: A contribution to modeling two-way communication with retrial queueing systems. In: Dudin, A., Nazarov, A., Moiseev, A. (eds.) ITMM/WRQ -2018. CCIS, vol. 912, pp. 236–247. Springer, Cham (2018). https://doi.org/10.1007/978-3-319-97595-5_19
17. Nazarov, A., Phung-Duc, T., Paul, S.: Heavy outgoing call asymptotics for MMPP/M/1/1 retrial queue with two-way communication. In: Dudin, A., Nazarov, A., Kirpichnikov, A. (eds.) ITMM 2017. CCIS, vol. 800, pp. 28–41. Springer, Cham (2017). https://doi.org/10.1007/978-3-319-68069-9_3
18. Nazarov, A.A., Paul, S., Gudkova, I., et al.: Asymptotic analysis of Markovian retrial queue with two-way communication under low rate of retrials condition. In: Proceedings 31st European Conference on Modelling and Simulation (2017)
19. Phung-Duc, T., Rogiest, W.: Two way communication retrial queues with balanced call blending. In: Al-Begain, K., Fiems, D., Vincent, J.-M. (eds.) ASMTA 2012. LNCS, vol. 7314, pp. 16–31. Springer, Heidelberg (2012). https://doi.org/10.1007/978-3-642-30782-9_2
20. Pustova, S.: Investigation of call centers as retrial queuing systems. Cybern. Syst. Anal. **46**(3), 494–499 (2010)
21. Sakurai, H., Phung-Duc, T.: Two-way communication retrial queues with multiple types of outgoing calls. Top **23**(2), 466–492 (2015). https://doi.org/10.1007/s11750-014-0349-5
22. Wolf, T.: System and method for improving call center communications, uS Patent App. 15/604,068. Accessed 30 Nov 2017

On the Geo/G/1 System

László Lakatos[✉]

Eötvös Loránd University, Budapest, Hungary
lakatos@inf.elte.hu

Abstract. The Pollaczek-Khinchin transform equation is usually derived by means of the embedded Markov chain technique considering the number of customers in an M/G/1 queueing system at moments just after having served customers. The probabilities contained in this generating function, in [6] we determined them by using results from the theory of regenerative processes. The approach was based on the determination of mean values of times spent in different states in a busy period. Using the same general idea, we get an analogous result in the case of discrete time, for the Geo/G/1 system and derive the discrete-time version of the Pollaczek-Khinchin transform equation.

Keywords: Pollaczek-Khinchin transform equation · Discrete queueing system · Geo/G/1

1 Introduction

The classical formula, the Pollaczek-Khinchin transform equation (e.g. [1])

$$P(z) = \sum_{i=0}^{\infty} p_i z^i = \frac{(1-\rho)(1-z)b(\lambda(1-z))}{b(\lambda(1-z)) - z}$$

(where λ is the arrival rate, τ the mean value of service time of a customer, $\rho = \lambda\tau$, $b(s)$ the Laplace-Stieltjes transform of distribution function of a customer's service time) is usually derived by means of the embedded Markov chain technique considering the number of customers in the M/G/1 queueing system at moments just after having served customers. The probabilities contained in this generating function, in [6] we have determined them by using results from the theory of regenerative processes. The approach was based on the determination of mean value of the regenerative cycle and the mean values of times spent in different states in a busy period, it is described in [5]. This method may be used even for more complicated service disciplines, e.g. the case of bulk arrivals appears in [2], the case of vacation in [3] and [4]. Using this general idea, we get an analogous result for the corresponding discrete time (Geo/G/1) system and derive the discrete-time version of the Pollaczek-Khinchin transform equation.

The standard approach for the discrete-time system based on the use of embedded Markov chain is presented in [9] or [8].

© Springer Nature Switzerland AG 2019
A. Dudin et al. (Eds.): ITMM 2019, CCIS 1109, pp. 285–295, 2019.
https://doi.org/10.1007/978-3-030-33388-1_23

We will consider a queueing system with geometrically distributed interarrival and generally distributed service time, one server and FCFS service discipline, i.e. the discrete analogue of the $M/G/1$ system. Introduce the notations:

r - the probability of arrival of a customer in a slot (time unit);

b_k - the probability the service time is equal to k slots;

$B(z)$ - the generating function of service time;

τ - the mean value of service time;

$\rho = r\tau$;

$A(z)$ - the generating function of number of customers arriving for the service time of a customer;

$\zeta = \dfrac{\tau}{1-\rho}$ - the mean value of length of busy period;

ζ_i - the mean value of time spent above the i-th level for a busy period;

ξ_i - the mean value of time spent on the i-th level in a busy period;

$P(z) = \sum\limits_{i=0}^{\infty} p_i z^i$ - the generating function of ergodic distribution in the Geo/G/1 system.

As in the case of $M/G/1$ system, we will consider the number of customers at moments after having served a customer, each service will be identified with the number of remaining in the system customers. This number is called the state of system, sometimes we will tell the system is at a certain level.

The use of embedded Markov chain means that the system is considered only at special moments, namely at moments $t_n + 0$ (when the n-th customer already left the system). So we have a modified model, the arriving customers are taken into account not at the moments of their real arrival, but at the moment when the service of actual customer is completed. The quantity of present in the system at this moment customers is called *the state of the system*. For our purposes it will be more convenient to regard the system at the starting moments of services and accept that it does not change till the completion of service. This will be called *the number of present customers in the system*. The such defined notions of state and present customers must be distinguished, the difference will be clear from the following reasoning. For the busy period there change intervals of staying on the first and above the first level. The services of customers when at the beginning there is one customer in the system corresponds to the state 1 excluding two cases. If the interval on the first level ends with a jump above the first level, then the service of this last customer will correspond to the new level. But coming down from the second level this service time will be returned to the first level. The second case is the service of last customer in the busy period. From the viewpoint of states it corresponds to the zero state (after service no customer remains in the system), from the viewpoint of number of present customers it is 1 (at the beginning of service there is one customer), so it is necessary to exclude from the number of customers served on the first level. In the case of other levels the general number of customers served on these levels does not change.

2 Ergodic Probabilities via Mean Values

We are going to consider a queueing system in which the service time has general distribution with generating function

$$B(z) = \sum_{k=1}^{\infty} b_k z^k,$$

i.e. the service time is equal to k slots (time units) with probability b_k; for a slot a new customer enters independently of other slots with probability r, i.e. the interarrival time has geometrical distribution

$$P\{\eta = k\} = (1-r)^{k-1}r \qquad (k = 1, 2, \ldots).$$

The generating function of arrival of a customer for a slot is $1 - r + rz$.
    We have

Theorem 1. *Let us consider a Geo/G/1 queueing system where in a time unit (slot) a new customer arrives with probability r (there is no entry with probability $1 - r$), the service time has general distribution with generating function $B(z)$. If the service time of a customer has a finite mean τ, $r\tau < 1$, then there exists an equilibrium distribution in the system. These equilibrium probabilities are determined by the fractions $p_i = \xi_i/\zeta$ $(i = 0, 1, 2, \ldots)$, where ζ is the mean value of the busy period and ξ_i is the mean value of time spent on the i-th level during the busy period.*

Proof. The proof of the theorem is a direct consequence of Theorem 4.40 [7] (or see [10] Theorems 1.3.2 and 1.3.3). The mean values appearing in the theorem are given by the lemma below.

    First we find the generating function of number of customers served in a busy period, let us denote it by $G(z)$. The generating function of arriving customers for a service time is

$$A(z) = \sum_{k=1}^{\infty} b_k(1 - r + rz)^k = B(1 - r + rz) = \sum_{i=0}^{\infty} a_i z^i.$$

Consider the structure of the busy period. After free state there appears a customer, during its service arrive k new ones. Each of them with customers arriving during their services generate periods with the same structure as the whole busy period, i.e. $G(z)$ satisfies the functional equation

$$G(z) = z \sum_{i=0}^{\infty} a_i G^i(z) = zA(G(z)) = zB(1 - r + rG(z)).$$

By means of derivation we obtain

$$G'(z) = B(1 - r + rG(z)) + zB'(1 - r + rG(z))rG'(z)$$

and

$$G'(1) = 1 + B'(1)rG'(1),$$

from which

$$G'(1) = \frac{1}{1 - rB'(1)} = \frac{1}{1 - r\tau}.$$

This result can also be found in [9].

Lemma 1. *In the Geo/G/1 system*

$$\xi_0 = \tau, \qquad \xi_1 = \frac{1 - a_0}{a_0}\tau, \qquad \xi_2 = \frac{1 - a_0 - a_1}{a_0}(\xi_0 + \xi_1), \qquad (1)$$

and ξ_k $(k \geq 3)$ satisfy the recurrence relation

$$\xi_k = \sum_{i=1}^{k-2} \frac{1 - a_0 - a_1 - \ldots - a_i}{a_0}\xi_{k-i} + \frac{1 - a_0 - a_1 - \ldots - a_{k-1}}{a_0}(\xi_0 + \xi_1). \quad (2)$$

Proof. Let j customers be present in the system. Then with probability a_1 we remain at this level, and with probability $1 - a_1$ we leave it, namely with probability $\dfrac{a_0}{1 - a_1}$ come to the $j - 1$-st level, and with probability $\dfrac{1 - a_0 - a_1}{1 - a_1}$ to a level above j.

The mean value of number of customers served for a period when only one customer is present in the system is

$$\sum_{k=1}^{\infty} ka_1^{k-1}(1 - a_1) = \frac{1}{1 - a_1}.$$

The mean value of number of customers served for a period above the first level is

$$\sum_{k=2}^{\infty} \frac{a_k}{1 - a_0 - a_1}(k - 1)\frac{1}{1 - \rho} = \frac{1}{(1 - \rho)(1 - a_0 - a_1)}[\rho - a_1 - (1 - a_0 - a_1)]$$

$$= \frac{\rho - 1 + a_0}{(1 - \rho)(1 - a_0 - a_1)},$$

where we used the equalities

$$\rho = \sum_{k=1}^{\infty} ka_k \qquad \text{and} \qquad \sum_{k=0}^{\infty} a_k = 1.$$

We mention that the mean values can be measured either in time or in the number of customers, the difference appears in the factor τ (the mean value of a customer's service time).

For a busy period we have some periods during which there is only one customer in the system, it can be finished in two different ways: either there is no entry (it means the end of the busy period), or more than one customer enter. We have $1, 2, \ldots, k, \ldots$ such periods with probabilities

$$\frac{a_0}{1 - a_1}, \quad \frac{1 - a_0 - a_1}{1 - a_1} \frac{a_0}{1 - a_1}, \ldots, \frac{(1 - a_0 - a_1)^{k-1}}{(1 - a_1)^{k-1}} \frac{a_0}{1 - a_1}, \ldots$$

The mean value of such periods is

$$\sum_{k=1}^{\infty} k \frac{(1 - a_0 - a_1)^{k-1}}{(1 - a_1)^{k-1}} \frac{a_0}{1 - a_1} \frac{\tau}{1 - a_1} = \frac{\tau}{a_0};$$

the mean value of periods above the first level will be

$$\sum_{k=1}^{\infty} k \frac{(1 - a_0 - a_1)^{k}}{(1 - a_1)^{k}} \frac{a_0}{1 - a_1} \frac{\rho - 1 + a_0}{(1 - \rho)(1 - a_0 - a_1)} \tau = \frac{\rho - 1 + a_0}{a_0(1 - \rho)} \tau.$$

The sum of two values is

$$\frac{\tau}{a_0} + \frac{\rho - 1 + a_0}{a_0(1 - \rho)} \tau = \frac{\tau}{1 - \rho},$$

it gives the mean value of the busy period.

Our above computations were based on the number of present customers, so the mean value of time on the first level means the sum of times in states 0 and 1, i.e. $\xi_0 + \xi_1$. To the zero state in a busy period belongs only the last customer, consequently $\xi_0 = \tau$ and

$$\xi_1 = \frac{\tau}{a_0} - \tau = \frac{1 - a_0}{a_0} \tau.$$

So, we were able to divide the busy period into two parts, to find the mean values of times spent on the zero plus first levels and above the first level. In the following we continue to divide the periods above the first level determining the mean values of times above the k-th ($k = 2, 3, \ldots$) level and finding on their base the mean values of times spent on the concrete levels. First we consider the second level. We have two possibilities:

1. from the first level we come to the second one;
2. from the first level we come at least to the third level.

Coming from the first level to the second one we are in the same situation as in case of the first level: serving a certain number of customers on the second level we come to the first one or above the second one. In the first case the sojourns on and above the second level change, and spending on average ζ_1 time above the second level we come to the first one. In the second case from the first level we jump above the second one, the return time to the second level is

$$\sum_{k=3}^{\infty} \frac{a_k}{1 - a_0 - a_1 - a_2} (k - 2) \frac{\tau}{1 - \rho} = \frac{\rho - 2 + 2a_0 + a_1}{(1 - \rho)(1 - a_0 - a_1 - a_2)} \tau = \varepsilon_2.$$

Now we are in the previous situation and spend ζ_1 time above the second level. The probabilities of two possibilities are

$$\frac{a_2}{1 - a_0 - a_1} \quad \text{and} \quad \frac{1 - a_0 - a_1 - a_2}{1 - a_0 - a_1},$$

so the average sojourn time above the second level for a period beginning and ending on the first level is

$$\frac{a_2}{1 - a_0 - a_1}\zeta_1 + \frac{1 - a_0 - a_1 - a_2}{1 - a_0 - a_1}(\zeta_1 + \varepsilon_2) = \zeta_1 + \varepsilon_2',$$

where

$$\varepsilon_2' = \frac{\rho - 2 + 2a_0 + a_1}{(1 - \rho)(1 - a_0 - a_1)}\tau.$$

During a busy period we will have i such intervals with probability

$$\frac{(1 - a_0 - a_1)^i}{(1 - a_1)^i}\frac{a_0}{1 - a_1},$$

so by using

$$\sum_{i=1}^{\infty} i\frac{(1 - a_0 - a_1)^{i-1}}{(1 - a_1)^{i-1}} = \frac{1}{\left(1 - \frac{1-a_0-a_1}{1-a_1}\right)^2} = \frac{(1 - a_1)^2}{a_0^2}$$

we get

$$\begin{aligned}
\zeta_2 &= \sum_{i=1}^{\infty} i\frac{(1 - a_0 - a_1)^i}{(1 - a_1)^i}\frac{a_0}{1 - a_1}(\zeta_1 + \varepsilon_2') \\
&= \frac{1 - a_0 - a_1}{1 - a_1}\frac{a_0}{1 - a_1}\sum_{i=1}^{\infty} i\frac{(1 - a_0 - a_1)^{i-1}}{(1 - a_1)^{i-1}}(\zeta_1 + \varepsilon_2') \\
&= \frac{1 - a_0 - a_1}{1 - a_1}\frac{a_0}{1 - a_1}\frac{(1 - a_1)^2}{a_0^2}\zeta_1 \\
&\quad + \frac{1 - a_0 - a_1}{1 - a_1}\frac{a_0}{1 - a_1}\frac{(1 - a_1)^2}{a_0^2}\frac{\rho - 2 + 2a_0 + a_1}{(1 - \rho)(1 - a_0 - a_1)}\tau \\
&= \frac{1 - a_0 - a_1}{a_0}\zeta_1 + \frac{1 - a_0 - a_1 - a_2}{a_0}\frac{\rho - 2 + 2a_0 + a_1}{(1 - \rho)(1 - a_0 - a_1 - a_2)}\tau \\
&= \frac{1 - a_0 - a_1}{a_0}\zeta_1 + \frac{1 - a_0 - a_1 - a_2}{a_0}\varepsilon_2.
\end{aligned}$$

The mean value of time spent on the second level is obtained as the difference of mean values of times spent above the first and second levels. Since

$$\frac{1 - a_0 - a_1 - a_2}{a_0}\varepsilon_2 = \frac{\rho - 2 + 2a_0 + a_1}{a_0(1 - \rho)}\tau = \frac{\rho - 1 + a_0}{a_0(1 - \rho)}\tau - \frac{1 - a_0 - a_1}{a_0(1 - \rho)}\tau$$

$$= \zeta_1 - \frac{1 - a_0 - a_1}{a_0}\frac{\tau}{1 - \rho},$$

consequently

$$\xi_2 = \zeta_1 - \zeta_2 = \zeta_1 - \frac{1 - a_0 - a_1}{a_0}\zeta_1 - \frac{1 - a_0 - a_1 - a_2}{a_0}\varepsilon_2$$

$$= \zeta_1 - \frac{1 - a_0 - a_1}{a_0}\zeta_1 - \zeta_1 + \frac{1 - a_0 - a_1}{a_0}\frac{\tau}{1 - \rho}$$

$$= \frac{1 - a_0 - a_1}{a_0}(\zeta - \zeta_1) = \frac{1 - a_0 - a_1}{a_0}(\xi_0 + \xi_1).$$

The next step is to compute the mean value of time spent on the third level. From the first level one can come to the second, the third and above the third level. In the three cases the mean values of times to stay above the third level are

$$
\begin{array}{ll}
- \zeta_2 & \text{(level 2),} \\
- \zeta_1 + \zeta_2 & \text{(level 3),} \\
- \varepsilon_3 + \zeta_1 + \zeta_2 & \text{(above 3).}
\end{array}
$$

In the first case from the first level we come to the second one, the mean value of time spent above the third level coincides with the case of second level from the viewpoint of first level, so the desired mean value is ζ_2. In the second case the service begins on the third level, spending ζ_1 time above the third level we come to the second level and are in the previous situation. The corresponding mean value is $\zeta_1 + \zeta_2$. In the third case the service starts above the third level, let it be k, having served $k - 3$ customers and the generated by them ones we reach the third level and the second case takes place. The mean value of time to return to the third level is

$$\varepsilon_3 = \sum_{k=4}^{\infty} \frac{a_k}{1 - a_0 - a_1 - a_2 - a_3}(k - 3)\frac{\tau}{1 - \rho} = \frac{\rho - 3 + 3a_0 + 2a_1 + a_2}{1 - a_0 - a_1 - a_2 - a_3}\frac{\tau}{1 - \rho}.$$

The probabilities of three cases are

$$\frac{a_2}{1 - a_0 - a_1}, \qquad \frac{a_3}{1 - a_0 - a_1} \quad \text{and} \quad \frac{1 - a_0 - a_1 - a_2 - a_3}{1 - a_0 - a_1},$$

so the mean value of time spent above the third level for an interval beginning and ending on the first level is

$$\frac{a_2}{1 - a_0 - a_1}\zeta_2 + \frac{a_3}{1 - a_0 - a_1}(\zeta_1 + \zeta_2) + \frac{1 - a_0 - a_1 - a_2 - a_3}{1 - a_0 - a_1}(\varepsilon_3 + \zeta_1 + \zeta_2)$$

$$= \zeta_2 + \frac{1 - a_0 - a_1 - a_2}{1 - a_0 - a_1}\zeta_1 + \frac{1 - a_0 - a_1 - a_2 - a_3}{1 - a_0 - a_1}\varepsilon_3.$$

For a busy period we have i intervals above the first level with probability $\frac{(1 - a_0 - a_1)^i}{(1 - a_1)^i}\frac{a_0}{1 - a_1}$, the mean value of time above the third level is

$$\zeta_3 = \sum_{i=1}^{\infty} i\frac{(1 - a_0 - a_1)^i}{(1 - a_1)^i}\frac{a_0}{1 - a_1}\left[\zeta_2 + \frac{1 - a_0 - a_1 - a_2}{1 - a_0 - a_1}\zeta_1 + \frac{1 - a_0 - a_1 - a_2 - a_3}{1 - a_0 - a_1}\varepsilon_3\right]$$

$$= \frac{1 - a_0 - a_1}{a_0}\zeta_2 + \frac{1 - a_0 - a_1 - a_2}{a_0}\zeta_1 + \frac{\rho - 3 + 3a_0 + 2a_1 + a_2}{a_0}\frac{\tau}{1 - \rho}.$$

We find the mean value of time on the third level as the difference of mean values above the second and third levels, it is

$$\xi_3 = \zeta_2 - \zeta_3,$$

where

$$\zeta_2 = \frac{1 - a_0 - a_1}{a_0}\zeta_1 + \frac{\rho - 2 + 2a_0 + a_1}{a_0}\frac{\tau}{1 - \rho},$$

$$\zeta_3 = \frac{1 - a_0 - a_1}{a_0}\zeta_2 + \frac{1 - a_0 - a_1 - a_2}{a_0}\zeta_1 + \frac{\rho - 3 + 3a_0 + 2a_1 + a_2}{a_0}\frac{\tau}{1 - \rho}.$$

Since

$$\frac{\rho - 2 + 2a_0 + a_1}{a_0}\frac{\tau}{1 - \rho} - \frac{\rho - 3 + 3a_0 + 2a_1 + a_2}{a_0}\frac{\tau}{1 - \rho}$$
$$= \frac{1 - a_0 - a_1 - a_2}{a_0}\frac{\tau}{1 - \rho} = \frac{1 - a_0 - a_1 - a_2}{a_0}\zeta,$$

so

$$\xi_3 = \frac{1 - a_0 - a_1}{a_0}(\zeta_1 - \zeta_2) - \frac{1 - a_0 - a_1 - a_2}{a_0}\zeta_1 + \frac{1 - a_0 - a_1 - a_2}{a_0}\zeta$$
$$= \frac{1 - a_0 - a_1}{a_0}\xi_2 + \frac{1 - a_0 - a_1 - a_2}{a_0}(\xi_0 + \xi_1),$$

i.e. in the case $k = 3$ the statement of the lemma is true.

We determine ζ_k (the mean value of time spent above the k-th level for a busy period). ζ_k has the structure

ζ_k :	ζ_{k-1}	(level 2)
	$\zeta_{k-2} + \zeta_{k-1}$	(level 3)
	....	
	$\zeta_{k-i+1} + \zeta_{k-i+2} + \ldots + \zeta_{k-1}$	(level i)
	....	
	$\zeta_1 + \zeta_2 + \ldots + \zeta_{k-1}$	(level k)
	$\zeta_1 + \zeta_2 + \ldots + \zeta_{k-1} + \varepsilon_k$	(above k)

From the first level we can come to the second, ..., $k-1$-st, k-th levels or to a level above the k-th one. In the case of second level we are in the same situation as in the case of $k - 1$-st level from the viewpoint of first one, so the mean value is ζ_{k-1}.

In case of third level we have a period starting with the presence of three customers and ending with the presence of two ones. This corresponds to the situation: starting from the first level we consider the time spent above the $k-2$-nd level, so the mean value is ζ_{k-2}. We are in the previous situation (there are two customers), so the mean value of remaining part is ζ_{k-1}. So, the desired mean value in this case $\zeta_{k-2} + \zeta_{k-1}$.

We consider the last possibility. Then we go to a level above the k-th one. The mean value of time spent above the k-th level at the beginning is

$$\sum_{i=k+1}^{\infty} \frac{a_i}{1 - a_0 - a_1 - \ldots - a_k}(i - k)\frac{\tau}{1 - \rho}$$
$$= \frac{\rho - k + ka_0 + (k - 1)a_1 + \ldots + 2a_{k-2} + a_{k-1}}{(1 - \rho)(1 - a_0 - a_1 - \ldots - a_k)}\tau = \varepsilon_k.$$

Now we are at the k-th level, spending ζ_1 above the k-th one we come to $k - 1$, spending ζ_2 above k we come to $k - 2$, ..., and, finally, starting from the second level spending ζ_{k-1} above the k-th one we reach the first level. The desired mean value is $\zeta_1 + \zeta_2 + \ldots + \zeta_{k-1} + \varepsilon_k$. The probabilities of these possibilities are $\dfrac{a_2}{1 - a_0 - a_1}, \dfrac{a_3}{1 - a_0 - a_1}, \ldots, \dfrac{1 - a_0 - a_1 - \ldots - a_k}{1 - a_0 - a_1}.$

Multiplying the conditional mean values with the corresponding probabilities we get

$$\zeta_{k-1} + \frac{1 - a_0 - a_1 - a_2}{1 - a_0 - a_1}\zeta_{k-2} + \ldots$$
$$+ \frac{1 - a_0 - a_1 - \ldots - a_{k-1}}{1 - a_0 - a_1}\zeta_1 + \frac{1 - a_0 - a_1 - \ldots - a_k}{1 - a_0 - a_1}\varepsilon_k$$

In a busy period we will stay i times above the first level with probability $\dfrac{(1 - a_0 - a_1)^i}{(1 - a_1)^i}\dfrac{a_0}{1 - a_1}$, so

$$\zeta_k = \sum_{i=1}^{\infty} i\frac{(1 - a_0 - a_1)^i}{(1 - a_1)^i}\frac{a_0}{1 - a_1}\left\{\zeta_{k-1} + \frac{1 - a_0 - a_1 - a_2}{1 - a_0 - a_1}\zeta_{k-2}\right.$$
$$+ \ldots + \frac{1 - a_0 - a_1 - \ldots - a_{k-1}}{1 - a_0 - a_1}\zeta_1$$
$$\left. + \ldots + \frac{1 - a_0 - a_1 - \ldots - a_k}{1 - a_0 - a_1}\varepsilon_k\right\}$$
$$= \sum_{i=1}^{k-1}\frac{1 - a_0 - a_1 - \ldots - a_i}{a_0}\zeta_{k-i} + \frac{1 - a_0 - a_1 - \ldots - a_k}{a_0}\varepsilon_k$$

Substituting here the similar value for ζ_{k-1} we obtain

$$\xi_k = \sum_{i=1}^{k-2}\frac{1 - a_0 - a_1 - \ldots - a_i}{a_0}\xi_{k-i} + \frac{1 - a_0 - a_1 - \ldots - a_{k-1}}{a_0}(\xi_0 + \xi_1).$$

The lemma is proved.

3   The Derivation of the Pollaczek-Khinchin Transform Equation

We give another method to derive the generating function.

Theorem 2. *The generating function of present customers in the Geo/G/1 system*

$$P(z) = \frac{(1-\rho)(1-z)B(1-r+rz)}{B(1-r+rz)-z},\tag{3}$$

may be obtained from (1), (2) by using the theory of regenerative processes.

Remark 1. We underline that in the Geo/G/1 system under the regenerative cycle one has to understand the busy period, according to the definition of embedded chain the service time of last customer corresponds to the free state.

Proof. Let us write the formulae of ξ_i for the first values. We have

$$\xi_0 = \xi_0,$$
$$\xi_1 = \frac{1-a_0}{a_0}\xi_0,$$
$$\xi_2 = \frac{1-a_0-a_1}{a_0}\xi_1 + \frac{1-a_0-a_1}{a_0}\xi_0,$$
$$\xi_3 = \frac{1-a_0-a_1}{a_0}\xi_2$$
$$\quad + \frac{1-a_0-a_1-a_2}{a_0}\xi_1 + \frac{1-a_0-a_1-a_2}{a_0}\xi_0,$$
$$\xi_4 = \frac{1-a_0-a_1}{a_0}\xi_3 + \frac{1-a_0-a_1-a_2}{a_0}\xi_2$$
$$\quad + \frac{1-a_0-a_1-a_2-a_3}{a_0}\xi_1 + \frac{1-a_0-a_1-a_2-a_3}{a_0}\xi_0,$$
$$\xi_5 = \frac{1-a_0-a_1}{a_0}\xi_4 + \frac{1-a_0-a_1}{a_0}\xi_4$$
$$\quad + \frac{1-a_0-a_1-a_2}{a_0}\xi_3 + \frac{1-a_0-a_1-a_2-a_3}{a_0}\xi_2$$
$$\quad + \frac{1-a_0-a_1-a_2-a_3-a_4}{a_0}\xi_1 + \frac{1-a_0-a_1-a_2-a_3-a_4}{a_0}\xi_0,$$

....

Let us multiply the expression for ξ_i by z^i and sum up them from the third row excluding the last term (containing ξ_0). We have

$$\frac{1-a_0-a_1}{a_0}z(\xi_1 z + \xi_2 z^2 + \ldots) + \frac{1-a_0-a_1-a_2}{a_0}z^2(\xi_1 z + \xi_2 z^2 + \ldots)$$
$$\quad + \frac{1-a_0-a_1-a_2-a_3}{a_0}z^3(\xi_1 z + \xi_2 z^2 + \ldots) + \ldots$$
$$= \left(\sum_{i=1}^{\infty}\xi_i z^i\right)\left\{\frac{1-a_0-a_1}{a_0}z + \frac{1-a_0-a_1-a_2}{a_0}z^2\right.$$
$$\left. + \frac{1-a_0-a_1-a_2-a_3}{a_0}z^3 + \ldots\right\}\tag{4}$$
$$= \left(\sum_{i=1}^{\infty}\xi_i z^i\right)\frac{1}{a_0}\left\{\frac{z}{1-z} - \frac{a_0 z}{1-z} - \frac{a_1 z}{1-z} - \frac{a_2 z^2}{1-z} - \frac{a_3 z^3}{1-z} - \ldots\right\}$$
$$= \left(\sum_{i=1}^{\infty}\xi_i z^i\right)\frac{1}{a_0(1-z)}\left\{z(1-a_0) - [A(z) - a_0]\right\}$$
$$= [P(z) - \xi_0]\tfrac{1}{a_0(1-z)}\left\{z(1-a_0) - [A(z) - a_0]\right\},$$

where $\overline{P}(z) = \sum\limits_{i=0}^{\infty} \xi_i z^i$. For the term containing ξ_0, we have

$$\xi_0 z \sum_{i=1}^{\infty} \frac{1 - a_0 - \ldots - a_i}{a_0} z^i = \xi_0 z \frac{1}{a_0(1 - z)} \{z(1 - a_0) - [A(z) - a_0]\}. \qquad (5)$$

Adding (4), (5), the first row and the second one multiplied by z, we obtain

$$\overline{P}(z) = [\overline{P}(z) - \xi_0] \frac{1}{a_0(1 - z)} \{z(1 - a_0) - [A(z) - a_0]\}$$

$$+ \xi_0 z \frac{1}{a_0(1 - z)} \{z(1 - a_0) - [A(z) - a_0]\} + \xi_0 + \frac{1 - a_0}{a_0} \xi_0 z.$$

From it

$$\overline{P}(z) = \frac{(1 - z)A(z)}{A(z) - z} \xi_0.$$

Dividing it by the mean value of busy period $\dfrac{\tau}{1 - \rho}$ and taking into account that $\xi_0 = \tau$, finally we obtain (3)

$$P(z) = \frac{(1 - \rho)(1 - z)A(z)}{A(z) - z} = \frac{(1 - \rho)(1 - z)B(1 - r + rz)}{B(1 - r + rz) - z}.$$

References

1. Gnedenko, B.V., Kovalenko, I.N.: Introduction to Queueing Theory, 2nd edn. Birkhäuser, Boston (1989)
2. Lakatos, L.: On the $M^X/G/1$ system. Ann. Univ. Sci. Bp. Sect. Comput. **18**, 137–150 (1999)
3. Lakatos, L.: Equilibrium distributions for the M/G/1 and related systems. Publ. Math. Debrecen **55**, 123–140 (1999)
4. Lakatos, L.: On the M/G/1 system with bulk arrivals and vacation. Theory Stoch. Process. 5(21), 127–136 (1999)
5. Lakatos, L.: The Pollaczek-Khinchin transform equation via regenerative processes. In: Udayabaskaran, S. (ed.) 2nd International Conference on Stochastic Modelling and Simulation, pp. 71–76. Vel Tech Technical University, Chennai (2012)
6. Lakatos, L., Čerić, V.: Another approach to the ergodic distribution in the M/G/1 system. In: Galambos, J., Kátai, I. (eds.) Probability Theory and Applications. Essays to the Memory of József Mogyoródi, pp. 289–298. Kluwer (1992)
7. Lakatos, L., Szeidl, L., Telek, M.: Introduction to Queueing Systems with Telecommunication Applications. Springer, Boston (2013). https://doi.org/10.1007/978-1-4614-5317-8
8. Lakatos, L., Szeidl, L., Telek, M.: Introduction to Queueing Systems with Telecommunication Applications, 2nd edn. Springer, Cham (2019). https://doi.org/10.1007/978-3-030-15142-3
9. Pechinkin, A.V., Razumchik, R.V.: Queueing Systems in Discrete Time. Fizmatlit, Moscow (2018). (in Russian)
10. Tijms, H.: Stochastic Models. An Algorithmic Approach. Wiley, New York (1994)

Methods to Reduce Loss Probability in Systems with Infinite Service Time Dispersion

Vladimir N. Zadorozhnyi and Tatiana R. Zakharenkova$^{(\boxtimes)}$

Omsk State Technical University, Omsk, Russia
zwn2015@yandex.ru, ZakharenkovaTatiana@gmail.com

Abstract. The problem of reducing loss probability in systems with finite buffer and infinite service time dispersion. Efficient method of loss reduction is developed for system $M/Pa/1/m$, in which Pareto distribution has infinite dispersion. The method is based on absolute priority discipline with afterservice. At infinite buffer $m = \infty$ the introduction of absolute priorities into the systems is proven to allow the transformation of waiting time distribution in a way that its average values becomes finite. This underlies the development of the numerical method for optimal requests separation into priority classes with respect to minimal average waiting time in a system with infinite buffer It is demonstrated that the resulting priority assignment decreases the loss probability significantly if the infinite buffer in a system is replaced by a finite one. The method remains highly efficient even when extended to other systems $GI/GI/1/m$ with infinite service time dispersion. The developed priority assignment method has a number of important advantages over other methods for dealing with requests losses. As a result, it can be efficiently used in practice, namely, to lower the messages loss probability in data networks with fractal traffic.

Keywords: Queueing systems · Fractal traffic · Loss probability in a finite buffer system · Service discipline · Optimal priority assignment

1 Introduction

At the beginning of the 1990 s, random processes describing data network traffic were found to have a fractal structure [1], which significantly affects the queueing properties in the network devises. The main problem in networks design was that of packet loss since fractal traffic is characterized by a large, in most cases infinite, dispersion. This results in unpredictable sizes of the formed queues. Consequently, "classical" models of the queues being inadequate for fractal traffic were replaced by queueing systems with power-law distribution tails [2], namely, queueing system with Pareto distribution, whose shape parameter lies in the region accounting for infinite dispersion [3]. These models adequately represent

© Springer Nature Switzerland AG 2019
A. Dudin et al. (Eds.): ITMM 2019, CCIS 1109, pp. 296–311, 2019.
https://doi.org/10.1007/978-3-030-33388-1_24

the properties of the packet queue and allow developing the methods to lower the packet loss probability in networks.

The following methods stand out as the main ones to lower the loss probabilities: increasing the size of the buffer capacity, increasing the node operation speed, increasing the number of channels operating in parallel and introducing the priority service meachnism.

In fractal traffic, increasing the buffer size is an inefficient means to lower to loss probability [4]. This can be explained by the fact that at infinite dispersion $\mathrm{Var}(x)$ of service time x average queue length in the system with an infinite buffer is infinite. In fact, let us consider the system M/Pa/1 with Pareto distribution of service time, the system commonly used for modelling network devices. Pareto distribution of random variable x is set by the following probability distribution function:

$$F(t) = 1 - \left(\frac{K}{t}\right)^{\alpha}, \qquad K > 0, \qquad t \geq K, \tag{1}$$

where K is the least value of the random variable x (scale parameter), $\alpha > 0$ is shape parameter.

It is easy to verify that at $\alpha > 1$ mathematical expectation (m.e.) $\mathrm{E}(x) < \infty$, and at $\alpha \leq 2$ the dispersion $\mathrm{Var}(x) = \infty$. Therefore, the range $1 < \alpha \leq 2$ of parameter α values, frequently used at network devices modeling, determined finite m.e. b and infinite time dispersion x. Consequently, the second moment $b^{(2)}$ of time x is infinite as well.

Applying Pollaczek-Khinchine formula to the considered system with $b^{(2)} = \infty$, stationary average waiting time for the system is found to be

$$W = \frac{\lambda b^{(2)}}{2(1 - \rho)} = \infty$$

at any load coefficient $\rho < 1$ and any non-zero rate λ of arriving requests flow. Using Little's formula $L = \lambda W$, one determines that stationary average queue length L in such system is also infinite (result $L = \infty$ is obtained by means of moment generating function in the article [5]). In a physical sense, infinite average waiting time explains not only inefficient increase of the buffer size as a means to lower loss probability, but inefficient increase of channel processing speed as well, since $W = \infty$ at any load coefficient $\rho < 1$. In [5] was established that when the buffer m is finite the loss probability P decreases with the growth m at asymptotical power rate: $P \sim Cm^{(1-\alpha)}$, where C is a constant.

A very effective method to reduce the loss probability is to increase the number of channels in the system [6]. This method is associated with reasonable expenses for additional equipment.

This article is devoted to the development of another effective method of dealing with losses (no longer focused on additional hardware costs). The developed method consists in the introduction of absolute priorities for incoming requests due to their service time. Special feature of the developed method is the use of an infinite number of priority classes, radically changing such systems' properties.

2   Formation of Priority Classes

Let us split the range $K \leq t < \infty$ of possible service time x values, using a sequence of points (markup)

$$\{t_k\} = t_0, t_1, \ldots, t_k, \ldots \tag{2}$$

into intervals $[t_0, t_1), [t_1, t_2), \ldots, [t_{k-1}, t_k), \ldots$, where $t_0 = K$. If the request entering the system has a service time belonging to the kth interval, i.e. if $x \in [t_{k-1}, t_k)$, this request is associated with the kth priority class. Moreover, the higher the number k of the request priority class, the lower its priority (thus, negative priorities are allowed, since the number of priority classes is infinite in general). Such absolute priority assignment will be called an assignment caused by service time axis marking. If the number of intervals $[t_{k-1}, t_k)$ is finite, in the last one $[t_{N-1}, t_N)$ we get $t_N = \infty$. Introducing absolute priorities, it is assumed that the discipline of absolute priorities with afterservice is used.

Let us determine the average waiting time W in the system M/Pa/1 at such division of arrival flow into priority components. Due to [7], average time U_k of staying in the system for the request with the kth priority class can be expressed as follows:

$$U_k = \frac{b_k}{1 - \sigma_{k-1}} + \frac{\sum_{i=1}^{k} \lambda_i b_i^{(2)}}{2(1 - \sigma_k)(1 - \sigma_{k-1})}, \qquad k = 1, 2, \ldots, \tag{3}$$

where b_k is average service time for requests of the kth priority class, λ_i is the arrival rate of requests with the ith priority class, $b_i^{(2)}$ is the second service moment for the requests of the ith priority class, $\sigma_k = \sum_{i=1}^{k} \rho_i$ is the sum of system load coefficients due to the requests of priority classes from the 1st to the kth, $\rho_i = \lambda_i b_i$, $\sigma_{k-1} = \sigma_k - \rho_k$.

Then average time U of requests staying in the system can be determined as the sum

$$U = \sum_k p_k U_k, \tag{4}$$

(where $p_k = P(t_{k-1} \leq x < t_k)$ is the probability of arriving request classified as the kth priority class), and average waiting time is a difference

$$W = U - b, \tag{5}$$

where $b = E(x)$.

Indicators (4) and (5) of the considered system depend on the marking of $\{t_k\} = t_0, t_1, \ldots, t_k, \ldots$ service time axis on intervals, defining priority classes of requests. Let us find for the given marking $\{t_k\}$ the parameter values for the system M/Pa/1, included in the right side of the formula (3).

Because the probability $p_k = P(t_{k-1} \leq x < t_k)$ of interval $[t_{k-1}, t_k)$ at service time x distribution (1) equals

$$p_k = \left(\frac{K}{t_{k-1}}\right)^\alpha - \left(\frac{K}{t_k}\right)^\alpha, \tag{6}$$

conditional distribution function for service time x belonging to the kth interval has the form:

$$F_k(t) = F(t|t_{k-1} \leq x < t_k) =$$
$$= \frac{F(t) - F(t_{k-1})}{p_k} = \frac{t_{k-1}^{-\alpha} - t^{-\alpha}}{t_{k-1}^{-\alpha} - t_k^{-\alpha}}, \qquad t_{k-1} \leq x < t_k. \tag{7}$$

Therefore, for formula (3) we obtain:

$$b_k = \int_{t_{k-1}}^{t_k} t\, dF_k(t) = \int_{t_{k-1}}^{t_k} t\, \frac{\alpha t^{-\alpha-1}}{t_{k-1}^{-\alpha} - t_k^{-\alpha}}\, dt = \frac{\alpha}{\alpha-1} \cdot \frac{t_{k-1}^{1-\alpha} - t_k^{1-\alpha}}{t_{k-1}^{-\alpha} - t_k^{-\alpha}}, \tag{8}$$

$$\lambda_k = p_k \lambda, \tag{9}$$

(all priority components of the arrival flow are Poisson),

$$b_k^{(2)} = \int_{t_{k-1}}^{t_k} t^2\, dF_k(t) = \begin{cases} \dfrac{\alpha}{\alpha-2} \cdot \dfrac{t_{k-1}^{2-\alpha} - t_k^{2-\alpha}}{t_{k-1}^{-\alpha} - t_k^{-\alpha}}, & \alpha \neq 2, \\[3ex] \dfrac{2}{t_{k-1}^{-2} - t_k^{-2}} \ln\left(\dfrac{t_k}{t_{k-1}}\right), & \alpha = 2, \end{cases} \tag{10}$$

(here the uncertainty of $0/0$ type at $\alpha = 2$ is disclosed by L'Hopital rule),

$$\rho_k = \lambda_k b_k.$$

$$\sigma_k = \sum_{i=1}^{k} \rho_i = \sum_{i=1}^{k} \lambda_i b_i,$$

$$\sigma_{k-1} = \sigma_k - \rho_k. \tag{11}$$

It is easy to verify that the obtained parameters of the priority components into which the arrival flow is divided meet the verification conditions

$$\sum_k p_k = 1, \quad \sum_k p_k F_k(t) = F(t), \quad \sum_k p_k b_k = b,$$

$$\sum_k \lambda_k = \lambda, \quad \sum_k p_k b_k^{(2)} = b^{(2)}, \quad \sum_k \rho_k = \rho. \tag{12}$$

Now let us investigate average waiting time W (5) at various markings $\{t_k\}$. First, let us note that at $1 < \alpha \leq 2$ any finite markings $\{t_k\} = \{t_k\}_1^N$ leave average time infinite since in the last, N^{th} semi-infinite interval

$[t_{N-1}, t_N) = [t_{N-1}, \infty)$ the second service time moment $b_N^{(2)}$ represented by Eq. (10) is infinite. Infinite moment $b_N^{(2)}$ results into infinite average staying time U_N – see (3) – for requests of the lower N^{th} priority class, into infinite unconditional average staying time U – see (4), and into infinite average waiting time W – see (5). We, however, are interested in possibility to obtain finite average time W due to absolute priorities at $1 < \alpha \le 2$. Therefore, further consideration will be given only of the infinite markings $\{t_k\} = \{t_k\}_1^\infty$, leaving no right semi-infinite interval on the axis of service time distributed due to Pareto principle. Let us start with the study of a regular marking.

3   Regular Service Time Axis Marking

Let us demonstrate that at regular marking

$$t_k - t_{k-1} = \Delta = \text{const}, \tag{13}$$

i.e. in which the lengths of all intervals are equal, introducing the discipline of absolute priorities with afterservice described above makes the average waiting time finite.

T h e o r e m. Introducing discipline of absolute priorities with afterservice into the system M/Pa/1 with $\rho < 1$, $1 < \alpha \le 2$, when priorities are determined by infinite regular marking (13) with positive step $\Delta < \infty$, makes the average waiting time finite.

P r o o f. The average staying time in the system, due to the expressions (3) and (4), is determined as follows:

$$U = \sum_k p_k U_k = \sum_{k=1}^\infty \frac{p_k b_k}{1 - \sigma_{k-1}} + \sum_{k=1}^\infty \frac{p_k \sum_{i=1}^k \lambda_i b_i^{(2)}}{2(1 - \sigma_k)(1 - \sigma_{k-1})}. \tag{14}$$

Here the first sum on the right is finite due to the following relations:

$$\sum_{k=1}^\infty \frac{p_k b_k}{1 - \sigma_{k-1}} \le \sum_{k=1}^\infty \frac{p_k b_k}{1 - \rho} = \frac{1}{1 - \rho} \sum_{k=1}^\infty p_k b_k = \frac{b}{1 - \rho}$$

(see (12)), $\rho < 1$ is the system load coefficient.

The second sum is limited by the expression above

$$\sum_{k=1}^\infty \frac{p_k \sum_{i=1}^k \lambda_i b_i^{(2)}}{2(1 - \sigma_k)(1 - \sigma_{k-1})} \le \sum_{k=1}^\infty \frac{p_k \sum_{i=1}^k \lambda_i b_i^{(2)}}{2(1 - \rho)(1 - \rho)} =$$

$$\frac{1}{2(1 - \rho)^2} \sum_{k=1}^\infty p_k \sum_{i=1}^k p_i \lambda b_i^{(2)} = \frac{\lambda}{2(1 - \rho)^2} \sum_{k=1}^\infty p_k \sum_{i=1}^k p_i b_i^{(2)},$$

and, therefore, only the finiteness of the sum $S = \sum_{k=1}^{\infty} p_k \sum_{i=1}^{k} p_i b_i^{(2)}$ remains to be proven.

By inserting the corresponding expressions (6) and (10), we obtain:

$$S = \sum_{k=1}^{\infty} \left[\left(\frac{K}{t_{k-1}} \right)^{\alpha} - \left(\frac{K}{t_k} \right)^{\alpha} \right] \sum_{i=1}^{k} \left[\left(\frac{K}{t_{i-1}} \right)^{\alpha} - \left(\frac{K}{t_i} \right)^{\alpha} \right] \left[\frac{\alpha}{\alpha-2} \cdot \frac{t_{i-1}^{2-\alpha} - t_i^{2-\alpha}}{t_{i-1}^{-\alpha} - t_i^{-\alpha}} \right] =$$

$$= \frac{\alpha K^{2\alpha}}{\alpha-2} \sum_{k=1}^{\infty} \left[\left(\frac{1}{t_{k-1}} \right)^{\alpha} - \left(\frac{1}{t_k} \right)^{\alpha} \right] \sum_{i=1}^{k} \left[\left(\frac{1}{t_{i-1}} \right)^{\alpha} - \left(\frac{1}{t_i} \right)^{\alpha} \right] \left[\frac{t_{i-1}^{2-\alpha} - t_i^{2-\alpha}}{t_{i-1}^{-\alpha} - t_i^{-\alpha}} \right] =$$

$$= \frac{\alpha K^{2\alpha}}{\alpha-2} \sum_{k=1}^{\infty} \left(t_{k-1}^{-\alpha} - t_k^{-\alpha} \right) \sum_{i=1}^{k} \left(t_{i-1}^{-\alpha} - t_i^{-\alpha} \right) \left(\frac{t_{i-1}^{2-\alpha} - t_i^{2-\alpha}}{t_{i-1}^{-\alpha} - t_i^{-\alpha}} \right) =$$

$$= \frac{\alpha K^{2\alpha}}{\alpha-2} \sum_{k=1}^{\infty} \left(t_{k-1}^{-\alpha} - t_k^{-\alpha} \right) \sum_{i=1}^{k} \left(t_{i-1}^{2-\alpha} - t_i^{2-\alpha} \right) <$$

$$< \frac{\alpha K^{2\alpha}}{\alpha-2} \sum_{k=1}^{\infty} \left(t_{k-1}^{-\alpha} - t_k^{-\alpha} \right) \sum_{i=1}^{k} \left(t_{i-1} - t_i \right) =$$

$$= \frac{\alpha K^{2\alpha}}{\alpha-2} \sum_{k=1}^{\infty} \left(t_{k-1}^{-\alpha} - t_k^{-\alpha} \right) \sum_{i=1}^{k} \Delta = \frac{\alpha K^{2\alpha}}{\alpha-2} \sum_{k=1}^{\infty} \left(t_{k-1}^{-\alpha} - t_k^{-\alpha} \right) k\Delta =$$

$$= \frac{\alpha K^{2\alpha} \Delta}{\alpha-2} \sum_{k=1}^{\infty} \left(t_{k-1}^{-\alpha} - t_k^{-\alpha} \right) k. \tag{15}$$

The last of the obtained sums is calculated, considering that the coordinates of the regular marking $(t_0, t_1, \ldots, t_k, \ldots)$ have the form $(K, K + \Delta, K + 2\Delta, \ldots, K + k\Delta, \ldots)$:

$$S_1 = \sum_{k=1}^{\infty} \left(t_{k-1}^{-\alpha} - t_k^{-\alpha} \right) k = \sum_{k=1}^{\infty} \left(k t_{k-1}^{-\alpha} - k t_k^{-\alpha} \right) =$$

$$= \left(1 t_0^{-\alpha} - 1 t_1^{-\alpha} \right) + \left(2 t_1^{-\alpha} - 2 t_2^{-\alpha} \right) + \left(3 t_2^{-\alpha} - 3 t_3^{-\alpha} \right) + \cdots +$$

$$+ \left(k t_{k-1}^{-\alpha} - k t_k^{-\alpha} \right) + \cdots = \sum_{k=0}^{\infty} t_k^{-\alpha} =$$

$$= \sum_{k=0}^{\infty} (K + k\Delta)^{-\alpha} = \Delta^{-\alpha} \sum_{k=0}^{\infty} \left(k + \frac{K}{\Delta} \right)^{-\alpha} = \Delta^{-\alpha} \zeta \left(\alpha, K/\Delta \right), \tag{16}$$

where $\zeta(s, q)$ is Hurwitz's zeta function [8]:

$$\zeta(s, q) = \sum_{k=0}^{\infty} (k + q)^{-s}. \tag{17}$$

Using integral test for convergence (Cauchy test) for series (17), we see that series converges for real $s > 1$ and $q > 0$ (zeta function is determined for complex s, q if $\mathrm{Re}(s) > 1$, $\mathrm{Re}(q) > 0$). Thus, sum S_1 (16) is finite for $\alpha > 1$ and, consequently, sum S (15), average staying time (14), and average waiting time W (5) are finite.

The theorem is proved.

The Hurwitz's zeta function can be well calculated by existing mathematical programs. For instance, calculating sum S_1 (16) when $\alpha = 8/7$, $K = 0.0625$, $\Delta = 0.2$ by means of the wolframalpha.com site, we obtain $S_1 = 68.767911\ldots$, moreover, the number of significant digits of the result easily increases to several hundred.

4   Optimization of Regular Marking Tick

When absolute priorities caused by infinite regular marking are introduced into the system M/Pa/1/m, at any Δ, for example, at $\Delta = \mathrm{E}(x)$, the loss probability P decrease is achieved. Moreover, as demonstrated by the simulation experiments, at increased buffer size m, $P(m)$ lowers with exponential speed. These qualitative changes can be explained by the fact that in queueing systems with infinite buffer the introduction of absolute priorities caused by infinite marking results in lower average waiting time W, and consequently, in lower queue length L from the infinite value to the finite one.

At the same time the exact value of W depends on the marking parameter Δ, therefore, it is natural to choose only the optimal Δ. The value of Δ will be called optimal, if it minimizes the value of W. In other words, to determine optimal Δ the following problem should be solved

$$W(\Delta) = U - b = \sum_k p_k U_k - b =$$

$$= \sum_{k=1}^{\infty} \frac{p_k b_k}{1 - \sigma_{k-1}} + \sum_{k=1}^{\infty} \frac{p_k \sum_{i=1}^{k} \lambda_i b_i^{(2)}}{2(1 - \sigma_k)(1 - \sigma_{k-1})} - b \to \min_{\Delta}, \qquad (18)$$

where the indicators in a minimized expression depend on the used regular marking $(t_0, t_1, \ldots, t_k, \ldots) = (K, K+\Delta, K+2\Delta, \ldots, K+k\Delta, \ldots)$, determined by the parameter Δ. Parameters λ, K, α of the system M/Pa/1/∞, for which problem (18) is solved, are considered as set. Parameters in (18) are determined by the marking $(t_0, t_1, \ldots, t_k, \ldots)$ with the formulae (6)–(11). Due to the service time of problem (18), it is reasonable to solve it with numerical methods. An example of numerical optimization is shown in Fig. 1.

Line 1 on the left figure is calculated for parameters $\alpha = 1.9$, $K = 9/38$, line 2 is for $\alpha = 1.5$, $K = 1/6$, line 3 is for $\alpha = 1.25$, $K = 0.1$ and line 4 is for $\alpha = 8/7$, $K = 0.0625$. For all four cases, $\mathrm{E}(x) = 0.5$, $\lambda = 1$, $\rho = 0.5$. The purpose of the calculation is to determine the dependence of the optimum on the parameter α.

The large red markers on the lines indicate the zones in which the optimal values of Δ are located, delivering the minimum average time W. Significant difficulties of calculating the dependences shown in the pictures are related to the fact that at considered Δ it is necessary to calculate the sums in (18) for a large number of summands.

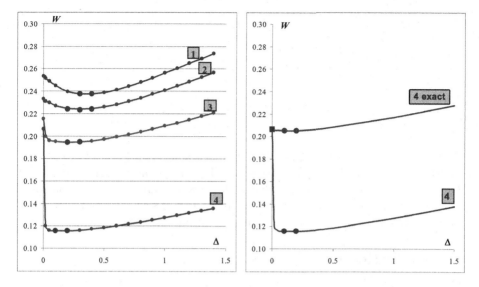

Fig. 1. Approximately calculated dependences of W on Δ at various α (on the left) and comparison of an approximate dependence with the exact one. The position of the found minimum zone does not change (Color figure online)

This is because the calculated sums converge very slowly. A special program had to be written to calculate 10 million of such summands, and the whole calculation of the figure took quite a long time. This, in particular, explains why the neighborhoods of the minima on the curves in this figure are defined with a rather large tolerance. Numerical calculation of W at $\Delta = 0$ was of course impossible. Therefore an analytical method had to be used to calculate $W(\Delta)$ at $\Delta = 0$.

The right side of Fig. 1 shows the most "difficult" dependence calculated approximately (at $\alpha = 8/7$, closest to one) which is compared to the exact one. The exact calculation of the dependence was performed by substituting parts of sums in (18), corresponding to values $k \in (10^6, \infty)$, with definite integrals. As expected, the values of W at redefined calculation have changed markedly, but the position the sought optimal Δ was preserved.

5   Calculation of $W(\Delta)$ at $\Delta \to 0$

At $\Delta \to 0$ we move from a discrete infinite marking to a continuous one, therefore, the numbering of the intervals becomes an inappropriate means. Let us turn to the representation of the intervals in the form of $[t, t + dt)$. The indicators needed to calculate W due to formula (18) transform into the indicators of the corresponding infinitesimal intervals, while the sums included in (18) transform into integrals, respectively. As a result of such transformations for sum (18), we obtain:

$$W(0) = \alpha K^\alpha \int\limits_K^\infty \frac{t^{-\alpha}}{\left[1 - \frac{\lambda \alpha K^\alpha}{\alpha - 1}\left(K^{1-\alpha} - t^{1-\alpha}\right)\right]}dt +$$

$$+\frac{\lambda \alpha^2 K^{2\alpha}}{2(2-\alpha)} \int\limits_K^\infty \frac{t^{-1-\alpha}\left(t^{2-\alpha} - K^{2-\alpha}\right)}{\left[1 - \frac{\lambda \alpha K^\alpha}{\alpha - 1}\left(K^{1-\alpha} - t^{1-\alpha}\right)\right]}dt - b. \qquad (19)$$

By calculating (19), the values of $W(0)$ are estimated for Fig. 1. The integrals in (19) are well calculated by mathematical packages at the given values of numerical coefficients. Calculating $W(0)$ allows performing additional control of the accuracy for the calculations of $W(\Delta)$ in the area of small Δ, where necessary number of calculated summands in sum (18) becomes especially large.

6   Introduction of Priorities at $\alpha > 2$

Regular marking can also be used to realize absolute priorities in such systems $M/Pa/1/\infty$, in which Pareto distribution has the parameter $\alpha > 2$. At the same time, the optimization of the regular marking tick can also be performed by a numerical method. For example, Fig. 2 shows the calculation of the dependence $W(\Delta)$ at $\lambda = 1$, $K = 0.48$, $a = 2.5$ (in this system $b = 0.8$, $\rho = 0.8$, $b^{(2)} = 1.152$).

The dependence calculated by formula (18) is shown by the red curve, the round markers on it show the corresponding values obtained by simulation. The upper horizontal green line corresponds to the value of W_0, which characterizes the original system (the system without priorities). The lower horizontal line corresponds to the value of $W(0)$, which the system has when introducing absolute priorities due to regular marking with infinitesimal intervals defining priority classes. The fact that the curve starts almost from the upper horizontal and ends almost with the lower horizontal indicates that a sufficiently wide range of Δ values is considered. It can also be noted that for small Δ the calculated curve slightly exceeds the exact limit value of $W(0)$, to which it seems to be asymptotically approaching from below. This indicates that 60 thousand summands which were used to calculate the values of $W(\Delta)$ according to formula (18), were not enough at low Δ. A relatively small number of summands was chosen in this case because at $\alpha > 2$ the sums in (18) converge relatively quickly.

Optimal value Δ comprises the value $\Delta_{opt} \approx 0.4653$ here, at such Δ the value $W(\Delta_{opt}) \approx 1.236$ is achieved. Evaluating the efficiency of introducing absolute priorities at $\alpha > 2$ by reduction coefficient $\xi = W_0/W(\Delta_{opt})$, we get $\xi = 2.88/1.236 \approx 2.33$.

Figure 3 shows the calculation of dependence $W(\Delta)$ in the system with the parameters $\lambda = 1$, $K = 0.48$, $\alpha = 3.5$ (in this system $b = \rho = 0.672$, $b^{(2)} = 0.5376$). Here $\Delta_{opt} \approx 0.4046$, $W(\Delta_{opt}) \approx 0.6541$, efficiency indicator $\xi = 0.8195/1.236 \approx 1.25$.

As seen in Fig. 3, at small Δ the values of W calculated by the finite sums are almost equal to the value of $W(0)$ calculated precisely by formula (18), though sums (18) were truncated to the same 60 000 summands at calculation, just like in the calculation shown in Fig. 2. This can be explained by the sums (18) convergence rate growing along with α.

The use of 60 000 summands in sums (18) allowed their calculation to be realized in Excel, and, consequently, Δ_{opt} and $W(\Delta_{opt})$ to be determined using add-in 'Solver'. Thus, a solution is searched within a couple of tens of seconds. Therefore, the calculation of dependences in Figs. 2 and 3 was performed only to verify the uniqueness of the sought local minimum.

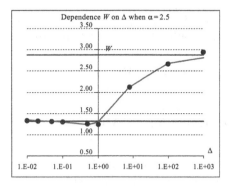

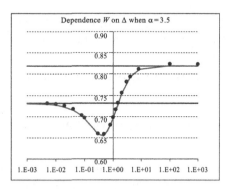

Fig. 2. The calculation of the dependence $W(\Delta)$ in the system M/Pa/1/∞ with the parameter $\alpha = 2.5$ (Color figure online)

Fig. 3. The calculation of the dependence $W(\Delta)$ in the system M/Pa/1/∞ with the parameter $\alpha = 3.5$

In numerical and simulation experiments, it was established that the efficiency indicator ξ grows quickly when α approaches the value $\alpha = 2$ from above, reaching the infinity at $\alpha = 2$ (and thus, it remains the same at the entire range $1 < \alpha \leq 2$).

Figure 1 (see above) shows the calculation of $W(\Delta)$ at $1 < \alpha \leq 2$, when in sums (18) dozens of millions of summands had to be used. Therefore, standard functions of Excel cannot be used efficiently. Optimization means were not included in a program specially written to calculate such sums. It was mainly caused by the fact that a more efficient marking than regular one had been found, so that the problem of optimizing the marking parameters, when calculating large sums, disappeared by itself.

7 Exponential Marking

Numerical experiment with a large number of other markings differing from a regular one showed that the most efficient marking is infinite exponential $(t_0, t_1, \ldots, t_k, \ldots)$, in which points t_k are distributed as follows:

$$t_0 = K, \qquad t_k = K + ce^{ak}, \qquad (k = 1, 2, \ldots), \tag{20}$$

where c, a are coefficients that can be optimized to minimize the average waiting time W at given λ, K, α.

In addition to this marking being able to significantly reduce the time W, its advantage is that due to the rapid growth of the intervals between the points t_k, sufficiently remote marking horizons, providing the proximity of finite sums to infinite ones in (18), overlap after only several tens of summands in these sums. Because of this, the calculation of W can be performed in a few rows of Excel spreadsheets, and at the same time it is possible to perform precise optimization of the marking parameters c, a due to the gradient method built into add-in 'Solver'.

A small number of practically realizable infinite exponential marking levels (20) are also beneficial for the practical implementation of the absolute priorities.

8   Experiments with Loss Probabilities at Finite Buffer Sizes

8.1   The Main Task Solved by the Introduction of Absolute Priorities

The ultimate goal of the research carried out in the article is a radical reduction of the failure probability in queueing systems with heavy distribution tails. The developed method consists in introducing absolute priorities caused by infinite markings in systems with a finite buffer. The created in the article method takes into account the results of well-known empirical studies of systems operating under fractal traffic [9] and achievements for the classical theory of priority disciplines optimization [10–12]. However, the proposed method differs significantly from well-known ones by considering of infinite markings. We propose to optimize the parameters of the used markings by numerical methods based on the exact formulae of the queueing theory. Since there are no exact formulae to calculate the considered systems with a finite buffer, we use the idea to optimize the marking parameters on systems with an infinite buffer, i.e. to choose such marking parameters that minimize the average service time (average queue length) in a system with an infinite buffer. Then, the subsequent buffer limit occurs under conditions where the queue becomes on average as short as possible, and therefore the loss probability should be reduced as much as possible. This heuristic justification of the developed method also explains the independence of the selected marking from the buffer length and, therefore, indirectly justifies the calculation of the rapid decrease in the loss probability with the increasing buffer size. This section describes the simulation experiments performed with different systems with a finite buffer, which most convincingly confirm the effectiveness of the method developed to reduce the loss probability.

8.2   Experiments with the Systems M/Pa/1/m at $1 < \alpha \leq 2$

Figure 4 shows the results of six simulation experiments in which the dependences of loss probabilities P on the buffer size m were determined for various systems

M/Pa/1/m at $1 < \alpha \leq 2$. The horizontal coordinate axis corresponds to the buffer size m, and the vertical axis corresponds to the loss probability P. In all experiments, 10 million requests were passed through the system.

Continuous red lines represent the graphs of the dependences $P(m)$, obtained at the optimal exponential markings. Blue round markers represent the dependences $P(m)$, obtained at the optimal regular markings.

Figure 4 illustrates that optimal exponential markings are as good as optimal regular markings. Exponential markings, however, are much more economical, since they require the use of only a few dozen priority classes. Figure 4 shows the results of the experiments, in which 68 priority classes on exponential markings were used, while optimal regular markings at $1 < \alpha \leq 2$ lead to using tens of millions of priority classes (see explanations to Fig. 1). Some of the best results for minimizing P at the exponential markings in Fig. 4 can be explained by the fact that regular markings are optimized less accurately.

Trajectory 1 in Fig. 4 corresponds to the experiments with the M/Pa/1/m system where $\alpha = 1.5$, $\rho = 0.5$. The optimal value of Δ for the corresponding regular marking, determined up to two decimal digits, is 0.30. The optimal values of the exponential marking parameters are as follows: $a = 1.055$, $c = 0.08424$.

Trajectory 2 is obtained for the system with the same $\rho = 0.5$, but with a much heavier Pareto distribution tail, here, $\alpha = 8/7 \approx 1.1428$. From the comparison of trajectories 1 and 2, it can be concluded that the heavier distribution tail only increases the efficiency of the proposed method, while in conventional approaches heavier tail (a simple increase in the buffer volume or an increase in the channel performance) leads to almost insurmountable problems. In the experiments, whose results are shown on trajectory 2, $\Delta_{opt} = 0.15$, optimal values of the parameters a and c for the exponential markings were 1.01007 and 0.03896, respectively.

Trajectory 3 corresponds to the system, where $\alpha = 1.5$, but the load on the system is increased, in this case $\rho = 0.8$. The parameters of the optimal markings used here are as follows: $\Delta_{opt} = 0.45$, $a = 0.8980$, $c = 0.1424$.

Another pair of experiments was carried out to compare the efficiency of the proposed method and the usually recommended method of reducing the loss probability, consisting in a simple increase in the buffer size without introducing absolute priorities. Simulation experiments were carried out with the same system, which was considered first in the previous series of experiments: this is the system M/Pa/1/m, in which $\alpha = 1.5$, $\rho = 0.5$. The results are shown in Fig. 5 (here coordinate axes are the same as in Fig. 4).

A continuous, sharply downward, red line is the result of the buffer build-up, using absolute priorities due to optimal exponential marking (this line is taken from trajectory 1 in Fig. 4). The line converges to a straight line that at the logarithmic scale of the ordinate axis indicates that the dependence $P(m)$ represented by this line is asymptotically exponential.

The solid green line at the top of the diagram is calculated by simulating the initial system that does not use priorities. Since the corresponding dependence is power-law, here the line goes down with deceleration. At the buffer size $m = 100$,

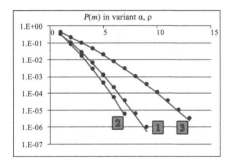

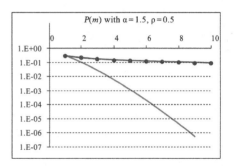

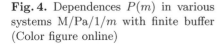

Fig. 4. Dependences $P(m)$ in various systems M/Pa/1/m with finite buffer (Color figure online)

Fig. 5. Dependences $P(m)$ in the normal and priority modes (Color figure online)

the probability P decreases here only to 0.029, at $m = 1000$ we get $P = 0.00764$, and at $m = 10\,000$ we get $P = 0.00125$. As you can see from Fig. 5, when using absolute priorities, such loss probability is achieved already at the buffer size $m = 5$. Moreover, using buffers designed to store 10 000 packets in real network devices is sure to make no sense because of the high cost of the corresponding equipment and the large delays occurring in the corresponding queues.

The markers on the top line indicate the results obtained when using relative priorities (with the same set of priority classes as with absolute priorities).

The effects observed at simulation should be somehow explained at the level of physical meaning. The extraordinary efficiency of absolute priorities both in terms of reducing the loss probability and in terms of transforming the average infinite waiting time into a finite one seems to be due to the fact that the service time x has an infinite dispersion in the systems under consideration. Absolute priorities "reschedule" the order of service for arriving messages, and to change the order of service is especially reasonable in cases when the service time of arriving requests is of a considerable diversity. However, when comparing high efficiency of absolute priorities with low efficiency of relative ones, it is necessary to clarify the explanation with the following remark. At infinite dispersion of service time x, the system occasionally receives requests with very high, "catastrophic" service time. If such a request enters the system and occupies a channel in non-priority mode or relative priority mode, a long queue is created at the system input during its service, which leads to an overflow of even a very large buffer. In the mode of absolute priorities, incoming "non-catastrophic" requests with a higher priority "do not notice" a catastrophic request and simply push it out into the queue, consequently they are served as if there were no catastrophic request. As a result, long queues accumulating does not occur.

These reasonings make us suppose that the proposed method will be as effective in any systems GI/GI/1/m, where service time x distribution is a heavy tail distribution (HTR) with infinites (or simply large) dispersion.

8.3   Other Systems with HTR

Figure 6 shows the results of modeling two systems Pa/Pa/1/m. The lower line is
the graph for the dependence $P(m)$ in the system with the following parameters.
Both Pareto distributions have the shape parameter $\alpha = 1.5$. The parameter K of
the first Pareto distribution is 4/15, while for the second one it is twice as small.
Therefore, the load coefficient of the systems is $\rho = 0.5$. The system uses the
absolute priorities with priority classes determined by an exponential marking.
The marking is not optimized; the first exponential marking is simply taken from
the precious six experiments (the marking with the parameters $a = 1.055$,
$c = 0.08424$). The upper line is the results of modeling the second system
Pa/Pa/1/m obtained from the first one by substituting both shape parameters
$\alpha = 1.5$ with $\alpha = 1.25$. The same exponential marking was used here. The
figure demonstrates the fact that, despite the use of non-optimized markings,
the proposed method provides low loss probabilities at small m. As m grows,
the loss probabilities decrease according to the asymptotically exponential law.

   If necessary, the markings in such systems can be optimized using simulation.

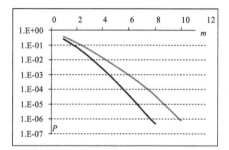

Fig. 6. The results of modeling two
systems Pa/Pa/1/m

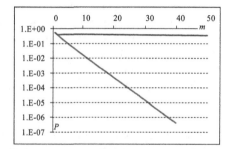

Fig. 7. The results of modeling system
gamma/Log$n$/1/m

   Figure 7 shows the results of modeling a system in which the arrival flow is set
by the gamma distribution of the intervals and the service time x has a lognormal
distribution. The gamma distribution has a mathematical expectation of 1 and
a relatively high dispersion of 16 (the parameters β and α of the distribution
are chosen to be 16 and 1/16, respectively). Lognormal distribution has the
parameters $\mu = -10$, $\sigma = 4.3$. At such parameters of the lognormal distribution,
for the service time x we get $E(x) = b = 0.47$, $Var(x) = 5.938 \cdot 10^{42}$.

   The upper almost horizontal line in Fig. 7 shows the dependence $P(m)$ in non-
priority mode: P decreases very slowly with the growth of m. At $m = 10\,000$,
the probability $P = 0.25436$ was obtained, at $m = 100\,000$, the probability
$P = 0.234502$. In Fig. 7 this sequence of probabilities begins with values of the
order 0.5...0.4.

   The lower line in the diagram corresponds to the dependence $P(m)$ obtained
in the experiment when using absolute priorities with infinite marking of service
time, i.e. the method developed in this article.

9   Conclusion

The article proposed a method to reduce loss probability for requests in systems GI/GI/1/m with infinite service time x dispersion Var(x). Dispersion Var(x) = ∞ in the systems M/G/1 results in infinite average waiting time and infinite average queue length, and in the systems M/G/1/m with finite buffer, the result is high loss probability P slowly decreasing with the growth of m. This problem is very relevant, for example, in the design of data networks, where the size of the transferred files and, accordingly, the time of their service by network devices are distributed according to the asymptotically power law, well approximated by the Pareto distribution with a shape indicator α lying within $1 < \alpha \leq 2$.

To solve the problems, which in practice result from an infinite dispersion Var(x), the article develops a method based on introducing absolute priorities for requests with a (formally) infinite number of priority classes.

In the course of the research carried out in the article, the following main results were obtained.

1. The regular marking in which all intervals $[t_{k-1}, t_k)$ have the same length Δ is investigated. The theorem is formulated and proved that if the absolute priorities determined by the regular marking are introduced into the system M/Pa/1/∞ with infinite average waiting time (i.e. at $1 < \alpha \leq 2$), its average waiting time W becomes finite.
2. The problem of optimizing regular marking (i.e. its defining parameter Δ) by criterion $W \to \min$ is set and solved by numerical methods.
3. The calculation formula to determine the average waiting time W at $\Delta \to 0$ is obtained.
4. Among a large number of irregular markings, the most effective one is found, i.e. the exponential marking, in which the length of successive intervals grows as an exponent with two constant coefficients. It is shown that in terms of reducing the average waiting time W the exponential marking not inferior to regular one. At the same time, exponential marking is much more economical: at $1 < \alpha \leq 2$, when tens of millions of priority classes are implemented by the optimal regular marking, the optimal exponential marking requires the implementation no more than 50 priority levels.
5. The efficiency analysis is performed for the developed method as a method of reducing the average waiting time W at $\alpha > 2$, where the original system M/Pa/1 has a finite W. It is shown that the efficiency of the method in this area (the reduction coefficient W) increases quickly as α approaches 2 from above and becomes infinite at the point $\alpha = 2$.
6. The efficiency of applying the developed method to the systems M/Pa/1/m with a finite buffer and Var(x) = ∞ is studied. It is established that the introduction of absolute priorities determined by the infinite marking drastically reduces the loss probability. In this case, the slow (with power speed) decrease of $P(m)$ turns into a decrease with an exponential speed.
7. It is assumed that the method will also be equally effective in other systems with infinite dispersion Var(x). In order to verify this assumption, a

series of experiments with other systems was carried out. Experiments have convincingly confirmed the high efficiency of the method in various systems with infinite (or very large) dispersion of service time.

References

1. Leland, W.E., Willinger, W., Taqqu, M.S., Wilson, D.V.: On the self-similar nature of Ethernet traffic. In: ACM SIGCOMM 1993, pp. 183–193 (1993)
2. Zwart, A.P.: Queueing Systems with Heavy Tails. Eindhoven University of Technology, Eindhoven (2001)
3. Paxon, V., Floyd, S.: Wide area traffic: the failure of Poisson modeling. IEEE/ACM Trans. Netw. **3**(3), 226–244 (1995)
4. Park, W., Willinger, W.: Self-Similar Network Traffic and Performance Evaluation. Wiley, New York (2000)
5. Likhanov, N. Tsybakov, B., Georganas, N.: Analysis of an ATM buffer with self-similar ("fractal") input traffic. In: IEEE INFOCOM 1995, vol. 3, pp. 985–992 (1995)
6. Zadorozhnyi, V.N., Zakharenkova, T.R.: Minimization of packet loss probability in network with fractal traffic. In: Dudin, A., Nazarov, A., Kirpichnikov, A. (eds.) ITMM 2017. CCIS, vol. 800, pp. 168–183. Springer, Cham (2017). https://doi.org/10.1007/978-3-319-68069-9_14
7. Kleinrock, L.: Queueing Systems: V. II - Computer Applications. Wiley, New York (1976)
8. Adamchik, V.S.: Some series of the zeta and related functions. Analysis **18**, 131–144 (1998)
9. Erramilli, A., Narayan, O., Willinger, W.: Experimental queueing analysis with long range dependent packet traffic. IEEE/ACM Trans. Netw. **4**(2), 209–223 (1996)
10. Zadorozhnyi, V.N., Zakharenkova, T.R., Tulubaev, D.A.: Estimation of prioritized disciplines efficiency based on the metamodel of multi-flows queueing systems. In: Dudin, A., Nazarov, A., Moiseev, A. (eds.) ITMM/WRQ -2018. CCIS, vol. 912, pp. 290–304. Springer, Cham (2018). https://doi.org/10.1007/978-3-319-97595-5_23
11. Bronstein, O.I., Rykov, V.V.: On optimal priority rules in queueing systems. Izv. AS USSR. Techn. Cibern **6**, 28–37 (1965). (in Russian)
12. Vishnevskiy, V.M.: Theoretical bases of designing computer networks. Technosphere, Moscow (2003). (In Russian)

On a Queue with Postponed Work Under N-Policy

Chembra Balan Ajayakumar$^{1(\boxtimes)}$ and Achyutha Krishnamoorthy^2

^{1} College of Engineering, Aranmula, Pathanamthitta, Kerala, India
ajayakumarcb@gmail.com
^{2} Centre for Research in Mathematics, CMS College, Kottayam, Kerala, India
achyuthacusat@gmail.com

Abstract. Models with postponement are an alternative to finite capacity queues in which overflow jobs are irrevocably lost. If the buffer is empty, an arriving customer enters in to it and his service starts immediately. In case the buffer is non-empty, but not full, with some probability depending on the number of works in the buffer direct entry to buffer is permitted. When a customer is rejected from the buffer, he is offered a pool of postponed work of infinite capacity. In this case the customer chooses to enter the pool with certain probability. On the contrary, if the buffer is full at a customer arrival epoch, the customer decides to join the orbit with certain probability; however the system may reject him with some probability. At a service completion epoch, if the number of customers in the buffer is less than a pre-assigned quantity, head of the pooled customers will be transferred to the buffer with a specified probability. In the N-policy introduced in this paper, the number of continuously served customers, taken from the buffer, is counted at each service completion epoch. When it reaches a pre-assigned number N, then the one ahead of all waiting in the pool gets transferred to the buffer for immediate service. We study its long run behaviour. Several system performance measures, and a few numerical illustrations are provided. A game theoretical approach to the queue is also introduced.

Keywords: Phase type distribution · Quasi birth death process · Matrix geometric solution · Postponed work · N-policy

1 Introduction

In many practical situations, finite capacity queues are more realistic than those with infinite capacity, considering the physical limitations of the system. But this will result in overflow of jobs and make considerable loss to the system. With this in view, Deepak et al. [3] introduced a concept called postponed work. Ajayakumar and Pramod [1] introduced N-policy into Deepak et al. [3]. They analysed such a system in great detail in the stationary case and provided a number of system performance measures. No further development in this is reported so

© Springer Nature Switzerland AG 2019
A. Dudin et al. (Eds.): ITMM 2019, CCIS 1109, pp. 312–326, 2019.
https://doi.org/10.1007/978-3-030-33388-1_25

far. Nevertheless this notion of postponement of work has been introduced into inventory by a few researchers (see Krishnamoorthy and Islam [5], Arivarignan et al. [2], Paul Manuel et al. [9], Sivakumar and Arivarignan [10]). In the present paper we modify the model described in [1] by introducing a selection rule for entry into buffer if there is a vacancy.

At a service completion epoch, if the number in the buffer is less than a pre-assigned quantity, a pooled customer is transferred to the buffer with some probability. In the N-policy introduced in this paper, the number of continuously served customers, taken from the buffer, is counted at each service completion epoch. When it reaches a pre-assigned number N, then the one ahead of all waiting in the pool gets transferred to the buffer for immediate service. A diagramatic representation of the model is given in Fig. 1.

Rest of this paper is arranged as follows. Section 2 provides mathematical formulation of the model and Sect. 3 analyses the system by describing stability criterion and stationary distribution. Section 4 describes computation of some expected values such as waiting time in buffer and pool, duration between two consecutive transfers under N-policy and FIFO violation. Section 5 gives some performance measures. Section 6 presents numerical results. A game theoretical approach is given in Sect. 7. Conclusion is given in Sect. 8.

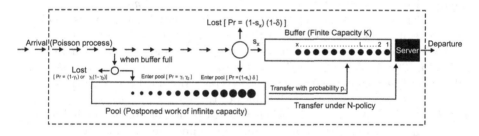

Fig. 1. Queue with postponed work under N-Policy

2 Mathematical Formulation

Consider an $M/PH/1$ queue with finite capacity K, called buffer and an infinite capacity pool for postponed work. If the system is empty, an arriving customer will join buffer and his service starts immediately. If the buffer has l persons where $1 \leq l \leq K - 1$, then a newly arriving customer will be allowed to enter buffer only with probability s_l. We assume that as l increases, the probability s_l decreases and that $s_0 = 1$ and $s_K = 0$. Customers who are denied admission to the buffer on arrival even when a vacant slot is available in the buffer, are directed to join the pool; at this epoch such customers decide to join the pool with probability δ or to leave the system with probability $1 - \delta$. If the buffer is full then the newly arriving customer may decide to join pool, with probability γ_1 or to leave with probability $1 - \gamma_1$. If he decides to join pool, then the server will permit him to join pool with probability γ_2 or not admit with probability $1 - \gamma_2$. So in this case, customers join pool with rate $\lambda\gamma_1\gamma_2$ or leave the system at rates $\lambda(1 - \gamma_1) + \lambda\gamma_1(1 - \gamma_2)$.

When at the end of a service, if there are customers in the pool, the system operates as follows. If the buffer is empty, the one ahead of all waiting in the pool gets transferred to the buffer for immediate service. If the buffer contains ξ jobs, where $1 \leq \xi \leq L - 1$; $2 \leq L \leq K - 1$ at a service completion epoch, then the customer at the head of the buffer starts getting service and with probability p, the head of the queue in the pool is transferred (we call this a p-transfer) to the buffer as the last among the waiting customers in it. With probability $q = 1 - p$, no such transfer takes place. If at a service completion epoch, the number of customers in the buffer is L or more but at most $K - 1$, then no customer transfer from pool to buffer is effected. At each service completion epoch, if the pool contains at least one customer, the number of continuously served customers taken from the buffer is counted. When it reaches N $(N > 0)$, the one ahead of all waiting in the pool gets transferred to the buffer for immediate service. Customers arrive according to a homogeneous Poisson process of rate λ [4]. Service time of customers are iid phase type distributed with representation (β, S) of order m. The vector $S^0 = -Se$ containing elements S_{h0} represents the absorption rate from the phase h, $h = 1, 2, ..., m$. Absorption (service completion) occurs with probability 1 from any phase i in $\{1, 2, ...m\}$ if and only if the matrix S is non singular. Then the mean time until absorption is $-\beta S^{-1}e$. Also the equilibrium distribution of the excess life is $PH(\pi^*, S)$ where π^* is the stationary probability vector satisfying $\pi^*Q^* = 0$ and $\pi^*e = 1$ where $Q^* = S + S^0\beta$. The model is studied as a Quasi Birth-Death(QBD) process and a solution of the classical matrix geometric type is obtained ([8] and [7]).

The state space consists of all tuples of the form (i, j, b, h) with $i \geq 1$, $1 \leq j \leq K$; $0 \leq b \leq N$; $1 \leq h \leq m$ where i is the number of postponed work, j is the number of work in the finite buffer including the unit in service, b is the number of continuously served customers from the buffer at a service completion and h is the phase of the service in progress at a time t. For a given value of i, $K(N + 1)m$ states constitute the level i of the QBD. Now consider the boundary level $i = 0$. Then we denote the empty system $(0, 0, 0, 0)$ by 0. Also there are Km states of the form $(0, j, 0, h)$, $1 \leq j \leq K$; $1 \leq h \leq m$. This is due to the fact that when the pool has no customers, N-policy is suspended. These have the same significance as before, except that in these states, no postponed jobs are present, but there are jobs in the finite buffer. These $Km + 1$ states make up the boundary level 0 of the QBD. The infinitesimal generator of the QBD describing the $M/PH/1/K$ queue with postponed customers under N-policy is of the form

$$Q = \begin{bmatrix} \mathcal{B}_1 & \mathcal{B}_0 & & & \\ \mathcal{B}_2 & Q_1 & Q_0 & & \\ & Q_2 & Q_1 & Q_0 & \\ & & Q_2 & Q_1 & Q_0 \\ & & & \ddots & \ddots & \ddots \end{bmatrix}$$

where the matrix $\mathcal{B}_0$ is of dimension $(Km+1) \times K(N+1)m$, $\mathcal{B}_1$ is square matrix of order $Km+1$ and $\mathcal{B}_2$ is of dimension $K(N+1)m \times (Km+1)$. Q_0, Q_1 and Q_2 are square matrix of order $K(N+1)m$. Each of these matrices is itself highly structured.

The matrix $\mathcal{B}_1$ corresponds to the transition from the level 0 to 0 in which $(l,l)^{th}$ entries are Δ_l for $l = 2, ..., K-1$, $(l,l+1)^{th}$ entries are Ω_l for $l = 2, ..., K-1$, $(1,2)^{th}$ entry is $\lambda\beta$, $(l,l-1)^{th}$ entries are $S^0\beta$ for $i = 3, ..., K$, $(2,1)^{th}$ element is S^0, $(K,K)^{th}$ element is Δ and $(1,1)^{th}$ element is $-\lambda$. where $\Delta = S - \lambda\gamma_1\gamma_2 I_m$, $\Delta_l = S - \epsilon_l I_m$, $\Omega_l = \lambda s_l I_m$, $\epsilon_l = \lambda(s_l + \delta - s_l\delta)$, $l = 2, ..., K-1$.

$$\mathcal{B}_0 = \begin{bmatrix} \bar{0} \\ diag(\Gamma_1, \Gamma_2, ..., \Gamma_{K-1}, \Gamma) \end{bmatrix}$$

where $\Gamma_l = \lambda(1-s_l)\delta t_5 \otimes I_m$, $l = 1, 2, ..., K-1$; $\Gamma = \lambda\gamma_1\gamma_2 t_5 \otimes I_m$ and t_5 is a row vector of order $N+1$ with first element 1 and all other elements zero with I_m representing identity matrix of order m. Also $\bar{0}$ is zero matrix of appropriate order and $diag(\Gamma_1, \Gamma_2, ..., \Gamma_{K-1}, \Gamma)$ represents a diagonal block matrix of order K.

$$\mathcal{B}_2 = \begin{bmatrix} \bar{0} \, diag \left(H_1, H_2, ..., H_L, H_{L+1}, ..., H_K, \right) \end{bmatrix}$$

where $diag(H_1, H_2, ..., H_L, H_{L+1}, ..., H_K)$ represents a diagonal block matrix of order K with diagonal block entries $H_1 = t_6 \otimes S^0\beta$, $H_2 = ... = H_L = t_7 \otimes S^0\beta$, $H_{L+1} = ... = H_K = t_8 \otimes S^0\beta$ and t_6 is a column vector of order $N+1$ with all entries are 1, t_7 is a column vector of order $N+1$ with all elements are p except 1 at $(N,1)^{th}$ position, t_8 is a column vector of order $(N+1)$ with $(N,1)^{th}$ element is 1 and all other elements zero. $Q_0 = diag(\omega_1, \omega_2, ..., \omega_{K-1}, \omega)$ where $\omega_l = \lambda(1-s_l)\delta I_{N+1} \otimes I_m$, $l = 1, 2, ..., K-1$; $\omega = \lambda\gamma_1\gamma_2 I_{N+1} \otimes I_m$ and I_{N+1} is the identity matrix of order $N+1$. Also $Q_2 = diag(\Lambda_1, \Lambda_2, ..., \Lambda_L, \Lambda_{L+1}, ..., \Lambda_K)$ where $\Lambda_1 = t_1 \otimes S^0\beta$, $\Lambda_2 = ... = \Lambda_L = t_2 \otimes S^0\beta$, $\Lambda_{L+1} = ... = \Lambda_K = t_3 \otimes S_0\beta$, t_1 is a square matrix of order $N+1$, given by

$$t_1 = \begin{bmatrix} \bar{0} & I_N \\ 1 & \bar{0} \end{bmatrix}$$

where I_N is identity matrix of order N. t_2 is a square matrix of order $N+1$ in which $(l,l+1)^{th}$ and $(N+1,1)^{th}$ entries are p for $l = 1, ..., N-1$, $(N,N+1)^{th}$ element is 1 and all other elements zero. t_3 is a square matrix of order $N+1$ with $(N,N+1)^{th}$ entry is 1 and all other entries are zero.

In matrix Q_1, $(l,l)^{th}$ entries are Θ_l for $l = 1, ..., K-1$, $(l,l+1)^{th}$ entries are Φ_l for $l = 1, ..., K-1$, $(l,l-1)^{th}$ entries are ζ for $l = 2, ..., L$, and η for

$l = L+1, ..., K$ and $(K, K)^{th}$ element is Θ where $\zeta = t_4 \otimes qS^0\beta$ which corresponds to the transition of the buffer size from j to $j-1$ for $j = 2, 3, ..., L$; $\eta = t_4 \otimes S^0\beta$ which corresponds to the transition of the buffer size from j to $j-1$ for $j = L+1, ..., K$; $\Theta = I_{N+1} \otimes (S - \lambda\gamma_1\gamma_2 I_m)$ which corresponds to the transition of the buffer size from K to K; $\Theta_j = I_{N+1} \otimes (S - \epsilon_j I_m)$ which corresponds to the transition of the buffer size from j to j for $j = 1, 2, ..., L, L+1, ..., K-1$ where $\epsilon_j = \lambda(s_j + \delta - s_j\delta)$; $\Phi_j = \lambda s_j I_{N+1} \otimes I_m$ which corresponds to the transition of the buffer size from j to $j+1$ where $j = 1, 2, ..., L, L+1, ..., K-1$. Also t_4 is a square matrix of order $N+1$ in which $(l, l+1)^{th}$ entries are 1 for $l = 1, ..., N-1$, $(N, 1)^{th}$ element is 1 and all other elements are zero.

3   Analysis of the System

3.1   Stability Criterion

Theorem 1. *The system is stable if and only if*

$$\lambda\gamma_1\gamma_2 \sum_{b=0}^{N}\sum_{h=1}^{m} \pi_{Kbh} + \lambda\delta \sum_{j=1}^{K-1}\sum_{b=0}^{1}\sum_{h=1}^{n} (1-s_j)\pi_{jbh} < \frac{1}{K(N+1)m \sum_{l=1}^{} m_{1_l}}.$$

Proof. Let $G_{ll'}(k, x)$ be the conditional probability that the QBD process starting in the state $l = (i, j, b, h)$ (for $i > 1$) where $1 \le j \le K, 0 \le b \le N, 1 \le h \le m$ at time $t = 0$ reaches the state $l' = (i-1, j', b', h')$ where $1 \le j' \le K, 0 \le b' \le N$, $1 \le h' \le m$ for the first time, involving exactly k transitions and completing before time x. Because of the structure of Q, the probability $G_{ll'}(k, x)$ does not depend on i. The matrix with elements $G_{ll'}(k, x)$ is denoted by $G(k, x)$.

Now introduce the transform matrix,

$$\hat{G}(z, \theta) = \sum_{k=1}^{\infty} z^k \int_0^{\infty} e^{-\theta x} dG(k, x)$$

for $|z| \le 1, \theta > 0$. The matrix $\hat{G}(z, \theta)$ satisfies the matrix equation

$$\hat{G}(z, \theta) = z(\theta I - Q_1)^{-1}Q_2 + (\theta I - Q_1)^{-1}Q_0\hat{G}^2(z, \theta).$$

Use the notations $C_0(\theta) = (\theta I - Q_1)^{-1}Q_2$ and $C_2(\theta) = (\theta I - Q_1)^{-1}Q_0$. Now the transform matrix $\hat{G}(z, \theta)$ is equal to the minimal non negative solution of the matrix quadratic equation $X(z, \theta) = zC_0(\theta) + C_2(\theta)X^2(z, \theta)$ and it is obtained by successive substitutions starting with the zero matrix. Also we have

$$\lim_{z \to 1, \theta \to 0} \hat{G}(z, \theta) = G(k, x) = [G_{ll'}(k, x)].$$

Suppose the matrix $A = Q_0 + Q_1 + Q_2$ is irreducible. Then the necessary and sufficient condition for the positive recurrence of the process is that the matrix

G is stochastic. For this, the condition $\pi Q_2 e > \pi Q_0 e$ must be satisfied where π is the stationary probability vector associated with $A = Q_0 + Q_1 + Q_2$. That is it is the unique solution to $\pi A = 0$ and $\pi e = 1$. The quantity $\rho = \frac{\pi Q_0 e}{\pi Q_2 e}$ is called the traffic intensity of the QBD process. G is obtained as the minimal non negative solution to the equation $G = C_0 + C_2 G^2$ where $C_0 = (-Q_1)^{-1} Q_2$ and $C_2 = (-Q_1)^{-1} Q_0$. That is, G is the minimal non negative solution of the matrix quadratic equation $Q_2 + Q_1 G + Q_0 G^2 = 0$.

Let $m_1 = [m_{1_l}]$ denotes the column vector of dimension $K(N+1)m$ where m_{1_l} denotes the mean first passage time from the level i ($i > 1$) to the level $i-1$ given that the first passage time started in the state l. Then,

$$m_1 = \left[-\frac{\partial}{\partial \theta} \hat{G}(z, \theta) e \right]_{\theta=0, z=1} = -(Q_1 + Q_0(I + G))^{-1} e.$$

For the system stability, the rate of drift from level i to level $i-1$ should be greater than that to level $i+1$. This means that the Markov Chain is stable if and only if $\pi Q_2 e > \pi Q_0 e$. The rate of drift from level i to the level $i+1$ is given by $\lambda \gamma_1 \gamma_2 \sum_{b=0}^{N} \sum_{h=1}^{m} \pi_{Kbh} + \lambda \delta \sum_{j=1}^{K-1} \sum_{b=0}^{1} \sum_{h=1}^{n} (1-s_j) \pi_{jbh}$. It follows that the condition $\pi Q_0 e < \pi Q_2 e$ is equivalent to

$$\lambda \gamma_1 \gamma_2 \sum_{b=0}^{N} \sum_{h=1}^{m} \pi_{Kbh} + \lambda \delta \sum_{j=1}^{K-1} \sum_{b=0}^{1} \sum_{h=1}^{n} (1-s_j) \pi_{jbh} < \frac{1}{\sum_{l=1}^{K(N+1)m} m_{1_l}}.$$

So by an appropriate choice of γ_1 and γ_2, that is by postponing a fraction of overflowing customers, one can obtain a stable system even if arrival rate is greater than service rate.

3.2   Stationary Distribution

Since the model is studied as a QBD process [6], its stationary distribution, if it exists, has a matrix geometric solution. Assume that the stability criterion is satisfied. Let the stationary vector x of Q be partitioned by the levels in to subvectors x_i for $i \geq 0$. Then x_i has the matrix geometric form

$$x_i = x_1 R^{i-1} \tag{1}$$

for $i \geq 2$ where R is the minimal non negative solution to the matrix equation $Q_0 + R Q_1 + R^2 Q_2 = 0$ and the vectors x_0, x_1 are obtained by solving the equtions

$$x_0 \mathcal{B}_1 + x_1 \mathcal{B}_2 = 0, \tag{2}$$

$$x_0 \mathcal{B}_0 + x_1 (Q_1 + R Q_2) = 0 \tag{3}$$

subject to the normalising condition $x_0 e + x_1 (I - R)^{-1} e = 1$.

From the above discussion it is clear that to determine x, a key step is the computation of the rate matrix R. We can partition x_i by sublevels as $x_0 = (x_{00}, x_{01}, x_{02}, ..., x_{0K})$ and $x_i = (x_{i1}, x_{i2}, x_{i3}, ..., x_{iK})$ where $i \geq 1$ and x_{00} is a scalar and x_{0j}, $1 \leq j \leq K$ are vectors of order m and $x_{ij} = (x_{ij0}, x_{ij1}, ..., x_{ijN})$ where $i \geq 1$, $1 \leq j \leq K$ and x_{ijb}, $0 \leq b \leq N$ are vectors of order m.

4   Computation of Expected Values

In this section we derive the expected waiting time of a tagged customer in buffer and pool, the expected duration between two consecutive transfers under N-policy, the expected duration for the first N-policy transfer in a busy cycle and the probability of FIFO violation.

4.1   Expected Waiting Time in Buffer

We denote the mean waiting time of customers who upon their arrivals enter the buffer by $E(W_1)$.

Case1: $N \geq K$; in this case the tagged customer is not affected by the new arrivals in buffer and in pool. So we can calculate the waiting time by considering the system state at which the tagged customer enters. Hence $E(W_1) = \sum_i \sum_j \sum_b \sum_h E$(waiting time of the customer who finds the system in state (i, j, b, h)) Pr(system is in state (i, j, b, h)).

$$E(W_1) = \sum_{j=1}^{k-1} \sum_{h=1}^{m} -\beta S^{-1} e(j-1) x_{0j0h}$$

$$+ \sum_{i=1}^{\infty} \sum_{j=1}^{K-1} \sum_{b=0}^{N-1} \sum_{h-1}^{m} -\beta S^{-1} e(j-1+\psi) x_{ijbh}$$

$$+ \sum_{i=1}^{\infty} \sum_{j=1}^{K-1} \sum_{h=1}^{m} -\beta S^{-1} e(j-1) x_{ijNh} - \pi^* S^{-1} e$$

where $\pi^* Q^* = 0$, $\pi^* e = 1$, $Q^* = S + S^0 \beta$ and

$$\psi = 1 + \left[\frac{j - (N-b)}{N} \right], 0 \leq b < N$$

where $[.]$ denotes the greatest integer of the value within paranthesis and $-\pi^* S^{-1} e$ is the additional time required to complete the service of the customer who is at the server when the tagged person enters the buffer.

In $M/M/1$ case with service rate μ,

$$E(W_1) = \sum_{j=1}^{K-1} \frac{1}{\mu} j x_{0j0} + \sum_{i=1}^{\infty} \sum_{j=1}^{K-1} \sum_{b=0}^{N-1} \frac{1}{\mu} (j+\psi) x_{ijb}$$

$$+ \sum_{i=1}^{\infty} \sum_{j=1}^{K-1} \frac{1}{\mu} j x_{ijN}.$$

Case2: $N < K$; in this case, the tagged customer in the buffer will be affected by the new arrivals in the pool and so the new arrivals in the buffer. So the waiting time of the tagged customer depends on the various susequent developments in the pool such as visits to zero level one or more, but a finite number in the pool joining after the tagged customer. Because of the complexity of calculation, we may turn to computing an upper bound on the waiting time, by keeping in mind, the fact that only a maximum finite number K of persons in the pool will affect the tagged person. In the worst case we have $N = 1$ which represents service alternating between buffer and pool. So an upper bound for the waiting time of a customer who upon his arrival enters the buffer in the state (i, j, b, h), is

$$UB(W_1) = \sum_{j=1}^{k-1} \sum_{h=1}^{m} -\beta S^{-1} e(j - 1 + [\frac{j}{N}]) x_{0j0h}$$

$$+ \sum_{i=1}^{\infty} \sum_{j=1}^{K-1} \sum_{b=0}^{N-1} \sum_{h=1}^{m} -\beta S^{-1} e(j - 1 + \psi) x_{ijbh}$$

$$+ \sum_{i=1}^{\infty} \sum_{j=1}^{K-1} \sum_{h=1}^{m} -\beta S^{-1} e(j - 1 + [\frac{j-1}{N}]) x_{ijNh} - \pi^* S^{-1} e$$

where $-\pi^* S^{-1} e$ is the excess time required to complete the service of the customer who is at the server when the tagged person enter buffer.

In $M/M/1$ case with service rate μ,

$$UB(W_1) = \sum_{j=1}^{K-1} \frac{1}{\mu}(j + [\frac{j}{N}]) x_{0j0} + \sum_{i=1}^{\infty} \sum_{j=1}^{K-1} \sum_{b=0}^{N-1} \frac{1}{\mu}(j + \psi) x_{ijb}$$

$$+ \sum_{i=1}^{\infty} \sum_{j=1}^{K-1} \frac{1}{\mu}(j + [\frac{j-1}{N}]) x_{ijN}.$$

4.2   Expected Waiting Time in Pool

We denote the expected waiting time of a customer who upon his arrival enters the pool, by $E(W_2)$. To find this, first we define the Markov process $\{X(t)\}$ as follows. $X(t) = (a, j, b, h)$ where a denotes the rank of the tagged customer entered pool, j denotes the number of customers in the buffer, b denotes the number of continuously served customers from buffer and h is the phase of the service process at time t. The rank a of the customer is assumed to be r if he joins as the rth customer in pool. His rank decreases to 1 with the customers ahead of him transferred from the pool to the buffer. Since the customers who arrive after the tagged customer cannot change his rank, level changing transitions in $\{X(t)\}$ can takeplace only to one side of the diagonal. We arrange the statespace of $\{X(t)\}$ as $\{r, r - 1, ..., 2, 1\} \times \{1, 2, ..., K\} \times \{0, 1, ..., N\} \times \{1, 2, ..., m\}$ with absorbing state 0 in the sense that the tagged customer is either selected to be served under N-policy or placed in the buffer with probability p or to the server with probability 1 if the buffer size reduces to 0 at the end of a service. The infinitesimal generator of the process is

$$\tilde{Q} = \begin{bmatrix} T & T^0 \\ \bar{0} & 0 \end{bmatrix}$$

where T has as entries transition rates in the transient part having dimension $rK(N+1)m$ and T^0 has entries, which are absorption rates. Now the expected absorption time of a particular customer is given by the column vector $E_w^{(r)} = -\tilde{I}T^{-1}e$ where $\tilde{I} = \begin{bmatrix} I_{K(N+1)m} & \bar{0} \end{bmatrix}$ having order $K(N+1)m \times rK(N+1)m$. So the expected waiting time of the customer is

$$W_L = \sum_{r=1}^{\infty} x_r E_w^{(r)}$$

where x_r is the steady state probability vector corresponding to $i = r$. Also W_L gives the waiting time of a customer in the pool up to the epoch of his transfer to the buffer.

Case1: $N \geq K$; in this case, expected waiting time in pool is

$$E(W_2) = \sum_{i=1}^{\infty} \sum_{j=1}^{K} \sum_{b=0}^{N} \sum_{h=1}^{m} x_{iKbh} W_L (x_{ij(N-1)h} s_{h0} + x_{i1bh} s_{h0})$$

$$+ \sum_{i=1}^{\infty} \sum_{b=0}^{N} \sum_{h=1}^{m} x_{iKbh} (W_L + W^{(1)} p(\sum_{j=1}^{L} x_{ijbh} s_{h0}))$$

where

$$W^{(1)} = \sum_{j=1}^{L} \sum_{h=1}^{m} -\beta S^{-1} e(j-1) x_{0j0h}$$

$$+ \sum_{i=1}^{\infty} \sum_{j=1}^{L} \sum_{b=0}^{N-1} \sum_{h=1}^{m} -\beta S^{-1} e(j-1+\psi) x_{ijbh}.$$

Case2: $N < K$; in this case we get an upperbound $UB(W_2)$ for the waiting time in pool.

$$UB(W_2) = \sum_{i=1}^{\infty} \sum_{j=1}^{K} \sum_{b=0}^{N} \sum_{h=1}^{m} x_{iKbh} W_L (x_{ij(N-1)h} s_{h0} + x_{i1bh} s_{h0})$$

$$+ \sum_{i=1}^{\infty} \sum_{b=0}^{N} \sum_{h=1}^{m} x_{iKbh} (W_L + UB(W^{(1)}) p(\sum_{j=1}^{L} x_{ijbh} s_{h0}))$$

where

$$UB(W^{(1)}) = \sum_{j=1}^{L} \sum_{h=1}^{m} -\beta S^{-1} e(j-1+\left\lceil \frac{j-1}{N} \right\rceil) x_{0j0h}$$

$$+ \sum_{i=1}^{\infty} \sum_{j=1}^{L} \sum_{b=0}^{N-1} \sum_{h=1}^{m} -\beta S^{-1} e(j-1+\psi) x_{ijbh}.$$

4.3   Expected Duration Between Two Consecutive Transfers Under N-Policy

For computing expected duration between two consecutive transfers under N-policy, we consider the Markov process $\{X(t)\}$ described as follows. $X(t) = (b, i, j, h)$ where b is the number of continuously served customers from the buffer, if pool has at least one person, at time t, measured from the service completion of the last customer who was transferred under N- policy. So $b = 0, 1, 2, ..., N$. Here we regard $0, 1, 2, ..., N-1$ as transient states and N as absorbing state (that is the state at which a new N-policy transfer occurs). i denotes the number of postponed jobs at time t. Even if pool is of infinite capacity, we restrict here it to be a finite value say V for sufficiently large V. So $i = 0, 1, 2, ..., V$. Now $j(= 0, 1, 2, ..., K)$ denotes the number of customers in the buffer at time t. Also $h = 1, 2, ..., m$ denotes the phase of the service in progress at a time t. The process $\{X(t)\}$ has the state space

$$\{0, 1, 2, ..., N-1, N\} \times \{0, 1, 2, .., V\} \times \{0, 1, 2, ..., K\} \times \{1, 2, 3, ..., m.\}$$

Infinitesimal Generator of the process is

$$\hat{Q} = \begin{bmatrix} U & U^0 \\ \bar{0} & 0 \end{bmatrix}$$

where U has as entries transition rates in the transient part having dimension $(NVKm+Km+1) \times (NVKm+Km+1)$ and U^0 has entries, which are absorption rates with dimension $(NVKm+Km+1) \times (KVm)$. The initial probability vector of $\hat{Q}$ is

$$\delta = \begin{bmatrix} \dfrac{1}{\overline{KVm+Km+1}} \begin{bmatrix} x_0 & x_1 & \cdots & x_{KVm+Km+1} \end{bmatrix} \bar{0} \\ \displaystyle\sum_{r=1} x_{r-1} \end{bmatrix}_{1\times(NKVm+Km+1)}$$

where $x_0 = x_{0000}$, $x_r = x_{ij0h}, 0 \le i \le V; 1 \le j \le K; 1 \le h \le m$ and r varies from 1 to $KVm + Km$ according to its lexicographic order. Then expected duration between two consecutive transfers under N-policy follows PH distribution with representation (δ, U) and it is given by $N_{ABSORB} = -\delta U^{-1} e$.

4.4   Expected Duration for the First N-Policy Transfer in a Busy Cycle

Here we compute the expected duration of the time elapsed from the epoch of the first arrival to an idle system until the first N-policy transfer is effected.

The time elapsed, starting with an arrival to an idle system, until the realization of the N-policy for the first time follows the PH-distribution with representation (α, U) where

$$\alpha = \frac{1}{\displaystyle\sum_{r=1}^{m} x_r} \begin{bmatrix} 0 & x_1 & x_2 & \cdots & x_m & \bar{0} \end{bmatrix}_{1\times(NKVm+Km+1)}$$

where $x_r = x_{010h}; 1 \leq h \leq m$, r varies from 1 to m and U is described in Sect. 4.3.

At the epoch of the first arrival to an idle system, process starts with the service in one of the m phases with steady state probability $x_r = x_{010h}; 1 \leq h \leq m$, r varies from 1 to m. This justifies the form of the initial probability vector α as given above. Then the expected duration for the realization of the above random variable is $N_{FIRST} = -\alpha U^{-1} e$.

4.5   Probability of FIFO Violation

It may be noted that the N-policy leads to violation of FIFO rule for customers in the pool. For example assume that there are two or more customers in the pool at a service completion epoch at which the number in the buffer droped to $L - 1$ or below and the number of continuously served customers reached $N - 1$. So the first in the pool may be selected under p-transfer and placed as the last in the buffer. When the next service is completed, the current head of the pool gets transferred to the buffer for immediate service there by violating the FIFO rule for pooled customers. Further it may be noted that this situation does not arise among the queued customers in the buffer. We compute the expectation of the indicator random variable defined as FIFO violation in pool. Its expectation is the probability for FIFO violation in pool which is given by

$$P_{\mathcal{FIFO}} = \sum_{i=1}^{\infty} \sum_{j=2}^{L} \sum_{b=N-j+1}^{N-1} x_{ijbh} p S_{h0}.$$

The FIFO may be violated by more than one successors if $N < L$. However this can be overcome by making N sufficiently large than L. If $N \geq K$, a customer joining the pool will not overtake any of the customers in the buffer who had joined before his entering to pool. At this time, FIFO is violated by atmost one successor in pool. Even this can be overcome by a slight modification by redefining the N-policy by resetting b in (i, j, b, h) as zero at the time of p-transfer.

5   Performance Measures

1. The mean number of pooled customers is

$$\mu_{POOL} = \sum_{i=1}^{\infty} i a_i = x_1 (I - R)^{-2} e.$$

2. The mean buffer size is $\mu_{BUFFER} = \sum_{j=1}^{K} j b_j$.

3. The rate at which the customer who leave the system without service is

$$\theta_{LOST} = \lambda(1 - \delta) \sum_{j=1}^{K-1} (1 - s_j) b_j + \lambda(1 - \gamma_1) b_K + \lambda \gamma_1 (1 - \gamma_2) b_K.$$

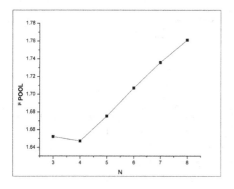
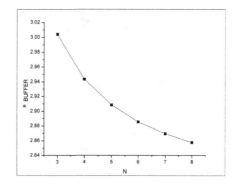

Fig. 2. N versus μ_{POOL} and μ_{BUFFER}

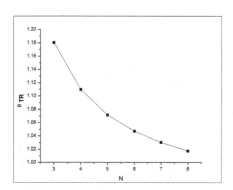
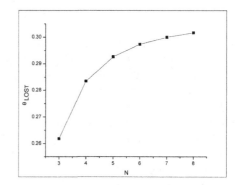

Fig. 3. N versus θ_{LOST} and θ_{TR}

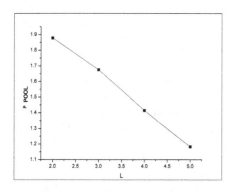
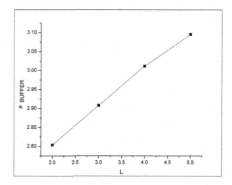

Fig. 4. L versus μ_{POOL} and μ_{BUFFER}

4. The rate at which pooled customers transfer in to the buffer is

$$\theta_{TR} = \sum_{i=1}^{\infty}\sum_{b=0}^{N}\sum_{h=1}^{m} x_{i1bh}S_{h0} + \sum_{i=1}^{\infty}\sum_{j=2}^{L}\sum_{b=0}^{N-2}\sum_{h=1}^{m} x_{ijbh}pS_{h0}$$

$$+ \sum_{i=1}^{\infty} \sum_{j=1}^{K} \sum_{h=1}^{m} x_{ij(N-1)h} S_{h0} + \sum_{i=1}^{\infty} \sum_{j=2}^{L} \sum_{h=1}^{m} x_{ijNh} p S_{ho}.$$

6   Numerical Results

We present some numerical results in order to illustrate the performance of the system. It is meaningful if we consider the following probabilites as a measure depending on the situations of the model as given below: γ_1: Customer's attraction to the pool when the buffer is full. γ_2: Server's interest on a new work to postpone when the buffer is full. s_x: Server's interest on a new work to accept in to the buffer by considering the buffer size. δ: Customer's special interest to the service station. Take

$$\gamma_1 = \frac{Lp}{K} + \frac{1}{N}$$

in order to bring out explicitly the dependence of γ_1 on the system parameters. This is justified as follows. Larger the L value, the customer encountering the buffer full, will be inclined to join the pool with higher probability. Also same is the relationship of γ_1 with p. On the other hand, γ_1 inversely varies with K. The additional term $\frac{1}{N}$ comes through N-policy. Here as N increases γ_1 decreases so that γ_1 and N vary inversely. But the relationship is feasible for those values of L, p, K and N such that $0 \le \gamma_1 \le 1$. This is possible if $N \ge K$ and such a selection is highly consistent. But N can be made less than K by suitably selecting other variables so that $0 \le \gamma_1 \le 1$, and that can be considered as an incentive to customers joining the pool.

The impact of N on various measures of descriptors with $K = 6, L = 3, m = 2, \lambda = 7, p = 0.5, \gamma_1 = \frac{Lp}{K} + \frac{1}{N}$,

$$\beta = \begin{bmatrix} 0.3 & 0.7 \end{bmatrix} \quad S = \begin{bmatrix} -12.5 & 6.0 \\ 6.0 & -12.5 \end{bmatrix} \quad S^0 = \begin{bmatrix} 6.5 \\ 6.5 \end{bmatrix}$$

is shown in Figs. 2 and 3. As N increases μ_{BUFFER}, θ_{TR} decrease monotonically whereas θ_{LOST} increases monotonically; but μ_{POOL} decreases at first and then increases. This is due to the fact that by our assumption γ_1 varies inversely as N. So as N increases, customer's attraction to the pool decreases when the buffer is full. So the pool size decreases. Also transfer rate decreases. This will make the buffer size decreasing and not full. So the influence of δ increases and which makes the pool size increasing.

By keeping $K = 6, p = 0.5, m = 2, \lambda = 7, N = 5, \gamma_1 = \frac{Lp}{K} + \frac{1}{N}$ the effect of L on various measures is shown in Fig. 4. Here also μ_{BUFFER} is monotonically increasing as L increases, as expected. This will gradually makes the buffer full. So the influence of δ decreases which makes the pool size decreasing and the effect of γ_1 increases which makes θ_{LOST} monotonically decreasing. The measures are numerically computed for various values of p by keeping $L = 3$ and shown in Table 1. Here also $\mu_{POOL}, \mu_{BUFFER}, \theta_{TR}$ are monotonically increasing and θ_{LOST} is monotonically decreasing as expected, in p. For a lower L, increase of

p does not much influence to make the buffer full, the effect of δ much influences the increasing of pool size. If $s_x = 1, \forall x, x = 1, 2, ..., K - 1$ and $\gamma_2 = 1$ then we get the model [1].

Table 1. $K = 6, L = 3, m = 2, \lambda = 7, N = 5, \gamma_1 = \frac{Lp}{K} + \frac{1}{N}$

p	μ_{POOL}	μ_{BUFFER}	θ_{TR}	θ_{LOST}
0.3	1.6284441	2.8322663	1.0228137	0.3299661
0.4	1.6445645	2.8706489	1.0458466	0.3119004
0.5	1.6752849	2.9082990	1.0713297	0.2926464
0.6	1.7203127	2.9454260	1.0992774	0.2722456
0.7	1.7798784	2.9821990	1.1297385	0.2507135

7   A Game Theoretical Approach

If the buffer is full with K customers, a newly arrived customer may join the pool with probability γ_1 or leave the system without joining the pool with probability $1 - \gamma_1$. At the same time the server may permit a customer to join pool with probability γ_2 or may sent out from the system without permitting to enter the pool with probability $1 - \gamma_2$. The situation can be modelled by a two-person zero-sum game as follows.

Let the customer be treated as the player 1 and the system as the player 2. The player 1 has two activities; (1) join pool (2) leave the system without join pool. The player 2 has also two activities; (1) allow the customer to enter pool (2) does not allow the customer to enter pool.

Let the mixed strategy of the player 1 be $(\gamma_1, 1 - \gamma_1)$ where $0 < \gamma_1 < 1$ and that of the player 2 be $(\gamma_2, 1 - \gamma_2)$ where $0 < \gamma_2 < 1$ where γ_1 is the probability of the customer to join the pool and γ_2 is the probability of the server to admit the customer to the pool. Then the pay-off matrix of player 1 is

$$\begin{bmatrix} \mathcal{C}_{11} & \mathcal{C}_{12} \\ \mathcal{C}_{21} & \mathcal{C}_{22} \end{bmatrix}$$

where $\mathcal{C}_{11}$ is the gain to the customer when he decides to join the pool and the server admit him to the pool; $\mathcal{C}_{12}$ is the gain to the customer when he decides to join the pool and at the same time the server does not admit him to the pool; $\mathcal{C}_{21}$ is the gain to the customer when he decides not to join the pool but the server is ready to admit him to the pool; $\mathcal{C}_{22}$ is the gain to the customer when he decides not to join the pool and the server decides not to admit him to the pool. Now a valid assumption can be $\mathcal{C}_{11} > \mathcal{C}_{12}, \mathcal{C}_{22} > \mathcal{C}_{12}$ and $\mathcal{C}_{22} > \mathcal{C}_{21}$. Then

$$\gamma_1 = \frac{\mathcal{C}_{22} - \mathcal{C}_{21}}{(\mathcal{C}_{11} + \mathcal{C}_{22}) - (\mathcal{C}_{21} + \mathcal{C}_{12})}, \gamma_2 = \frac{\mathcal{C}_{22} - \mathcal{C}_{12}}{(\mathcal{C}_{11} + \mathcal{C}_{22}) - (\mathcal{C}_{21} + \mathcal{C}_{12})}$$

and the value of the game is $\gamma = \frac{\mathcal{C}_{11}\mathcal{C}_{22} - \mathcal{C}_{12}\mathcal{C}_{21}}{(\mathcal{C}_{11} + \mathcal{C}_{22}) - (\mathcal{C}_{21} + \mathcal{C}_{12})}$.

8   Conclusion

This paper suggested modes of reducing loss of overflow customers in finite capacity queues by introducing "a pool of postponed work" and certain policies of transferring customers from this to the finite buffer. For several queueing situations this can be employed to reduce customer loss through impatience. We also introduced a game theoretic approach of customer joining strategy when he is denied admission to the pool at his arrival epoch.

References

1. Ajayakumar, C.B., Pramod, P.K.: An $M/PH/1$ queue with postponed work under N-policy. Bull. Kerala Math. Assoc. **9**(2), 395–419 (2012)
2. Arivarignan, G., Sivakumar, B., Jayaraman, R.: Stochastic modelling of inventory systems with postponed demands and multiple server vacations. Int. J. Appl. Math. **1**, 1–19 (2009)
3. Deepak, T.G., Joshua, V.C., Krishnamoorthy, A.: Queues with postponed work. Sociedad de Estdistica e Investigacion Operativa, Top **12**(2), 375–398 (2004)
4. Gross, D., Harris, C.M.: Fundamentals of Queueing Theory, 3rd edn. Wiley, New York (1998)
5. Krishnamoorthy, A., Islam, M.E.: Inventory system with postponed demands. Stoch. Anal. Appl. **22**(3), 827–842 (2004)
6. Latouche, G., Ramaswami, V.: A logarithmic reduction algorithm for quasi-birth-and-death processes. J. Appl. Probab. **30**, 650–674 (1993)
7. Latouche, G., Ramaswami, V.: Introduction to Matrix Analytic Methods in Stochastic Modellings. ASA-SIAM, Philadelphia (1999)
8. Neuts, M.F.: Matrix Geometric Solutions in Stochastic Models-An Algorithmic Approach. Dover Publications, New York (1994)
9. Manuel, P., Sivakumar, B., Arivarignan, G.: Perishable inventory system with postponed demands and negative customers. J. Appl. Math. Decis. Sci. **2007**, 1–12 (2007)
10. Sivakumar, B., Arivarignan, G.: An inventory system with postponed demands. Stoch. Anal. Appl. **22**(3), 827–842 (2008)

Author Index

Printed in the United States
By Bookmasters